PRO/ENGINEER WILDFIRE 4.0 INSTRUCTOR

Titles in the McGraw-Hill Graphics Series

PRO/ENGINEER WILDFIRE 4.0 INSTRUCTOR

David S. Kelley
Central Michigan University

Boston   Burr Ridge, IL   Dubuque, IA   New York   San Francisco   St. Louis
Bangkok   Bogotá   Caracas   Kuala Lumpur   Lisbon   London   Madrid   Mexico City
Milan   Montreal   New Delhi   Santiago   Seoul   Singapore   Sydney   Taipei   Toronto

Higher Education

PRO/ENGINEER WILDFIRE 4.0 INSTRUCTOR

Published by McGraw-Hill, a business unit of The McGraw-Hill Companies, Inc., 1221 Avenue of the Americas, New York, NY 10020. Copyright © 2009 by The McGraw-Hill Companies, Inc. All rights reserved. Previous editions © 2007, 2005 and 2002. No part of this publication may be reproduced or distributed in any form or by any means, or stored in a database or retrieval system, without the prior written consent of The McGraw-Hill Companies, Inc., including, but not limited to, in any network or other electronic storage or transmission, or broadcast for distance learning.

Some ancillaries, including electronic and print components, may not be available to customers outside the United States.

 This book is printed on recycled, acid-free paper containing 10% postconsumer waste.

1 2 3 4 5 6 7 8 9 0 QPD/QPD 0 9 8 7

ISBN 978–0–07–352266–1
MHID 0–07–352266–X

Global Publisher: *Raghothaman Srinivasan*
Executive Editor: *Michael Hackett*
Senior Sponsoring Editor: *Bill Stenquist*
Director of Development: *Kristine Tibbetts*
Developmental Editor: *Lora Kalb*
Executive Marketing Manager: *Michael Weitz*
Project Manager: *Joyce Watters*
Senior Production Supervisor: *Laura Fuller*
Associate Design Coordinator: *Brenda A. Rolwes*
Cover Designer: *Studio Montage, St. Louis, Missouri*
Compositor: *Laura Hunter, Visual Q*
Typeface: *10/12 Times Roman*
Printer: *Quebecor World Dubuque, IA*

Library of Congress Control Number: 2007937014

www.mhhe.com

This book is dedicated to my wonderful sister, Cindy Ross.
Thanks for being a great big sister.

BRIEF CONTENTS

CONTENTS

SECTIONS AND ADVANCED DRAWING VIEWS   335

SWEPT AND BLENDED FEATURES   369

Appendix A

SUPPLEMENTAL FILES   **543**

Appendix B

CONFIGURATION FILE OPTIONS   **545**

PREFACE

PURPOSE

This is my fifth installment of Pro/ENGINEER Instructor. My decision to write the original textbook was based on the lack of a comprehensive textbook on this popular mechanical computer-aided design package. My objectives for this project remain the same:

1. To write a textbook for an introductory course in engineering graphics.

2. To meet the needs of institutions teaching a course on parametric design and constraint-based modeling.

3. To create a book that would serve as a self-paced, independent study guide for the learning of Pro/ENGINEER for those who do not have the opportunity to take a formal course.

4. To incorporate a tutorial approach to the learning of Pro/ENGINEER in conjunction with detailed reference material.

5. To include topics that make the text a suitable supplement for an upper division course in mechanical design.

APPROACHES TO USING THE TEXTBOOK

This textbook is designed to serve as a tutorial, reference, and lecture guide. Chapters start by covering selected topics in moderate detail. Following the reference portion of each chapter are one or more tutorials covering the chapter's objectives and topics. At the end of each chapter are practice problems used to reinforce concepts covered in the chapter and previously in the book.

I had several ideas in mind when developing this approach to the book:

1. Since Pro/ENGINEER is a procedure-intensive computer-aided design application, the most practical pedagogical method to cover Pro/ENGINEER's capabilities (that would be the most beneficial both to students and instructors) would be to approach this book as a tutorial.

2. The book provides detailed reference material. A typical approach to teaching Pro/ENGINEER would be to provide a tutorial exercise followed by a nontutorial practice or practical problem. Usually students can complete the tutorial, but they may run into problems on the practice exercise. One of the problems that Pro/ENGINEER students have is digging back through the tutorial to find the steps for performing specific modeling tasks. The reference portion of each chapter in this text provides step-by-step guides for performing specific Pro/ENGINEER modeling tasks outside of a tutorial environment.

3. Supplemental model files (part, assembly, drawing, etc.) are available for download from the book's website (www.mhhe.com/kelley4). These files are used to enhance reference material. Each chapter contains reference guides for performing specific Pro/ENGINEER tasks. When appropriate in the reference guide, Pro/ENGINEER part files have been provided. This serves two purposes. First, the provided model provides a good starting point for instructors lecturing on specific topics of Pro/ENGINEER. Second, a student of Pro/ENGINEER can use a reference guide and part file to practice specific tasks.

4. The book is flexible in the order that topics are covered. Chapters 1 and 2 are primarily reference material. The first in-depth modeling tutorial starts with Chapter 3 (sketching fundamentals), and the first three-dimensional modeling begins in Chapter 4 (extruding features). Using this textbook in a course, an instructor may

decide to start with any of Chapters 1, 2, 3, or 4. Additionally, many of the chapters after Chapter 4 can be reordered to meet the needs of an individual instructor.

STUDENTS OF PRO/ENGINEER

One of the objectives of this book is to serve as a stand-alone text for independent learners of Pro/ENGINEER. This book is approached as a tutorial to help meet this objective. Since Pro/ENGINEER is menu-intensive, tutorials in this book use numbered steps to guide the selection of menu options. The following is an example of a tutorial step:

STEP 6:   **Place Dimensions according to design intent.**

Use the Dimension icon to match the dimensioning scheme shown in Figure 4–24. Placement of dimensions on a part should match design intent. With Intent Manager activated (Sketch >> Intent Manager), dimensions and constraints are provided automatically that fully define the section. Pro/ENGINEER does not know what dimensioning scheme will match design intent, though. Because of this, it is usually necessary to change some dimension placements.

MODELING POINT   If possible, a good rule of thumb to follow is to avoid modifying the section's dimension values until your dimension placement scheme matches design intent.

The primary menu selection is shown in bold. In this example, you are instructed to use the dimension option (portrayed by the Dimension icon) to create dimensions that match the part's design intent. Following the specific menu selection, when appropriate, is the rationale for the menu selection. In addition, Modeling Points are used throughout the book to highlight specific modeling strategies.

CHAPTERS

The following is a description and rationale for each chapter in the book.

CHAPTER 1 INTRODUCTION TO PARAMETRIC DESIGN

This chapter covers the basic principles behind parametric modeling, parametric design, and constraint-based modeling. Discussed is how Pro/ENGINEER can be used to capture design intent and how it can be an integral component within a concurrent engineering environment.

CHAPTER 2 PRO/ENGINEER'S USER INTERFACE

This chapter covers basic principles behind Pro/ENGINEER's interface and menu structure. The purpose is to serve as a guide and reference for later modeling activities. A tutorial is provided to reinforce the chapter's objectives.

CHAPTER 3 CREATING A SKETCH

Parametric modeling packages such as Pro/ENGINEER rely on a sketching environment to create most features. This chapter covers the fundamentals behind sketching in Pro/ENGINEER's sketcher mode. Two tutorials are provided along with practice problems.

CHAPTER 4 EXTRUDING, MODIFYING, AND REDEFINING FEATURES

Chapter 4 is the first chapter covering Pro/ENGINEER's solid modeling capabilities. Pro/ENGINEER's protrusion and cut commands are introduced and the extrude option is covered in detail. In addition, modification and datum plane options are introduced. Two tutorials are provided.

CHAPTER 5 FEATURE CONSTRUCTION TOOLS

While the protrusion and cut commands are Pro/ENGINEER's basic tools for creating features, this chapter covers additional feature creation tools. Covered in detail are the hole, round, rib, and chamfer commands; creating draft surfaces; shelling a part; cosmetic features; and creating linear patterns. Two tutorials are provided.

CHAPTER 6 REVOLVED FEATURES

Many Pro/ENGINEER features are created by revolving around a center axis. Examples include the revolve option found under the protrusion and cut commands, the sketched hole option, the shaft command, the flange command, and the neck command. These options, along with creating rotational patterns and datum axes, are covered in this chapter. Two tutorials are provided.

CHAPTER 7 FEATURE MANIPULATION TOOLS

Pro/ENGINEER provides tools for manipulating existing features. Manipulation tools covered include the group option, copying features, user-defined features, creating relations, family tables, and cross sections. In addition, using the model tree to manipulate features is covered.

CHAPTER 8 CREATING A PRO/ENGINEER DRAWING

Since Pro/ENGINEER is primarily a modeling and design application, the creation of engineering drawings is considered a downstream task. Despite this, there is a need to cover the capabilities of Pro/ENGINEER's drawing mode. This chapter covers the creation of general and projection views. Other topics covered include sheet formatting, annotating drawings, and creating draft entities. Two tutorials are provided, one of which covers the creation of geometric and dimensional tolerances.

CHAPTER 9 SECTIONS AND ADVANCED DRAWING VIEWS

Because of the length and depth of Chapter 8, the creation of section and auxiliary views is covered in a separate chapter.

CHAPTER 10 SWEPT AND BLENDED FEATURES

The protrusion and cut commands have options for creating extruded, revolved, swept, and blended features. Extruded and revolved features are covered in previous chapters. This chapter covers the fundamentals behind creating sweeps and blends. Three tutorials are provided.

CHAPTER 11 ADVANCED MODELING TECHNIQUES

The protrusion and cut commands have options for creating advanced features. Covered in this chapter are helical sweeps, variable section sweeps, and swept blends. Three tutorials are provided.

Chapter 12 Assembly Modeling

This chapter covers the basics of Pro/ENGINEER's assembly mode. Other topics covered include the creation of assembly drawings using report mode, the control of assemblies through layout mode, top-down assembly design, and mechanism design. Three tutorials are provided.

Chapter 13 Surface Modeling

This chapter covers the basics behind creating surface features within Pro/ENGINEER. Two tutorials are provided.

More Information on the Web

Please visit our web page, www.mhhe.com/kelley4. Ancillary materials are available for reading and download. For instructors, solutions to end-of-chapter Questions and Discussion are available.

Acknowledgments

I would like to thank many individuals for contributions provided during the development of this text. I would like to thank faculty, friends, and students at Purdue University and Western Washington University. I also want to thank the following individuals their efforts reviewing and testing over the years:

Holly K. Ault, *Worcester Polytechnic Institute*

Douglas H. Baxter, *Rensselaer Polytechnic Institute*

John R. Baker, *University of Kentucky—Paducah*

Dan Beller, *University of Wisconsin—Milwaukee*

Patrick Connolly, *Purdue University*

Robert Conroy, *Cal Poly State University*

Malcolm Cooke, *Case Western Reserve University*

Rollin C. Dix, *Illinois Institute of Technology*

Ismail Fidan, *Tennessee Tech University*

Boris Fritz, *Northrop Grumman Corporation*

Hector Gutierrez, *Florida Institute of Technology*

Quentin Guzek, *Michigan Technological University*

J. W. Hansberry, *University of Massachusetts Dartmouth*

Lawrence K. Hill, *Iowa State University*

Shawna Lockhart, *Montana State University*

Thomas Malmgren, *University of Pittsburgh at Johnstown*

Vojin R. Nikolic, *Indiana Institute of Technology*

Ronald J. Pederson, *New Mexico State University*

Mike Pierce, *Oklahoma State University—Okmulgee*

Mike L. Philpott, *University of Illinois*

Marie Planchard, *Massachusetts Bay Community College*

Sally Prakash, *University of Missouri—Rolla*

Jeff Raquet, *University of North Carolina at Charlotte*

John Renuad, *University of Notre Dame*

Larry F. Stikeleather, *North Carolina State University*

C. Steve Suh, *Texas A&M University*

Mary Tolle, *South Dakota State University*

Donald Wright, *Duke University*

Tao Yang, *California Polytechnic State University—San Luis Obispo*

The editorial and production groups of McGraw-Hill have also been wonderful to work with.

TRADEMARK ACKNOWLEDGMENTS

The following are registered trademarks of Parametric Technology Corporation®: PTC, Pro/ENGINEER, Pro/INTRALINK, Pro/MECHANICA, Windchill, and most other applications in the Pro/ENGINEER family of modules. Windows, Windows XP, Windows 2000, and Notepad are registered trademarks of Microsoft Corporation.

ABOUT THE AUTHOR

David Kelley, an Associate Professor in the Department of Engineering and Technology, joined Central Michigan University in January, 2004. He received his B.S. and M.S. degrees from the University of Southern Mississippi (1990 and 1992) and his Ph.D. from Mississippi State University (1998). Prior to joining CMU, he was an Assistant Professor in the Department of Computer Graphics Technology at Purdue University in West Lafayette, IN. Before his tenure at Purdue, he served on the faculty in the Engineering Technology Department at Western Washington University. He has also served as a faculty member at Oklahoma State University–Okmulgee, Northeastern State University (Tahlequah, Oklahoma), Northwest Mississippi Community College, and Itawamba Community College (Fulton, Mississippi).

Dr. Kelley's primary teaching responsibilities are in the areas of computer-aided design and mechanical design. In addition to his CAD teaching background, he has taught courses in computer-aided manufacturing, quality control, animation, and engineering design graphics. His research and scholarly interests include parametric design, collaborative engineering, and technology education.

1

INTRODUCTION TO PARAMETRIC DESIGN

Introduction

This chapter introduces the basic concepts behind parametric design and modeling. Parametric design is a powerful tool for incorporating design intent into computer-aided design models. Parametric models, often referred to as feature-based models, can be intuitively created and modified. Within this chapter, engineering graphics and three-dimensional modeling concepts will be explored. Additionally, parametric modeling principles will be covered. Upon finishing this chapter, you will be able to:

- Describe the utilization of computer-aided design within engineering graphics.
- Compare three-dimensional modeling techniques.
- Describe concepts associated with parametric modeling and design.
- Describe the use of parametric design within a concurrent manufacturing environment.

DEFINITIONS

Associativity	The sharing of a component's database between its application modes.
Design intent	The intellectual arrangement of assemblies, parts, features, and dimensions to meet a design requirement.
Parametric design	The incorporation of component design intent in a graphical model by means of parameters, relationships, and references.
Parametric modeling	A computer model that incorporates design parameters.

INTRODUCTION TO COMPUTER-AIDED DESIGN

Engineering design graphics has made significant changes since the early 1980s. For the most part, these changes are a result of the evolution of computer-aided design (CAD). Before CAD, design was accomplished by traditional board drafting utilizing paper, pencil, straightedges, and various other manual drafting devices. Used concurrently with manual drafting were sketching techniques, which allowed a designer to explore ideas freely without being constrained within the boundaries of drafting standards.

Many of the drafting and design standards and techniques that existed primarily because of the limitations of manual drafting still exist today. Popular midrange CAD packages still emphasize two-dimensional orthographic projection techniques. For example, these techniques allow a design to be portrayed on a computer screen by means once accomplished on a drafting table. Drafting standards have changed little since the beginning of CAD. These standards still place an emphasis on the two-dimensional representation of designs.

Many engineering fields continue to rely on orthographic projection to represent design intent. Some fields, such as manufacturing and mechanical engineering, foster a paper-less environment that does not require designs to be displayed orthographically. In this theoretically paperless environment, products are designed, engineered, and produced without a hard-copy drawing. Designs are modeled within a CAD system and the electronic data is utilized concurrently in various departments, such as manufacturing, marketing, quality control, and production control. Additionally, CAD systems are becoming the heart of many product data management systems. With a computer network, CAD designs can be displayed throughout a corporation's intranet. With Internet capabilities, a design can be displayed by using the World Wide Web.

ENGINEERING GRAPHICS

The fundamentals of engineering graphics and the displaying of three-dimensional (3D) designs on a two-dimensional surface have changed little since the advent of CAD. Despite the explosion of advanced 3D modeling packages, many design standards and techniques that once dominated manual drafting remain relevant today.

Sketching is an important tool in the design process. Design modeling techniques using two-dimensional CAD, three-dimensional CAD, or manual drafting can restrict an individual's ability to work out a design problem. It takes time to place lines on a CAD system or to construct a solid model. Sketching allows a designer to work through a problem without being constrained by the standards associated with orthographic projection or by the time required to model on a CAD system.

There are two types of sketching techniques: artistic and technical. Many individuals believe that artistic sketching is a natural inborn ability. This is not always the case. There are techniques and exercises that engineering students can perform that will improve their ability to think in three dimensions and solve problems utilizing artistic sketching skills. Despite this, few engineering students receive this type of training.

If engineering or technology students receive training in sketching, it is often of the technical variety. Technical sketching is similar to traditional drafting and two-dimensional computer-aided drafting. This form of sketching allows a design to be displayed orthographically or pictorially through sketching techniques.

The design process requires artistic sketching and technical sketching to be utilized together. Conceptual designs are often developed through artistic sketching methods. Once a design concept is developed, technical sketches of the design can be developed that will allow the designer to display meaningful design intent information. This information can then be used to develop orthographic drawings, prototypes, and/or computer models.

The traditional way to display engineering designs is through orthographic projection. Any object has six primary views (Figure 1–1). These views display the three primary dimensions of any feature: height, width, and depth. By selectively choosing a combination of the primary views, a detailer can graphically display the design form of an object. Often, three or fewer views are all that is necessary to represent design intent (Figure 1–2). A combination of views such as the front, top, and right-side will display all three primary dimensions of any feature. By incorporating dimensions and notes, design intent for an object can be displayed.

Orthographic projection is not a natural way to display a design. The intent of orthographic drawings is to show a design in such a way that it can be constructed or manufactured. Pictorial drawings are often used to represent designs in a way that non-technically trained individuals can understand. Pictorial drawings display all three primary dimensions (height, width, and depth) in one view (Figure 1–3). There are many forms of pictorial

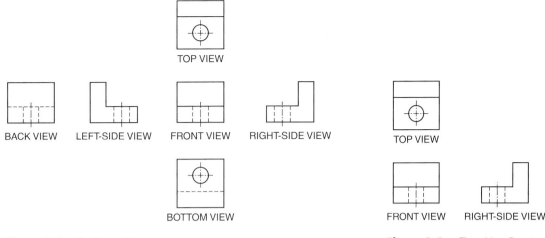

Figure 1-1 Six Primary Views

Figure 1-2 Three View Drawing

drawings. The most common are isometric, diametric, and trimetric. Naturally, objects appear to get smaller as one moves farther away. This is known as perspective. Perspective is another form of pictorial drawing. Orthographic, isometric, diametric, and trimetric projections do not incorporate perspective. Perspective drawings are often used to display a final design concept that can be easily understood by individuals with no technological training.

Before an object can be manufactured or constructed, technical drawings are often produced. Technical drawings are used to display all information necessary to properly build a product. These drawings consist of orthographic views, dimensions, notes, and details. Details are governed by standards that allow for ease of communication between individuals and organizations.

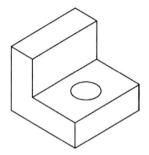

Figure 1-3 Isometric Pictorial View

PARAMETRIC MODELING CONCEPTS

Parametric modeling is an approach to computer-aided design that gained prominence in the late 1980s. A commonly held assumption among CAD users is that similar modeling techniques exist for all CAD systems. To users who follow this assumption, the key to learning a different CAD system is to adapt to similar CAD commands. This is not entirely true when a two-dimensional CAD user tries to learn, for the first time, a parametric modeling application. Within parametric modeling systems, though, you can find commands that resemble 2D CAD commands. Often, these commands are used in a parametric modeling system just as they would be used in a 2D CAD package. The following is a partial list of commands that cross over from 2D CAD to Pro/ENGINEER:

LINE

The Line option is used within Pro/ENGINEER's sketcher mode (or environment) as a tool to create sections. Within a 2D CAD package, precise line distances and angles may be entered by using coordinate methods, such as absolute, relative, and polar. Pro/ENGINEER does not require an entity to be entered with a precise size. Feature size definitions are established after finishing the geometric layout of a feature's shape.

CIRCLE

As with the line command, the Circle option is used within Pro/ENGINEER's sketcher environment. Precise circle size is not important in sketching the geometry.

ARC

As with the line and circle options, the Arc command is used within Pro/ENGINEER's sketching environment. Pro/ENGINEER's arc command also includes a fillet command for creating rounds at the intersection of two geometric entities.

DELETE

The Delete command is used within a variety of Pro/ENGINEER modes. Within the sketcher environment, Delete removes geometric entities such as lines, arcs, and circles; within Part mode, Delete removes features from a part. For assembly models, the Delete command deletes features from parts and parts from assemblies.

OFFSET

Offset options can be found within various Pro/ENGINEER modes. Within the sketcher environment, existing part features can be offset to form sketching geometry. Additionally, planes within Part and Assembly modes can be offset to form new datum planes.

TRIM

The Trim command is used within Pro/ENGINEER's sketching environment. Geometric entities that intersect can be trimmed at their intersection point.

MIRROR

The Mirror option is used within Pro/ENGINEER's Sketch and Part modes. Geometry created as a sketch can be mirrored across a centerline. Also, part features can be mirrored across a plane by executing the Copy option.

COPY

The Copy option is used within Part mode to copy existing features. Features can be copied linearly, mirrored over a plane, or rotated around an axis. Within Assembly mode, parts can be copied to create new parts.

ARRAY

Polar and rectangular array commands are common components among 2D CAD packages. Pro/ENGINEER's Pattern command serves a similar function. Features may be patterned using existing dimensions. Selecting an angular dimension will create a circular pattern.

Parametric modeling represents a different approach to CAD, compared to 2D drafting and to Boolean-based 3D modeling. Often, an experienced CAD user will have trouble learning a parametric modeling package. This is especially true when users try to approach three-dimensional parametric modeling as they would approach Boolean solid modeling. There are similar concepts, but the approaches are different.

FEATURE-BASED MODELING

Parametric modeling systems are often referred to as feature-based modelers. In a parametric modeling environment, parts are composed of features (Figure 1–4). Features can contain either positive or negative space. Positive space features are composed of actual mass. An example of a positive space feature is an extruded boss. A negative space feature is where a part has a segment cut away or subtracted. An example of a negative space feature is a hole.

Parametric modeling systems such as Pro/ENGINEER incorporate an intuitive way of constructing features. Often, the feature is first sketched in two dimensions, then either extruded, revolved, or swept to form the three-dimensional object. When sketching the

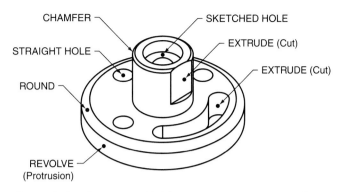

Figure 1-4   Features in a Model

feature, design intent is developed in the model by dimensioning and constraining the sketch.

Features can be predefined or sketched. Examples of predefined features include holes, rounds, and chamfers. Many parametric modeling packages incorporate advanced ways of modeling holes. Within a parametric modeling package, predefined holes can be simple, counterbored, countersunk, or drilled. Pro/ENGINEER's Hole command allows users the opportunity to sketch unique hole profiles, such as may be required for a counterbore. Sketched features are created by sketching a section that incorporates design intent. Sections may be extruded, revolved, or swept to add positive or negative space features.

Compared to Boolean modeling, feature-based modeling is a more intuitive approach. In Boolean modeling, a common way to construct a hole is to model a solid cylinder and then subtract it from the parent feature. In a parametric design environment, a user can simply place a hole either by using a predefined hole command or by cutting a circle through the part. With most Boolean-based modelers, in order to change a parameter of the hole, such as location or size, a user has to plug the original hole, then subtract a second solid cylinder. With a feature-based hole, the user can change any parameter associated with the hole by modifying a dimension or parameter. Similarly, a feature's sketch can be redefined or modified.

SKETCHING

As previously mentioned, parts consist of features. As shown in Figure 1–5, features are normally created by first sketching a section of the feature's profile. Sketch construction

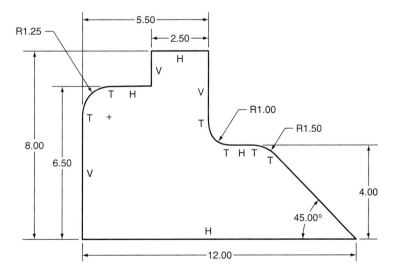

Figure 1-5   A Section Created in the Sketcher Environment

techniques are similar to 2D CAD drawing methods. In 2D CAD, a user has to use precision CAD techniques to draw a design. Parametric modeling sketcher environments do not require this. The sketching component within parametric design systems was developed to allow a user to quickly construct a design feature without having to be concerned with time-consuming precision. As an example, a user who wanted to draw a 4-in-square polygon in 2D CAD might draw four lines each exactly 4 in long and all perpendicular to one another. In a parametric design system, the user would sketch four lines forming roughly the shape of a square. If it is necessary to have a precise 4-in-square object, the user could dimension the object, constrain all lines perpendicular to one another, and then modify the dimension values to equal 4 in. Modifying the size of a feature (or sketch) in a parametric design system is as simple as modifying the dimension value and then updating the model.

CONSTRAINT MODELING

Design intent is incorporated in a feature by applying dimensions and constraints that meet the intent of the design. If the intent for a hole is to be located 2 in from a datum edge, the user would locate the hole within the modeling or sketching environment by dimensioning from the desired edge. Constraints are elements that further enhance design intent. Table 1–1 is a list of common parametric constraints. Once elements are constrained, they typically remain constrained until the user deletes them or until a dimension overrides them.

Some parametric modeling packages, such as Pro/ENGINEER, require a sketch to be fully constrained before it can be protruded into a feature. Other packages do not require this. There are advantages and disadvantages to packages that require fully constrained sketches. One advantage is that a fully constrained sketch requires all necessary design intent to be incorporated into the model. Additionally, a fully constrained sketch does not present as many surprises later in the modeling of a part. Many times, when a sketch is not completely constrained, the model will solve regenerations in a way not expected by the CAD user. Underconstrained sketches can have advantages though. Complicated sketches are often hard to fully constrain. Additionally, fully constraining a sketch may inhibit the design process. Often, designers do not want to fully define a feature.

DIMENSIONAL RELATIONSHIPS

Design intent can be incorporated in a model by establishing relationships between dimensions. A dimension can be constrained to be equal to another dimension (e.g., length = width). Additionally, mathematical relationships can be set between dimensions. As an example, the length of a feature can be set to twice its width (length = width x 2). Or the width could be set to equal half the length (width = length/2). Most forms of algebraic equations can be used to establish relationships between dimensions. An example of this would be to set the length of a feature equal to half the sum of the width and depth of a second feature [length = (width + depth)/2].

Table 1–1   Common Constraints Used in the Sketcher Environment

Constraint	Meaning
Perpendicular	Two lines are perpendicular to each other
Parallel	Two lines are parallel to each other
Tangent	Two elements are tangent to each other
Coincident	Two elements lie at the same location
Vertical	A line is vertical
Horizontal	A line is horizontal

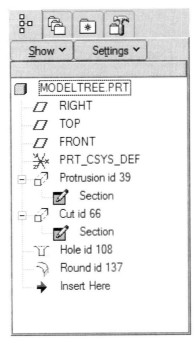

Figure 1–6   Pro/ENGINEER's
Model Tree

FEATURE REFERENCES

Within a part, features are related to each other in a hierarchical relationship. The first feature of a part is the base feature, which is the parent of all features that follow. Once a base feature is constructed, child features can be added to it. Child features may contain positive or negative space. A part can have an elaborate and complicated family tree. This tree is graphically displayed within Pro/ENGINEER as a model tree (Figure 1–6). It is important for the user to understand the relationship between a parent and a child. For any feature, if a parent is modified, it can have a devastating effect on its child features. Also, if a parent is deleted, all child features will be deleted. The user should be aware of this when constructing features.

MODEL TREE

Parametric modeling packages utilize a graphical model (or history) tree to list in chronological order all features that make up a part or assembly (Figure 1–6). The model tree does more than just list features, though. The following is a list of some of the uses of Pro/ENGINEER's model tree:

REDEFINING A FEATURE

A feature's definition can be modified by selecting it from the model tree. Some definitions that can be modified by using the Edit Definition command are:

- **Section** A feature's section can be modified.
- **Placement Refs** A feature's placement plane and references can be modified.
- **Depth** A feature's extrusion depth can be modified.
- **Hole Attributes** Attributes associated with a hole, such as diameter and direction, may be modified.

DELETING A FEATURE

A feature can be deleted or redefined by selecting the feature in the work screen. Sometimes a part can become so complicated that selecting a feature in this matter is difficult. To ensure that the correct feature is deleted, the feature can be selected on the model tree.

REORDERING FEATURES

Features that rely on other features in the modeling process are referred to as child features. A child feature must follow a parent feature in the order of regeneration. Usually, this means it should come after it on the model tree. Sometimes it is necessary to reorder a feature on the Model Tree to place it after a potential parent feature.

INSERTING FEATURES

Normally, new features are placed chronologically at the end of the model tree. At times, it may be necessary to place a new feature before an existing feature. When the Insert Feature command is used, Pro/ENGINEER will allow features to be inserted before or after existing features in the model tree.

SUPPRESSING FEATURES

Being able to remove a feature from the Model Tree can be a useful tool. The Suppress option is used to remove a feature temporarily from the order of regeneration.

ASSOCIATIVITY

Modules of Pro/ENGINEER are fully integrated. These modules share a common database. Often referred to as associativity, this feature allows modifications and redefinitions made in one module, such as Part mode, to be reflected in other modules, such as Drawing and Manufacturing. As an example, an object can be created in Part mode. This model can be used directly to create an orthographic drawing in Drawing mode. Dimensions used to create the model can be displayed in Drawing mode. If a dimension value is changed in Part mode, the same dimension and feature is changed in Drawing mode. Additionally, dimension values can be changed in Drawing mode, and Part mode will reflect these changes.

DATUM FEATURES

A datum is not a new idea to engineering design. A datum plane is a theoretically perfectly flat surface. It is used often in quality control for the inspection of parts. Geometric dimensioning and tolerancing practices utilize datum surfaces to control the size, location, and orientation of features. Many (but not all) parametric modeling systems use datum surfaces (or datum planes) to model features. Within Pro/ENGINEER, datum planes are often the parent features of all geometric features of a part. Datum planes are used as sketching surfaces, especially for sketching the first feature of a part. Datum planes are also used in parametric packages to locate features or to create new features. As an example, within Pro/ENGINEER, features can be mirrored about a datum plane to create new features (see Figure 1–7). In this case, the new features would be children of both the original features and the datum plane. The following is a description of the types of datums available within Pro/ENGINEER.

DATUM PLANE

A datum plane is a theoretically perfectly flat surface. Within Pro/ENGINEER, datum planes are used as sketching surfaces and references. Features may be

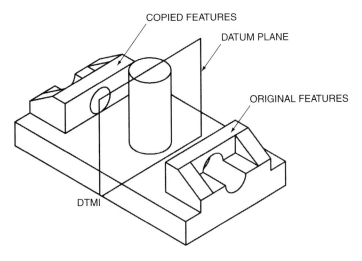

COPIED FEATURES

DATUM PLANE

ORIGINAL FEATURES

DTMI

Figure 1-7   A Feature Mirrored about a Datum Plane

sketched on any plane. Often, a suitable part plane does not exist. Datum planes can be created that will serve this purpose.

Many feature creation and construction processes require the use of a plane. A datum plane can be created for use as a reference in creating or constructing a feature. As an example, the Radial-Hole option requires a hole to be located at an angle from an existing plane. A datum plane can be created for this purpose. Another example would be patterning a feature around an axis. This construction technique requires that an angular dimension exist. A datum plane can be created that forms an angle with an existing plane. This new datum can then be combined with the feature to be patterned using the Group option. By grouping a datum plane with a feature, the angle parameter used to create the datum plane can be used within the Group >> Pattern option.

Datum Axis

A datum axis is similar to the centerline required at the center of holes and cylinders on orthographic drawings. Pro/ENGINEER provides an option for creating a datum axis. When the Create >> Datum >> Axis option sequence is used to create an axis, the axis is represented as a feature on the model tree. Pro/ENGINEER will also create a datum axis at the center of revolved features, holes, and extruded circles. These datum axes are not considered part features.

Datum axes are useful modeling tools. As an example, the Radial option under the Hole command requires the user to select an axis to reference the hole location. Another example is the Move >> Rotate option under Copy. A feature can be rotated around an axis or coordinate system.

Datum Curve

Datum curves are useful for the creation of advanced solid and surface features. Datum curves are considered part features and can be referenced in the sketching environment. They can be used as normal or advanced sweep trajectories or as edges for the creation of surface models. There are several techniques available for creating datum curves. One method is to sketch the curve using normal sketching techniques. Another construction methodology is to select the intersection of two surfaces.

Datum Point

Datum points are used in the construction of surface models, to locate holes, and to attach datum target symbols and notes. They are required for the creation of Pipe features. Datum points are considered features of a part, and Pro/ENGINEER labels the first point created PNT0. Each additional point is sequentially increased one numeric value.

COORDINATE SYSTEM

Pro/ENGINEER does not utilize a Cartesian coordinate system as most midrange computer-aided design packages do. Midrange two-dimensional drafting and three-dimensional Boolean-based modeling applications are based on a Cartesian coordinate world. Most parametric modeling packages, including Pro/ENGINEER, do not model parts on the basis of this system. Because of this, many users fail to understand the importance of establishing a datum coordinate system. Coordinate systems are used for a variety of purposes within Pro/ENGINEER. Many analysis tasks such as mass properties and finite-element analysis utilize a coordinate system. Coordinate systems are also used in modeling applications. As an example, the Copy-Rotate command provides an option to select a coordinate system to revolve around. Coordinate systems are also used frequently in Assembly mode and Manufacturing mode.

CONCURRENT ENGINEERING

The engineering design process was once linear and decentralized. Modern engineering philosophies are integrating team approaches into the design of products. As shown in Figure 1–8, team members may come from a variety of fields. Teaming stimulates a nonlinear approach to design, with the CAD model being the central means of communicating design intent.

Concurrent engineering has many advantages over a traditional design process. Individuals and groups invest significant resources and time in the development of products. Each individual and group has needs that have to be met by the final design. As an example, a service technician wants a product that is easy to maintain while the marketing department wants a design that is easy to sell. Concurrent engineering allows everyone with an interest in a design to provide input.

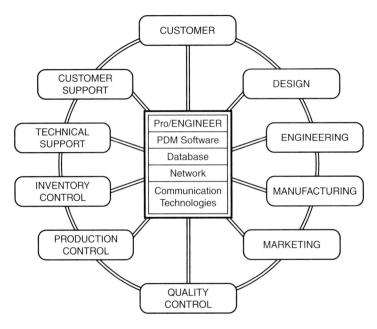

Figure 1–8   Concurrent Engineering Members

Modern engineering and communication technologies allow the easy sharing of designs among team members. CAD three-dimensional models graphically display designs that can be interpreted by individuals not trained in blueprint reading fundamentals. Because of this accessibility, the CAD system has become the heart of many product-data-management systems. Internet capabilities allow design to be shared over long distances. Most CAD applications have Internet tools that facilitate the sharing of design data. The name of Pro/ENGINEER's latest release, Wildfire, was derived from its newly integrated Internet tools. Built into Wildfire are conferencing tools that allow designers, engineers, and manufacturers to share desktops. Unlike most commercially available conferencing tools, which submit screen captures of computer screens, Wildfire's Conferencing Center first downloads the user's model then submits Pro/ENGINEER commands across the connection. This technique allows true real-time collaboration.

DESIGN INTENT

A capability unique to parametric modeling packages, compared to other forms of CAD, is the ability to incorporate design intent into a model. Most computer-aided design packages have the ability to display a design, but the model or geometry does not hold design information beyond the actual vector data required for construction. Two-dimensional packages display objects in a form that graphically communicates the design, but the modeled geometry is not a virtual image of the actual shape of the design. Traditional three-dimensional models, especially solid models, display designs that prototype the actual shape of the design. The problem with solid-based Boolean models is that parameters associated with design intent are not incorporated. Within Boolean operations, when a sketch is protruded into a shape or when a cylinder is subtracted from existing geometry to form a hole, data associated with the construction of the part or feature is not readily available.

Parameters associated with a feature in Pro/ENGINEER exist after the feature has been constructed. An example of this is a hole. A typical method used within Pro/ENGINEER to construct a straight hole is to locate the hole from two edges. After locating the hole, the system provides the hole diameter and depth. The dimensional values used to define the hole can be retrieved and modified at a later time. Additionally, parametric values associated with a feature, such as a hole diameter, can be used to control parameters associated with other dimensions.

With most Boolean operations, of primary importance is the final outcome of the construction of a model. In modeling a hole, the importance lies not in parameters used to locate a hole but where the hole eventually is constructed. When the subtraction process is accomplished, the cylinder location method is typically lost. With parametric hole construction techniques, these parameters are preserved for later use.

The dimensioning scheme for the creation of a feature, such as a hole, is important for capturing design intent. Figure 1–9 shows two different ways to locate a pair of holes. Both

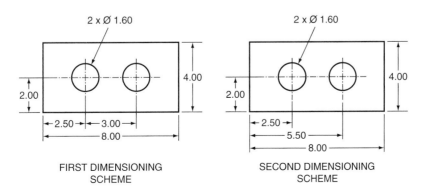

FIRST DIMENSIONING
SCHEME

SECOND DIMENSIONING
SCHEME

Figure 1-9   Dimensioning Scheme Differences

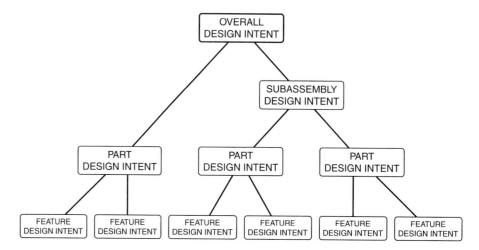

Figure 1-10   The Hierarchical Order of Design Intent

examples are valid ways to dimension and locate holes. Which technique is the best? The answer depends on the design intent of the part and feature. Does the design require that each hole be located a specific distance from a common datum plane? If it does, then the first example might be the dimensioning scheme that meets design intent. If the design requires that the distance between the two holes be carefully controlled, the second example might prove to be the best dimensioning scheme.

Designs are created for a purpose. Design intent is the intellectual arrangement of assemblies, parts, features, and dimensions to solve a design problem. Most designs are composed of an assembly of parts. Each part within a design is made up of various features. Design intent governs the relationship between parts in an assembly and the relationship between features in a part. As shown in Figure 1–10, a hierarchical ordering of intent can be created for a design. At the top of the design intent tree is the overall intent of the design. Below the overall design intent is the component design intent. Components are composed of parts and subassemblies. The intent of each component of a design is to work concurrently with other components as a solution to the design problem. Features comprise parts. Features must meet the design intent of the parts of which they are composed.

Parametric modeling packages provide a variety of tools for incorporating design intent. The following is a list of these tools:

DIMENSIONING SCHEME

The placement of dimensions is extremely important for the incorporation of design intent into a model. During sketching within Pro/ENGINEER, dimensions that will fully define a feature are placed automatically (when intent manager is activated). These dimensions may not match design intent, though. Dimensions within a section or within the creation of a feature should match the intent of a design.

FEATURE CONSTRAINTS

Constraints are powerful tools for incorporating design intent. If a design requires a feature's element to be constrained perpendicular to another element, a perpendicular constraint should be used. Likewise, design intent can be incorporated with other constraints, such as parallel, tangent, and equal length.

ASSEMBLY CONSTRAINTS

Assembly constraints are used to form relationships between components of a design. Within an assembly, if a part's surface should mesh with the surface of another part, a Mate constraint should be used. Examples of other common assembly constraints include Align, Insert, and Orient.

DIMENSIONAL RELATIONSHIPS

Dimensional relationships allow the capture of design intent between and within features while in Part mode, and between parts while in Assembly mode. A dimensional relationship is an explicit way to relate features in a design. Mathematical equations are used to relate dimensions. An example of a dimensional relationship would be to make two dimensions equal in value. Within Pro/ENGINEER, for this example, the first dimension would drive the second. Most algebraic and trigonometric formulas can be included in a relationship. In addition, simple conditional statements can be incorporated.

REFERENCES

Feature references can be created within Part and Assembly modes of Pro/ENGINEER. An example of a reference within Part mode is to use existing feature edges to create new geometry within the sketcher environment. A Parent-Child relationship is established between the two features. If the reference edge is modified, the child feature is modified correspondingly.

Within Assembly mode, an external reference can be established between a feature on one part and a feature on a second. Pro/ENGINEER allows for the creation of parts and subassemblies within Assembly mode. By creating a component using this technique, relationships can be established between two parts. Modification of the parent part reference will modify the child part.

PRO/ENGINEER MODES

Pro/ENGINEER is an integrated, fully associative package. Integrated parametric design packages such as Pro/ENGINEER share data with its various other operating modules. Pro/ENGINEER is the fundamental application in a powerful suite of tools capable of an integrated and concurrent environment. Objects created within Pro/ENGINEER can be shared with other applications. Because of a part's parametric associativity, changes made to an object in one mode will be reflected in other modes. As an example, a part can be modeled in Part mode. Following the modeling process, an orthographic drawing can be created in Drawing mode. Additionally, the part can be assembled with other components in Assembly mode. In Part, Drawing, or Assembly mode, a change can be made to a parameter of the part. Upon regeneration, this change will be reflected in the other modes in which the part resides.

The following is a description of the basic modules found with Pro/ENGINEER; they are included in Pro/ENGINEER's foundation package. Additionally, modules such as Pro/Designer, Pro/NC, Pro/Mechanica, Pro/Fly-Through, and Pro/Layout are available.

SKETCH MODE

A fundamental technique within most parametric modeling packages is to sketch feature entities and then to invoke a three-dimensional construction operation, such as Extrude, Revolve, or Sweep. Most features created by a sketching technique are constructed within Part or Assembly mode. Sketches can be created outside the part and assembly environment by utilizing Sketch mode. When this is done, the sketch can be saved for use in later modeling situations.

PART MODE

Part mode is the primary environment for the creation of solid and surface models. For many manufacturing enterprises, Part mode is the center of the design and production environment. Objects created in Part mode can be utilized in downstream applications, such as Pro/ENGINEER's Drawing and Manufacturing modes.

Additionally, part design intentions can be shared concurrently over Internet or intranet networks by using Pro/Fly-Through and/or Pro/Web-Publish.

DRAWING MODE

Drawing mode is the primary means within Pro/ENGINEER for constructing documentation drawings. While technical drawings were once considered the primary tool in engineering graphics, it is now a downstream application in the parametric modeling design process. In a true "paperless" manufacturing environment, an orthographic drawing is no longer required. A design can be developed in Part and Assembly mode, analyzed in Pro/Mechanical, and have its manufacturing code generated in Pro/NC. Despite this integrated philosophy, companies still need documentation.

Drawing mode can take an existing part or assembly and produce an orthographic drawing. It can produce detailed drawings with a variety of section and auxiliary view capabilities. Dimensioning tools, including geometric tolerancing, are available. Additionally, through Pro/Detail, drawing and construction tools are available. Figure 1–11 shows an example of a detailed drawing produced in Drawing mode.

ASSEMBLY MODE

Assembly mode allows for the combining of design components into a final design solution. A variety of tools exist for the integration of design intent. When parts are placed within an assembly, relationships can be established with existing parts, features, and subassemblies. Parts can be created within Assembly mode or placed from preexisting parts.

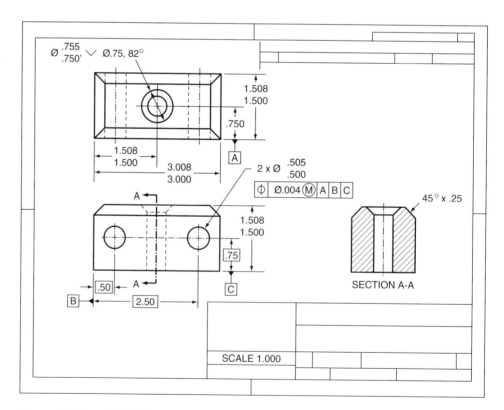

Figure 1-11   A Pro/ENGINEER Drawing

Pro/ENGINEER allows for bottom-up or top-down assembly design. Bottom-up design requires components to be modeled in Part mode and then assembled in Assembly mode. In top-down design, a skeleton model creates an assembly that starts at the overall design intent level and works down to the individual part level. Parts can be created within top-down design by modeling within Assembly mode.

SUMMARY

Parametric design packages such as Pro/ENGINEER are revolutionizing the engineering design process. Early CAD systems were capable of producing technical drawings in an electronic format, but added little to the actual design process beyond what could be accomplished with paper and pencil. Three-dimensional CAD applications, especially solid modeling systems, have design tools that allow a designer to model a design as a virtual prototype.

Parametric design fundamentals have increased the design capabilities of three-dimensional modeling CAD systems. As with solid modeling applications, parametric modeling systems can construct a design as an electronic prototype. Parametric modeling objects have intelligence that not only display a design as a graphic image but also incorporates parameters that can describe the intent of a design.

Integrated design applications such as Pro/ENGINEER are being used in companies not simply as a modeling tool but as the center of the product data management system. Design data from a Pro/ENGINEER modeling file can be viewed and retrieved throughout a company's intranet. Additionally, Internet tools such as Pro/Web-Publish allow data to be viewed by individuals external to a company's localized network. This powerful capability has increased the collaborative tools of computer-aided design and enhanced Pro/ENGINEER's concurrent engineering capabilities.

QUESTIONS AND DISCUSSION

1. Describe two types of sketching done in engineering graphics.

2. How many primary views are possible in orthographic projection?

3. Define the term Feature within a parametric design package.

4. List and describe five types of geometric constraints used during the sketch construction process of a parametric design package.

5. Explain what is meant by the Parent-Child relationship that exists between parametric features.

6. Describe uses of Pro/ENGINEER's Model Tree.

7. Explain what is meant when a parametric modeling package is fully associative.

8. What is a datum surface? Name ways that datums are used in Pro/ENGINEER?

9. Describe how Pro/ENGINEER can be used within a concurrent manufacturing environment.

10. List and describe the uses of the various modes found within Pro/ENGINEER's foundation package.

2

PRO/ENGINEER'S USER INTERFACE

Introduction

Pro/ENGINEER has UNIX, Windows, and Linux versions. With its Wildfire release, Pro/ENGINEER introduces a significantly new and more intuitive interface. As in older versions of Pro/ENGINEER, this interface resembles a typical Windows application, but unlike previous versions, the new release offers commonly used Windows interface techniques. However, in manipulating Pro/ENGINEER, it is important to remember that it does not always function like a true Windows application. This chapter will introduce the fundamentals of Pro/ENGINEER's interface. Upon finishing this chapter, you will be able to:

- Describe the purpose behind each menu on Pro/ENGINEER's Menu bar.
- Use Pro/ENGINEER's file management capabilities to save object files.
- Set up a Pro/ENGINEER object to include units, tolerances, and materials.
- Customize Pro/ENGINEER through the use of the Configuration file.
- Customize Pro/ENGINEER commands using Mapkeys.
- Organize items using the Layers option.

DEFINITIONS

Configuration file	A Pro/ENGINEER file used to customize environmental and global settings. Configuration options can be set through the Utilities >> Preferences option.
Mapkeys	Keyboard macros used to define frequently used command sequences.
Model	An object that represents the actual sculptured part, assembly, or work piece.
Nominal dimension	A dimension with no tolerance.
Object	A file representing an item, part, assembly, drawing, layout, or diagram created in Pro/ENGINEER.
Tolerance	The allowable amount that a feature's size or location may vary.

MENU BAR

The following is a description of many of the options available on Pro/ENGINEER's Part mode menu bar (see Figure 2–1). While this interface may appear to make Pro/ENGINEER a true Windows application, many typical Windows functions are not available (Copy, Paste, etc.).

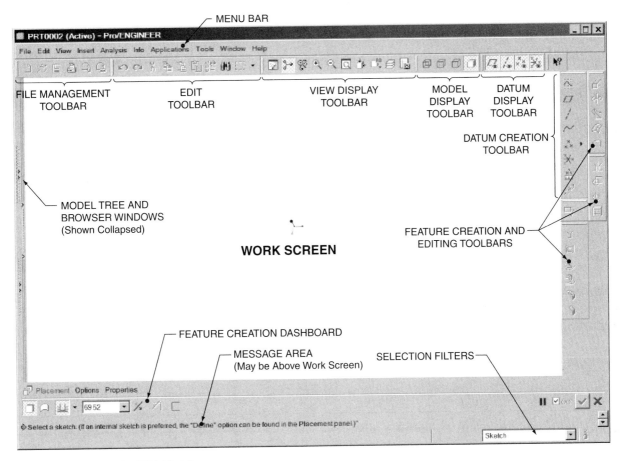

Figure 2-1   Pro/ENGINEER's Work Screen

FILE MENU

The File menu is Pro/ENGINEER's interface for the manipulation of files and objects. Found under the File menu are options for saving and opening objects. Also, options are available for printing and exporting objects.

EDIT MENU

The edit menu provides options for the modification of geometric elements. Within Part mode, commands are available for performing feature manipulation and modification techniques such as Edit Definition, Reroute, Suppress, and Delete. Within sketch mode, options are available for moving, copying, and trimming sketched entities.

VIEW MENU

The View menu is used to change the appearance of models and Pro/ENGINEER's work screen. Many of the view options that are available exist as shortcut keys, or can be found on the Toolbar. Commonly used view manipulation options under the View menu include exploding an assembly view, repainting the view, and retrieving the default view. Options are available for orienting a model and saving a view. Under the View menu, additional options are available for modifying a model's color and appearance, and for changing the lighting of Pro/ENGINEER's work screen.

INSERT MENU

The Insert menu provides selections for the creation of traditional Pro/ENGINEER features (e.g., Protrusion, Hole, Datum Plane, Cosmetic Thread).

ANALYSIS

Options for finding assembly and part properties can be found under the Analysis menu. As an example, the mass of a part can be obtained through the Model Analysis option.

INFO MENU

The Info menu is used to obtain information about Pro/ENGINEER objects. Information can be found on Parent-Child relationships, features, references, and geometry. Messages, such as error messages created during regeneration failures, can be displayed by using the Message Log option. Additional information about failed regenerations can be found by using the Geometry Check option. A commonly used option under the Info menu is Switch Dims. Dimensions may be displayed with numeric values or with dimension symbols. This option toggles between the two dimension display modes.

APPLICATIONS MENU

The Applications menu will allow a user to switch between Pro/ENGINEER modes and applications. As an example, a user may switch between Part mode and Manufacturing mode.

TOOLS MENU

The Tools menu is available for the customization of Pro/ENGINEER's interface. The Tools\Environment option is a common tool for temporarily changing the work screen appearance. Many of the features found under the environment menu, such as datum and model display, can now be found on the Toolbar. The Tools menu provides access to Pro/ENGINEER's configuration file (config.pro). Additional options are available for customizing toolbar selections and for creating Mapkeys.

WINDOW MENU

The Window menu is used to manipulate Pro/ENGINEER windows. Windows can be activated, opened, or closed. Within Pro/ENGINEER, multiple windows of multiple parts can be open at once. To work in one menu, a user has to first **Activate** it. Opened windows are displayed under the Window menu, thus allowing a user to easily switch from one object to another.

HELP MENU

Pro/ENGINEER utilizes a Web browser to access help information. The Pro/Help CD has to be loaded before all help options can be utilized, but some help options do have Web links. The help option provides search capabilities for Pro/ENGINEER options and a context-sensitive help option is also available. Use context-sensitive help to find information on individual Pro/ENGINEER menus and options.

PRO/ENGINEER'S TOOLBAR

As shown previously in Figure 2–1, Pro/ENGINEER provides toolbars for easy access to frequently used options through the use of icons. By default, Pro/ENGINEER's initial toolbar is divided into five groups. The Customize dialog box (Tools >> Customize Screen) allows additional toolbars to be added to Pro/ENGINEER's work environment.

FILE MANAGEMENT TOOLBAR

The file management toolbar is available for manipulating files.

Figure 2-2   File Management Icons

- **New** The New option is used to start a new Pro/ENGINEER file.
- **Open** The Open option is used to open a Pro/ENGINEER file.
- **Save** The Save option is used to save a Pro/ENGINEER file.
- **Print** The Print option is used to print or plot a Pro/ENGINEER object.
- **Email** The Email option is used to send model files to a mail recipient as either an attachment or as a link.

EDIT TOOLBAR

The Edit toolbar is available for modifying the display of Pro/ENGINEER objects on the work screen.

Figure 2-3   Edit Toolbar

- **Undo** The Undo option is used to remove the last operation.
- **Redo** The Redo is used to redo the last operation.
- **Cut** The Cut option is used to remove a text or annotation and place it on the clipboard.
- **Copy** The Copy option places an object on the clipboard.
- **Paste** The Paste option will place an object from the clipboard.
- **Paste Special** The Paste Special option will place an object from the clipboard dependently.
- **Regenerate** The Regenerate option will perform a regeneration of the features of a model.
- **Search** The Search icon will open the Search Tool dialog box.

VIEW DISPLAY

The view display group of icons is available for modifying the display of Pro/ENGINEER objects on the work screen.

Figure 2-4   View Toolbar

- **Repaint** The Repaint option is used to redraw the work screen.
- **Spin Center** The Spin Center option is used to turn off the spin center tripod.
- **Orient Mode** The Orient Mode icon is used to activate Orient mode on and off.
- **Zoom In** The Zoom In icon is used to zoom in to a user-defined window.
- **Zoom Out** The Zoom Out icon is used to zoom out from the work screen.
- **Refit** The Refit icon is used to fit the extent of a Pro/ENGINEER object into the work screen.
- **Reorient** The Reorient icon is used to orient a Pro/ENGINEER object on the work screen.

- **Saved View List** The Saved View List icon is used to access saved views.
- **Layers** The Layers icon will allow for the creation and manipulation of layers.
- **View Manager** The View Manager icon provides access to the View Manager dialog box.

MODEL DISPLAY

Figure 2–5
Model Display Toolbar

The model display group of icons is available for changing the display of Pro/ENGINEER objects. Only one of the four available icons under this group may be activated at a time.

- **Wireframe** The Wireframe icon displays a Pro/ENGINEER object as a wireframe.
- **Hidden Line** The Hidden Line icon displays a Pro/ENGINEER object with hidden lines.
- **No Hidden** The No Hidden icon displays a Pro/ENGINEER object without hidden lines.
- **Shade** The Shade icon shades a Pro/ENGINEER object.

DATUM DISPLAY

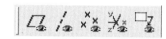

Figure 2–6
Datum Display Icons

The datum display group of icons is used to control the display of datums.

- **Datum Planes** The Datum Plane icon is used to turn on or off the display of Pro/ENGINEER datum planes.
- **Datum Axes** The Datum Axes icon is used to turn on or off the display of Pro/ENGINEER datum axes.
- **Point Symbols** The Point Symbols icon is used to turn on or off the display of Pro/ENGINEER datum points.
- **Coordinate Systems** The Coordinate Systems icon is used to turn on or off the display of Pro/ENGINEER coordinate systems.
- **Annotations** The Annotations icon turns on the display of 3D annotations and 3D elements.

CONTEXT-SENSITIVE HELP

The Context-Sensitive Help icon is used to display help information on individual menu or dialog box options. To use this help function, select the context-sensitive help icon, then select the menu item. Pro/Help will launch a Web browser displaying information about the selected item.

FILE MANAGEMENT

Various options are available for manipulating Pro/ENGINEER files. Pro/ENGINEER's file management capabilities provide a wide range of functions for managing projects and models. On first appearance, Pro/ENGINEER's file opening and saving commands resemble a Windows application. However, there are some significant differences between Pro/ENGINEER's file management and a Windows application.

- Pro/ENGINEER filename requirements are more restricted than Windows application filenames.
- Saving a Pro/ENGINEER object creates a new version of the object each time the object is saved. It does not override older versions.
- Pro/ENGINEER will not allow an object to be saved to a specific filename if that filename already exists. Pro/ENGINEER will not save on top of an existing file.

Table 2-1   File Extensions for Pro/ENGINEER Modes

Mode	Extension
Sketch	*.sec.*
Part	*.prt.*
Assembly	*.asm.*
Manufacturing	*.mfg.*
Drawing	*.drw.*
Format	*.frm.*

Table 2-2   Invalid and Valid Filenames

Invalid Filename	Problem	Valid Filename
part one	Space in filename	part_one
part@11	Nonalphanumeric character	part_11
Part[1_10]	Brackets used in filename	Part_1_10

FILENAMES

Pro/ENGINEER has different file extensions according to the mode being utilized. Table 2–1 shows file extensions based on five common Pro/ENGINEER modes.

Notice the extra asterisk at the end of each file extension. This asterisk represents the version of the file. The first time Pro/ENGINEER saves a file, this extra extension has a value of 1. The second time a file is saved, a new file is created with 2 as this value. The third time a file is saved, a new file is created with 3 as this value. Pro/ENGINEER creates a new object file each time a file is saved. If an object file is saved 10 times, 10 Pro/ENGINEER files will be created. To delete the previous Pro/ENGINEER files, select File >> Delete >> Old Versions.

Pro/ENGINEER file and directory names cannot be longer than 31 characters. Brackets, parentheses, periods, nonalphanumeric characters, and spaces cannot be used in a filename. An underscore (_) may be used in a file name, though. Table 2–2 shows examples of invalid and valid filenames.

MEMORY

When an object is opened, referenced, or created in Pro/ENGINEER, it is placed in memory. It remains there until it is erased, or until Pro/ENGINEER is exited. Also, when an assembly is opened, every part referenced by the assembly is placed in memory. Parts in active memory are displayed in a window. Multiple parts, assemblies, and drawings can be in active memory at once. This allows for ease of access between objects. Objects may also be in session memory. Session memory is the condition where the object is in memory, but not displayed in a graphics window.

WORKING DIRECTORY

Pro/ENGINEER utilizes a Working Directory to help manage files. The working directory is usually the modeling point for all Pro/ENGINEER objects. When a new file is saved, it is saved in the current working directory, unless a new directory is specified. To change the current working directory, from the File menu, select Working Directory. Select the desired directory as the working directory.

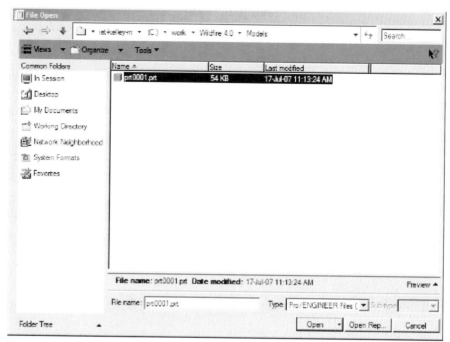

Figure 2–7 The File Open Dialog Box

OPENING AN OBJECT

As shown in Figure 2–7, Pro/ENGINEER objects are retrieved using the File Open dialog box. This dialog box may be retrieved by selecting Open from the File menu or by selecting the Open icon on the Toolbar menu.

By utilizing the File Open dialog box, Pro/ENGINEER objects, neutral file formats (e.g., STEP, IGES), raster images, and vector images. When an object is opened, Pro/ENGINEER defaults to one of the following directories:

- Directory associated with the active object (first alternative)
- Working Directory (second alternative)
- Directories contained in the search path (third alternative)

Perform the following steps to open a Pro/ENGINEER object:

STEP 1: **Select FILE >> OPEN to reveal the File Open dialog box.**

The File Open dialog box can also be revealed by using the **Open** icon.

STEP 2: **Select the directory in which the object is located.**

As shown in Figure 2–7 and as shown below, options are available for changing the viewing directory.

STEP 3: **Select the object to open.**

Objects may be selected by picking the object from the file list, or by typing in the file name. Older versions of an object may be opened also. If a version number is not known, you can type in the complete file name with an extension number relative to the older version. As an example, if you want to open a part file with the name *part_one*, but open the file three versions earlier, you would enter the filename *part_one.prt.-3*.

Wildcards are available for opening objects. An asterisk can replace multiple characters (e.g., *part*.prt*) while a question mark can replace a single character (e.g., *part?a.prt*).

STEP 4: **Select OPEN on the Dialog Box.**

MODELING POINTS Finding an object in a directory with many files can be difficult. There are several options available to lower the number of objects shown on the file list. One option is to select the type of file to open by using the Type option. Other view manipulation options are available under the Tools menu.

Pro/ENGINEER Wildfire has numerous Internet enhancements. Objects can be dragged and dropped from Wildfire's Web browser and from Web pages created with Windchill's Parts Link application. Wildfire also supports HTTP and FTP address in its File >> Open option.

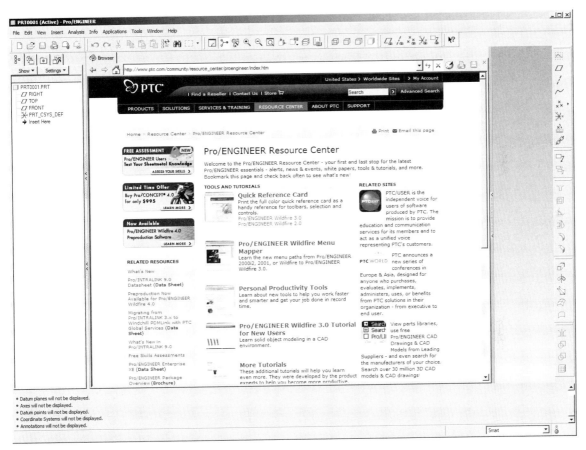

Figure 2–8   Pro/ENGINEER's Navigator Windows

PRO/ENGINEER NAVIGATOR

Pro/ENGINEER's Navigator windows contains tools for accessing, manipulating, and navigating through design data. As shown in Figure 2–8, the navigator environment consists of multiple tabs comprised within two windows. The following list describes available options:

Model Tree Navigator The model tree tables allows for the manipulation of Pro/ENGINEER features within the context of Part mode and assembly components within Assembly mode.

Folder Browser Navigator The folder browser option allows for the browsing, opening, and manipulation of Pro/ENGINEER objects through directory folders. Sub-options are available for creating new folders, deleting folders, and setting a folder as the current working directory.

Favorites Navigator Commonly used directory folders can be saved in the favorites tab. Web sites can also be saved to a favorites list.

Connections Navigator The connections navigator provides access to Parametric Technology Corporation services and to other user-defined sites.

The browser window is a unique environment that allows for traditional Internet browsing and file manipulation options. The following is a limited list of its capabilities:

- File system manipulation and browsing to include the previewing and opening of Pro/ENGINEER objects (see Figure 2–9).
- Bill of material (BOM) viewing.
- Web site viewing and interaction to include the manipulation of Pro/ENGINEER-related resources (i.e., Windchill's PartsLink, Projectlink, and Pro/COLLABORATE).
- Access FTP sites.

CREATING A NEW OBJECT

Files for most basic modes of Pro/ENGINEER are created by using the File >> New option from the Menu bar or using the New icon on the Toolbar. Perform the following steps to create a new object:

STEP 1:   **Select FILE >> NEW to open the New dialog box.**

New Pro/ENGINEER files can be created using the **File** icon also.

STEP 2:   **Select a Mode Type and Sub-type of Pro/ENGINEER.**

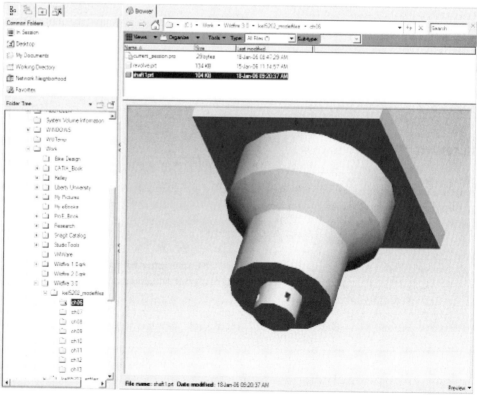

Figure 2–9   *Browser Window with Folder Preview*

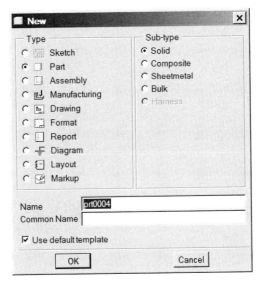

Figure 2-10   New Dialog Box

Available Mode Types are shown in Figure 2–10. Pro/ENGINEER defaults to Part mode with Solid as the sub-type.

Step 3:   **Enter a filename.**

You can enter a filename or take the default name.

MODELING POINT If you forget to enter a filename, or if you want to change a filename, use the Rename option found under the File menu. This option will rename all versions of a filename.

Step 4:   **Select OK from the dialog box.**

Saving an Object

Various options are available for saving objects. New objects are saved by default in the current working directory. If an object is retrieved from a directory other than the working directory, the object is saved in its original directory. Additionally, selecting Save in a sketcher environment will save the section (*.sec.*) and not the object being modeled. The following options are available for saving objects:

Save

This option saves an object to disk. When an assembly is saved, all individual parts that make up the assembly are saved. When a drawing is saved, the model used to create the drawing is saved only when changes have been made to the object. In sketching in a sketcher environment, the section under modification or creation is saved and not the object file. Pro/ENGINEER objects may be saved to a computer hard drive, floppy disk, Zip disk, or product data management system.

Save a Copy

The Save As option is used to either save an object as a new filename or to save an object to a new directory. When an object is saved using Save As, the original filename is not deleted and is still the active model. Save As practically creates a copy of the object being modeled. Any changes made to the original object are not reflected in the copied object.

BACKUP

The Backup option creates a copy of the object being modeled. The name of the object cannot be changed with this option. Any saves of an object conducted after a backup will be to the directory of the backup.

RENAME

The Rename option changes the name of a Pro/ENGINEER object. A suboption is available for renaming the object on disk and in memory, or just in memory. When an object that already exists is renamed all previous versions of the object are saved.

DELETE

Saving an object multiple times can create many versions of the object on disk. The Delete option is available to purge old versions. Options are available for deleting Old Versions of an object or All Versions of an existing object.

ERASE

Closing a window that contains a Pro/ENGINEER object does not remove it from memory. The Erase command is used to remove an object from memory. An object that is referenced by another opened object cannot be erased from memory. The Erase dialog box shows all objects referenced by a selected object. Options are available for erasing referenced objects from memory or for keeping them in memory.

ACTIVATING AN OBJECT

Multiple objects can be open at once within Pro/ENGINEER. Additionally, multiple windows can be opened. To modify an object, its associated window must be activated. To make a window active, use the Activate option found under the Window menu.

VIEWING MODELS

There are many different ways to view a Pro/ENGINEER object and to view Pro/ENGINEER's work screen. Options are available for panning, rotating, and zooming an object dynamically. Other options are available for changing the display of a model.

DYNAMIC VIEWING

A useful feature of Pro/ENGINEER that enhances its model-building capabilities is its dynamic viewing functions. A model can be dynamically zoomed, rotated, or panned through the use of the mouse or a mouse/key combination (Figure 2–11).

DYNAMIC ROTATE

A user may dynamically rotate a model by using Pro/ENGINEER's dynamic rotate option. Dynamic rotate is activated when the middle mouse button is selected. While the middle mouse button is held down, moving the cursor within the work screen will rotate the model around a specified spin center. The center of spin will be based on the location of the mouse at the time of selection.

DYNAMIC ZOOM

A user may dynamically zoom in or out on a model by using Pro/ENGINEER's dynamic zoom option. Dynamic zoom is activated when the keyboard's Control key

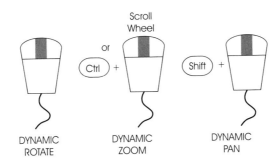

Figure 2-11   Dynamic Viewing Options

is selected at the same time as the middle mouse button. While the Control key and the middle mouse button are held down simultaneously, moving the cursor from the top of the work screen to the bottom will zoom in on a model. Correspondingly, moving the cursor from the bottom of the work screen to the top will zoom out on a model. For mice equipped with a scroll wheel, scrolling can be utilized for zooming a model.

Dynamic Pan

A user may dynamically pan a model by using Pro/ENGINEER's Dynamic Pan option. Dynamic pan is activated when the keyboard's Shift key is selected at the same time as the middle mouse button. While the Shift key and the middle mouse button are held down simultaneously, moving the cursor within the work screen will pan the model.

Model Display

As shown in Figure 2–12, there are four styles used to display a model in part, assembly, and manufacturing modes. Similarly, within drawing mode, there are three styles. There are situations when each display style is the most practical. The Model Display dialog box is located under the View menu and contains other display options. Each display style can be selected dynamically from the Toolbar menu.

Wireframe Display

Within all relevant modes of Pro/ENGINEER, the Wireframe style displays all edges of a model as a wireframe. Edges that would be hidden from view during a true representation of the model are displayed, just as edges that would not be hidden.

Hidden Display

With the Hidden display style, lines that would be hidden from view during a true representation of a model are shown in gray. Within Drawing mode, these gray lines represent hidden lines and will be printed as hidden lines.

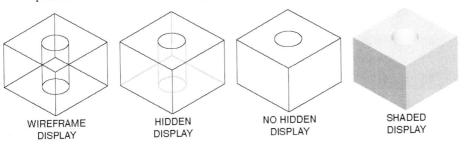

WIREFRAME DISPLAY   HIDDEN DISPLAY   NO HIDDEN DISPLAY   SHADED DISPLAY

Figure 2-12   Model Display Options

NO HIDDEN DISPLAY

With the No Hidden display style, lines that would be hidden from view during a true representation of a model are not shown.

SHADED DISPLAY

With the Shaded display, all solids and surfaces are displayed shaded. Hidden lines are not shown. This option is not available in drawing mode.

There are three possible default views within Pro/ENGINEER: Trimetric, Isometric, and User-Defined. When selecting Default View from the View menu, the model returns to this viewpoint. The initial setting within Pro/ENGINEER is Trimetric. This setting can be permanently changed by the configuration file option *orientation*, or temporarily changed in the Environment menu or in the Orientation dialog box.

VIEW ORIENTATION

For modeling in Part or Assembly mode, it can be advantageous to orient the model to one of the six primary orthographic views. Also, within Drawing mode, when a view is first established, the model is initially placed as the Default view. A correct view orientation can be created with the Orientation dialog box.

The Reorient option is available under the View menu and on the toolbar. This option opens the Orientation dialog box (Figure 2–13). Two orientation types are available under this dialog box:

DYNAMIC ORIENTATION

The Dynamic Orient option is similar to Pro/ENGINEER's dynamic view options. The Dynamic Orient option provides a dialog box for zooming, rotating, and panning a model.

ORIENTATION BY REFERENCE

The Orient by Reference option is used to create an orthographic view of a model. Figure 2–13 shows the Orientation dialog box with this option selected. Two references are required with this option. These references correspond to a selected primary orthographic view of the model (Front, Top, Right, etc.). The initial references are Front for the first reference and Top for the second. These references can be changed. In selecting a reference with a selected orthographic view, selecting a planar surface will orient the model toward that direction. As an example, when Top is the first reference and Front is the second, selecting as shown in Figure 2–14 will orient the model in the direction specified.

NAMING AND SAVING VIEWS

Often, it is necessary to return to a user-defined orientation of a model. View orientations can be obtained by using dynamic viewing or using the Reorient option. Once a view has been set, the Orientation and the View Manager dialog boxes have options for saving the view's orientation; the Saved View List icon on the toolbar is used to retrieve views. The View Manager dialog box (Figure 2–15) is assessable through an icon on the toolbar and is used to manage the display of views to include orientations and simplified representations. As shown in the figure, the New option is used to save a view, while the Edit option is available for redefining or removing saved views.

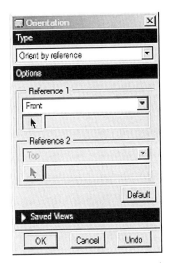

Figure 2-13   Orientation Dialog box

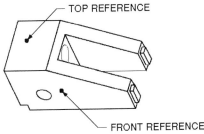

Figure 2-14   Orienting a Model

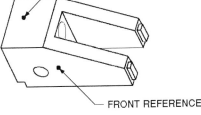

Figure 2-15   View Manager Dialog Box

SETTING UP A MODEL

Pro/ENGINEER provides several options for setting up a model. Properly establishing an object's parameters is an often overlooked, but important, step in the modeling process. This section will describe how to establish units for a model and how to define and assign materials. Additionally, setting tolerances and renaming features will be introduced.

UNITS

Within Pro/ENGINEER, there exist four Principle Unit categories: Length, Mass or Force, Time, and Temperature. Each category has a full range of possible units. As an example, available within the Length category are inches, feet, millimeters, centimeters, and meters.

SETTING A SYSTEM OF UNITS   Pro/ENGINEER utilizes a system of units to group the four principle categories. There are six predefined systems of units available within Pro/ENGINEER. These systems may be accessed through the System of Units tab found on the Units Manager dialog box.

- Meter Kilogram Second (MKS)
- Centimeter Gram Second (CGS)
- Millimeter Newton Second (mmNs)
- Foot Pound Second (FPS)
- Inch Pound Second (IPS)
- Inch lbm Second (Pro/E Default)

As shown from the above list of predefined systems of units, ***Inch-lbm-Second*** is the default system. To set a different system of units, select the specific system on the System of Units tab, then select the Set button.

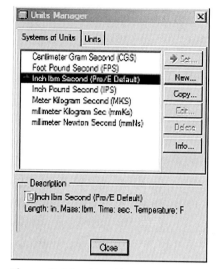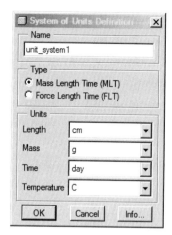

Figure 2–16   Units Menus

CREATING A SYSTEM OF UNITS   A user-defined system of units can be created. This allows a user to establish units that meet design intent for a given product. Perform the following steps to create a new system of units.

STEP 1:   **Select EDIT >> SETUP >> UNITS on the Menu bar**

When selecting the Units option, the Units Manager dialog box will appear (Figure 2–16). Options available under this dialog box include: (*a*) Creating a new system, (*b*) copying an existing system to create a new system, (*c*) reviewing a system's units, (*d*) deleting a user-defined system, and (*e*) setting a System of Units.

STEP 2:   **Select NEW on the Units Manager dialog box.**

STEP 3:   **In the System of Units Definitions dialog box, enter a NAME for the user-defined system (Figure 2–16).**

STEP 4:   **Select UNITS from each principle category (Figure 2–16).**

Notice the Type option for choosing between Mass and Force. Each subcategory has its own available units.

STEP 5:   **Select OK on the System of Units Definition dialog box then CLOSE the Units Manager dialog box.**

MODELING POINT   It is good modeling practice to set your units before creating a feature. If you forget to set them before creating the first feature, units can be converted from one system to another. Pro/ENGINEER will interpret your existing dimensions as another defined unit (e.g., 1 in will be interpreted as 1 mm). To allow for this interpretation, use the Interpret Existing Numbers option after setting a new unit.

MATERIALS

Material parameters may be defined for a part. Additionally, a defined material may be written to a file and used with other parts. A recommended procedure is to create a library of materials. Figure 2–17 shows the materials dialog box for assigning materials to a model.

Also shown is the Materials Definition dialog box for defining or modifying a material.

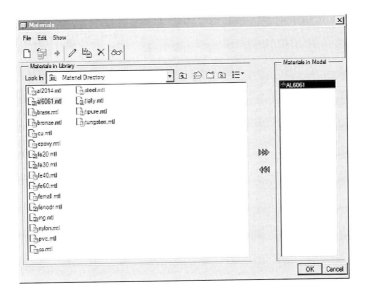

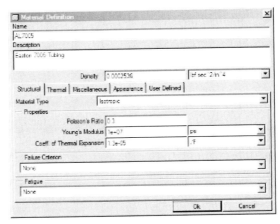

Figure 2-17   Available Material Parameters

CREATING A MATERIAL   When a material is defined, the information is stored in the database of the part being modeled. To create a material, perform the following steps:

STEP 1:   Select EDIT >> SETUP >> MATERIAL on the Menu bar.

STEP 2:   Select the NEW material option on the Materials dialog box.

STEP 3:   Enter a name for the material.

Any name can be used. It is recommended that the name be descriptive of the material being defined.

STEP 4:   Enter parameters associated with the material.

Pro/ENGINEER will display a default Materials file. Figure 2–17 shows an Figure 2–17 shows an example of parameters that can be created for a material. Materials created for a part are not automatically saved to disk. To save a material specification so other parts can use it, the material has to be saved to a permanent location. It is recommended that all materials be stored in a common Materials Library directory. Material files are in ASCII format and have a *.mat* extension.

DIMENSIONAL TOLERANCE SETUP

Pro/ENGINEER provides various methods for the display of dimensional tolerances. A tolerance is the allowable amount that a feature's shape or size may vary from exact. Tolerances are a necessity for most precision parts.

By default, Pro/ENGINEER displays tolerances as nominal values. There are two ways to display tolerance values. The first way is to set the configuration file option *tol_display* to YES. Additionally, when setting this option to yes, a table is displayed at the bottom of the work screen with the default tolerance values. The second way to display tolerances is to select the Dimension Tolerances check box within the Environment dialog box (Tools >> Environment). There are four formats for displaying tolerances.

- **Nominal** The Nominal option does not display tolerance values. This is Pro/ENGINEER's initial format.
- **Limits** The Limits option displays tolerances as an upper limit value and a lower limit value.
- **Plus-Minus** The Plus-Minus option displays tolerance values as a positive and negative deviation from the nominal value. The positive value and negative value may be different.
- **Plus-Minus Symmetric** The Plus-Minus Symmetric option displays tolerance values as a positive and negative deviation from the nominal value. The positive value and negative value are the same.

Tolerance formats can be changed by using the Dimension Properties dialog box (Figure 2–18). This dialog box can be accessed through the model tree's pop-up menu's Edit option (right-mouse select a feature on the model tree then pick Edit) or by selecting a dimension on the model with the Properties option from the dimension pop-up menu. The default format can be changed in the configuration file option *tol_mode.*

Within Pro/ENGINEER, a user may select ANSI or ISO for the tolerance display standard. The ANSI standard is selected initially. A specific *tolerance standard* can be selected through the Standard option found under the Tol Setup menu (Edit >> Set Up). Use the configuration file option *tolerance_standard* to change the default value.

ANSI TOLERANCE STANDARD

Tolerance values using the ANSI standard are initially set based on values found in the configuration file. The **linear_tol** and **angular_tol** options are used to set linear and angular tolerance values, respectively. The tolerance values are assigned to

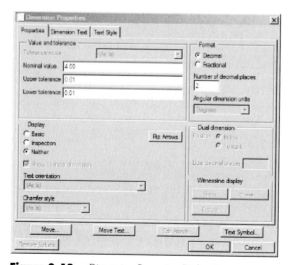

Figure 2-18   Dimension Properties Dialog Box

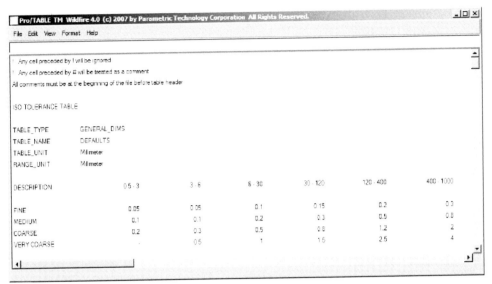

Figure 2-19   ISO Tolerance Standard Table

dimensions based on the number of decimal places. To change variational values and number of decimal places for dimensions, use the Dimension Properties dialog box.

ISO TOLERANCE STANDARD

Tolerance tables drive dimensional tolerances with the ISO tolerance standard. Tables are loaded when a model is created with an ISO standard, or when the standard is changed to ISO from ANSI. Figure 2–19 shows an example of a tolerance table. The TABLE_TYPE range is used to specify the type of tolerance to use.

Within the ISO standard, tolerances have an additional variation referred to as the Tolerance Class. The tolerance class is the relative looseness or tightness of a tolerance. There are four general classes of fit: Fine, Medium, Coarse, and Very Coarse. The coarser the fit, the more variation there is in the tolerance. Medium is Pro/ENGINEER's default class. Tolerance classes are specified in each tolerance table. To change classes of fit for individual dimensions, select Model Class from the TOL SETUP menu.

> **MODELING POINT**   Tables (referred to as Pro/TABLE) are used throughout Pro/ENGINEER to input information. A Tolerance table is just one example of a use of Pro/TABLE. Many Pro/TABLE cells have established values. As an example, the possible values for the Table_Unit cell (Figure 2–19) within a Tolerance table are inch, foot, micrometer, millimeter, centimeter, and meter. Instead of typing one of the possible units into the cell, selecting the **F4** key on the keyboard will display all of the available options.

Tolerances are based on the nominal size of the dimension. Figure 2–19 shows an example of a tolerance table with metric units. The following is a description of information found in a tolerance table:

- **Table_Type** Available options include General, Broken Edge, Holes, and Shafts. Use the F4 key to select between available options.
- **Table_Name** This cell represents the name of the table.
- **Table_Unit** Examples of tolerance units include inches, millimeters, and meters. Use the F4 key to select between available options.

- **Range_Unit** Examples of tolerance ranges include inches, millimeters, mci, and meters. Use the F4 key to select between available options.
- **Description** Description represents the nominal sizes of the table. The full range of available sizes of a unit should be covered. The table shown in Figure 2–19 ranges from less than 0.5 to 4000 mm.
- **Tolerance Class** Tolerances are provided for one or more of the available classes of fit. Providing all four classes will ensure that a change in a class will not be invalid.
- **Tolerance Value** Each tolerance value represents a symmetrical plus/minus variation from the nominal value. As an example, a value entered as 0.2 would provide a tolerance value of plus and minus 0.2 mm from a dimension's nominal value.

MODIFYING A TOLERANCE TABLE   Perform the following steps to modify an ISO tolerance table. The tolerance standard has to first be set to ISO/DIN:

STEP 1:   Select EDIT >> SETUP >> TOL SETUP from the menu bar.

STEP 2:   Select the TOL TABLES option on the Tolerance Setup menu.

STEP 3:   Select the MODIFY VALUE >> GENERAL DIMS options from the Tolerance Tables menu.

STEP 4:   Modify values in the table.

GEOMETRIC DIMENSIONING AND TOLERANCING   Geometric Dimensioning and Tolerancing (GD&T) controls the shape of geometric features of a part. Unlike dimensional tolerances, a geometric tolerance applied to a Pro/ENGINEER model does not affect the actual shape of the model.

Traditional tolerancing controls the location and size of part features, but does not control the geometric shape. As an example, a cylinder may have a tolerance applied to its diameter dimension. This dimension controls the size of the cylinder but not variations in the cylinder's shape. If extreme variations in the shape, such as a lack of cylindricity, will have negative affects on the design, GD&T can be used to control this form variation.

There are five classes of features that are controlled: form, profile, runout, orientation, and location. Cylindricity is a type of form control. Other form types include straightness, flatness, and circularity.

Another use of GD&T is to control the location of features. Traditional location tolerances create a square tolerance zone. A square tolerance zone may inadvertently lead to design problems. A geometric position tolerance creates a circular tolerance zone. Position is a type of location control. Other location types include concentricity and symmetry.

Many geometric tolerance values can be modified on the basis of a material condition. There are three possible material condition modifiers: maximum material condition (MMC), least material condition (LMC), and regardless of feature size (RFS). The MMC of a feature exists when a feature has the most material possible. The MMC for a shaft is the largest shaft possible within its established limits. Since a part will have more material if it has a smaller hole, the MMC for a hole is the smallest hole possible. The GD&T symbol for MMC is a circled letter M. The LMC for a part is the condition with the least amount of material present. The GD&T symbol for LMC is a circled letter L. Regardless of feature size is interpreted to mean that a tolerance value is not modified by a material condition. The symbol for RFS is a circled letter S. The absence of a material condition modifier, by definition, means RFS.

Datum references may be datum planes, axes, or points. Some types of feature controls do not require a datum reference. As an example, no Form type control (e.g., Flatness or Straightness) requires a datum reference. With many other types of controls, a datum reference is an absolute necessity. As an example, a Position tolerance might be used to control the location of a hole. It is common to locate a hole with references to two edges. In applying a Position tolerance to a hole, the hole has to be located with respect to two datum references. Datum references are also common with orientation controls.

When a geometric tolerance controls a feature's size or location, the size or location dimension becomes a basic dimension. Basic dimensions are theoretically perfect and are represented on a drawing as a nominal value enclosed in a rectangle. Since the geometric tolerance is controlling the dimension, there is no need to apply a traditional tolerance.

Figure 2–20 shows an example of a Feature Control Frame. Feature Control Frames are used in GD&T to control geometric features. The following is a description of each compartment:

TYPE

There are 14 types of GD&T controls available. Examples are Position, Perpendicular, Parallel, and Cylindricity.

TOLERANCE VALUE

The tolerance value determines the amount of variation allowed in a feature. Many GD&T types produce a cylindrical tolerance zone. If this occurs, a phi (∅) symbol representing diameter is placed before the tolerance value.

TOLERANCE MATERIAL CONDITION MODIFIER

A feature's tolerance value may be modified based on a material modifier. Three possible conditions exist: maximum material (MMC), least material (LMC), and regardless of feature size (RFS).

DATUM REFERENCE

Datum references are used within a feature control frame only when necessary. As an example, a Flat form control does not need a datum reference, but a perpendicular orientation control does. In the case of the latter example, a feature will be controlled, within a tolerance, to be perpendicular to a datum reference. With Perpendicularity, only one datum reference is needed.

CREATING A GEOMETRIC TOLERANCE   Geometric tolerances can be created within Part, Assembly, or Drawing mode. Since Pro/ENGINEER models have high accuracy (default value of 0.0012), a geometric tolerance applied to a feature does not affect the actual shape or form of the feature. Geometric dimensions and tolerances serve to provide graphical information only. Perform the following steps to create a geometric tolerance within Part mode:

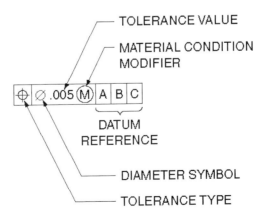

Figure 2–20   Feature Control Frame

> **MODELING POINT** Geometric dimensions and tolerances are important considerations for the intent of a design. Care should be taken in applying geometric tolerances to a part. Of considerable importance is the location of datum references. Too often, datum planes are placed without consideration for the design of a part. Many geometric tolerances require one or more datum references. These datums should be created before applying a geometric tolerance. Also, each datum reference must be set by using the Set Datum option before it can be utilized as a reference.

STEP 1:   Select INSERT >> ANNOTATIONS >> GEOMETRIC TOLERANCE on the menu bar.

STEP 2:   Select SPECIFY TOL on the Geometric Tolerance menu.

As shown in Figure 2–21, the Geometric Tolerance dialog box will appear. This dialog box is used to create and place Feature Control Frames.

STEP 3:   Select the type of feature control.

There are 14 feature control types used for geometric tolerancing. Each is available at the left end of the Geometric Tolerance dialog box.

STEP 4:   Select the model to apply feature control.

As shown in Figure 2–21, there are five tabs available under the Geometric Tolerance dialog box. The first tab specifies model references. The first model reference is the model to which the feature control is applied. By default, the active model is selected.

STEP 5:   Select the reference to apply feature control.

An understanding of GD&T principles is needed to select the proper reference for applying a feature control. As an example, when applying a Position tolerance to a hole, the tolerance is actually controlling the axis of the hole. The axis of the hole should be selected as the feature reference. With a Circularity tolerance, the tolerance is controlling the surface of the feature. This surface should be selected as the reference.

STEP 6:   Select a placement for Feature Control Frame.

There are several methods available for attaching a Feature Control Frame to a model. An understanding of GD&T drafting standards is needed to properly apply a frame. As an example, when attaching a feature control frame to a diameter dimension, the axis of the hole or cylinder is the element being controlled by the geometric tolerance.

When Leader is selected as the type of placement, a leader menu will appear. Under the leader menu is the option of attaching the leader to an Entity or to a Surface. Entity is selected by default. Often it is necessary to change this to Surface.

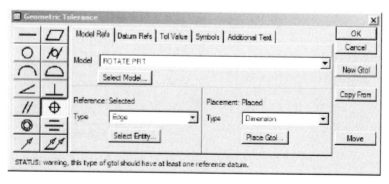

Figure 2–21   Geometric Tolerance Dialog Box

After selecting the Placement for the Feature Control Frame, notice how the frame appears on the model. During the addition of references and symbols throughout the Geometric Tolerance dialog box, notice how this symbol is updated.

STEP 7: **Under the DATUM REFS tab, select datum references to add to the Feature Control Frame.**

As shown in Figure 2–22, it is possible to utilize a Primary, Secondary, and Tertiary datum reference. Only use the datum references necessary for a feature control. Datums have to exist on the model before selecting a datum reference. Additionally, datums have to be set using the Set Datum option found under the Geom Tol menu before they can be utilized as a reference. Datum material condition modifiers also exist under each subtab.

STEP 8: **Under the TOL VALUE tab, enter the tolerance value to apply.**

As shown in Figure 2–23, tolerance values and tolerance material condition modifiers are entered through the Tol Value tab.

STEP 9: **Under the SYMBOLS tab, select Symbols and Modifiers to apply.**

As shown in Figure 2–24, symbols and/or modifiers such as Statistical Tolerance, Diameter Symbol, and Free State are added under this tab.

STEP 10: **If necessary, add any addition text to the note (Figure 2-25).**

STEP 11: **Select OK to apply the Feature Control Frame.**

MODELING POINT Before selecting OK to apply the Feature Control Frame, check the model to observe that the frame is being applied correctly. Values for the Feature Control Frame are added dynamically to the model.

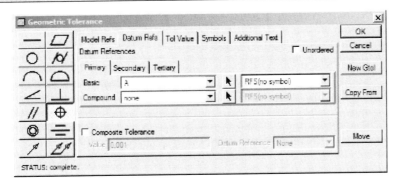

Figure 2–22   Datum Refs Tab

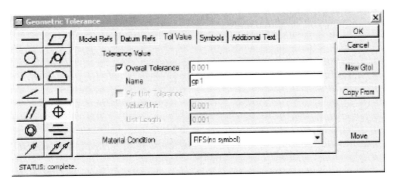

Figure 2–23   Tolerance Value Tab

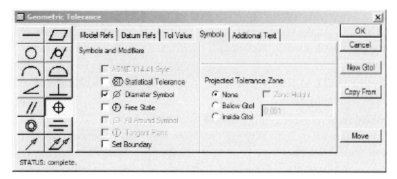

Figure 2–24   Symbols Tab

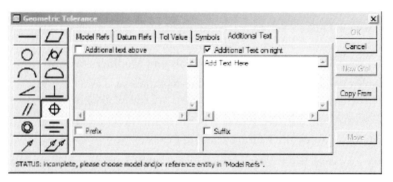

Figure 2-25   Additional Text Tab

NAMING FEATURES

Features created within Pro/ENGINEER are assigned a name based on the feature type. As an example, an extruded feature might be named on the model tree *Extrude 1.* One power of Pro/ENGINEER is its capability for providing a user with the ability to select a feature or part from the model tree. Finding a feature or part on the model tree is relatively easy for simple models, but can become difficult with complicated objects. It is recommended that a user rename features to better discriminate them from other components. The Setup menu has a Name option for renaming features, but the easiest method is to use the model tree's pop-up menu's Rename command (The pop-up menu is available by selecting the feature on the model tree).

Datums can be renamed by using the Name option. They can also be named by using the Set Datum option. Before a datum can be referenced by using geometric tolerancing, it must first be set. Options are available under Set Datum for renaming a datum, setting a datum, and unsetting a datum.

OBTAINING MODEL PROPERTIES

Pro/ENGINEER provides several options for obtaining information about an object. Most of these options can be found under the Info and Analysis menus. Functions available range from obtaining Parent-Child information to analyzing model properties.

PARENT/CHILD RELATIONSHIPS

Because of Pro/ENGINEER's nature as a parametric, feature-based modeling package, features are related to other features through references. When a feature references another

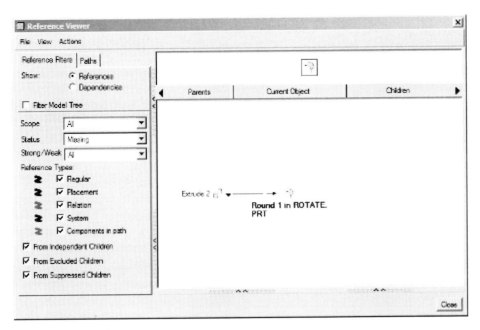

Figure 2-26   Reference Viewer

feature during the modeling process, the referenced feature becomes a parent of the part being modeled. Local references are references between features of a part, while external references occur between features of different parts. As a result of this dependency, parts composed of parent features within an external relationship must be present in memory when a child part is open.

Information about parent/child relationships can be found with the Reference Viewer dialog box (Figure 2–26). The following filter options are available:

- **Scope - External** Used to view external references (outside the current part model).
- **Scope - Local** Used to view internal references (within the current part model).
- **References** Child features of the selected feature will be displayed.
- **Dependencies** The children of the selected parent feature will be displayed.

MODEL ANALYSIS

A variety of model analysis types exist for obtaining model information. The Model Analysis dialog box is available under the Analysis menu.

MASS PROPERTIES

Mass properties can be obtained on parts and assemblies. The following properties are available:

- Volume
- Surface area
- Density
- Mass
- Center of gravity
- Inertia tensor
- Principal moments of inertia
- Rotation matrix and rotation angles
- Radii of gyration

X-Section Mass Properties

Mass properties can be obtained on a predefined X-section. The following properties are available:

- Area
- Inertia tensor
- Principal area moments of inertia
- Polar moment of inertia

- Rotation matrix and rotation angle
- Radii of gyration
- Section moduli

One-Sided Volume

The One-Sided Volume type is used to obtain the volume of a model on one side of a specified plane.

Pairs Clearance

The Pairs Clearance type is used to determine the clearance between two entities of a part or assembly. This option is available in Part, Assembly, and Drawing modes.

Global Clearance

The Global Clearance type is used to determine the clearance between two components of an assembly. This option is available in Assembly and Drawing modes.

Short Edge

The Short Edge type is used to determine the number of model edges shorter than a specific length.

Global Interference

The Global Interference type is used to calculate the amount of interference that exists between two components in an assembly.

Thickness

The thickness type is used to determine the minimum and maximum thickness of a part. This type is available in Part and Assembly modes.

Printing in Pro/ENGINEER

Pro/ENGINEER has the capability to print to a variety of printers and plotters. Additionally, objects may be printed to a file. The Print dialog box can be opened by selecting Print from the File menu or from the Toolbar.

The Print Dialog box Destination is used to select a specific printer. To change printers, select the Add Printer icon.

MODELING POINT   The Windows Printer Manager can be used to print objects. To use the default Windows printer, in the configuration file, enter **Windows Printer Manager** as the value for the Plotter option.

CONFIGURING THE PRINTER

To configure a printer, select the Configure button located on the Printer dialog box. Note that your model should be on a display mode other than shading. Multiple print configurations may be set. To save a configuration, select the Save button. The following tab options are available under Configure:

PAGE TAB

As shown in Figure 2–27, the Page tab is used to configure the sheet size. Standard sheet sizes are available (A, B, C, D, E, F, A0, A1, A2, A3, and A4), or a user may specify a variable sheet size.

PRINTER TAB

Printer tab is used to specify printer options that might be available. Not all options are available with every printer.

MODEL TAB

The Model tab (Figure 2–28) is used to adjust the way an object is printed on a sheet. The Plot field is used to adjust the area to be plotted. Options are available for creating a full plot, a clipped plot, a plot based on zoom, a plot of an area, and a plot based on the model size. The Based-on-Zoom option is the default value. For full sized plots, this should be changed to Full Plot.

PRO/ENGINEER'S ENVIRONMENT

Various selections are available for controlling Pro/ENGINEER's working environment. Figure 2–29, on the next page, shows Pro/ENGINEER's Environment dialog box. This dialog box is available under the Tools menu. Many of the Environment selection options are also available through other avenues, such as the Toolbar menu. The following is a description of each available selection:

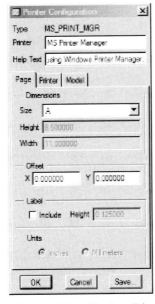

Figure 2-27   The Page Tab

Figure 2-28   The Printer Tab

- **Dimension Tolerance** This selection specifies whether an object's dimensions are displayed as a tolerance or as a nominal value. This option is also controllable with the configuration file option *Tolerance Display*. A dimension's tolerance mode can be set with the Dimension Properties dialog box, or by default with the configuration file option *Tolerance Mode*.

- **Datum Plane** This selection option controls the display of datum planes. This option is readily accessible through Pro/ENGINEER's Toolbar menu.

- **Datum Axes** This selection option controls the display of datum axes. This option is readily accessible through Pro/ENGINEER's Toolbar menu.

- **Point Symbols** This selection option controls the display of datum points. This option is readily accessible through Pro/ENGINEER's Toolbar menu.

- **Coordinate System** This selection option controls the display of coordinate systems. This option is readily accessible through Pro/ENGINEER's Toolbar menu.

- **Spin Center** The Spin Center is used to show the center of a spin when an object is rotated. This option is activated by default. This selection can be changed permanently using the configuration file option *spin_center_display*.

- **Notes as Names** This option displays the name of a note instead of the actual note text.

- **Reference Designators** This option displays reference designators such as cabling, ECAD, and Piping components.

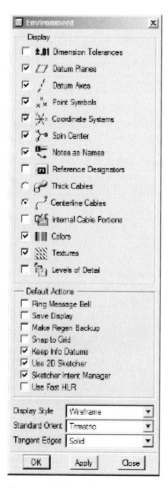

Figure 2-29   The Environment
Dialog Box

- **Thick Cables** This selection controls the three-dimensional display of cables. The Centerline Cables option and this option cannot be selected at the same time.

- **Centerline Cables** This selection displays the centerlines of cables. The Thick Cables option and this option cannot be selected at the same time.

- **Internal Cable Portions** This selection option allows for the display of cables that are hidden by other geometry.

- **Colors** This option will display a model in color.

- **Textures** This option will display surface textures on a shaded model.

- **Levels of Detail** This selection will allow for the use of levels of detail while a shaded model is dynamically viewed.

- **Ring Message Bell** When selected, a bell will sound when Pro/ENGINEER provides a message. This selection can be changed permanently with the configuration file *Bell* option.

- **Save Display** This option will save the display of an object, reducing the recalculations needed when the object is reopened at a future time.

- **Make Regen Backup** This selection creates a backup copy of the object before regenerations. This allows for the retrieval of a previously valid model. Backups are deleted when ending the object's session.

- **Snap to Grid** When Grid is activated, elements will snap to them, particularly within the sketcher environment.

- **Keep Info Datums** This selection option controls the display of datums created on the fly with the Info functionality. When this option is selected, these datums will be considered as features within the model.

- **Use 2D Sketcher** By default, sketching is set up within a two-dimensional environment. With this setting, the sketching plane is oriented parallel to the screen. When this setting is not selected, the sketcher environment remains in a three-dimensional orientation.

- **Sketcher Intent Manager** This option will allow the utilization of intent manager in sketching (see Chapter 3 for information on intent manager).

- **Use Fast HLR** This option allows for the faster acceleration of hardware while a model is dynamically viewed.

- **Display Style** There are four display styles available for viewing a model: Wireframe, Hidden Line, No Hidden, and Shading. These options are also available on Pro/ENGINEER's Toolbar menu.

- **Standard Orient** There are three settings available for Pro/ENGINEER's default standard orientation: Isometric, Trimetric, and User Defined. Trimetric is the initial setting. This can be changed with the configuration file option *Orientation*.

- **Tangent Edges** This selection option controls the display of tangent edges. There are five options available: Solid, No Display, Phantom, Centerline, and Dimmed (dimmed menu color).

CONFIGURATION FILE

Pro/ENGINEER's Options dialog box (Figure 2–30) is used to customize a variety of environmental and global settings. It is accessed through the Tools >> Options command. Options such as model orientation, system geometry color, background color, tolerance mode, and sketcher grid display can be set by default. As shown in the figure, the left-most column of the table is used to establish a specific option and the next column is used to define the value for the option. Available options can be found through Pro/ENGINEER's online help. A user can set an option by entering the option's name and value (e.g., *orientation* and *isometric* as shown in the illustration), followed by selecting the Add/Change icon.

> **MODELING POINT** The configuration file is used to permanently set environmental and global settings. Most settings can be changed temporarily using other options, such as under the Environment dialog-box.

Configuration options are saved in configuration files (with the default name *config.pro*). Configuration files can be defined in a variety of directories. When Pro/ENGINEER is first launched, it reads configuration files in order from the locations listed below. The last settings read from a configuration file are the ones that Pro/ENGINEER utilizes. Pro/ENGINEER has default values for unspecified options which are stored in a protective file name *config.sup*.

1. **Pro/ENGINEER LOADPOINT** The loadpoint directory is the location where Pro/ENGINEER is installed. A configuration file saved here is loaded first.

2. **LOGIN DIRECTORY** The login directory is the home directory for a login ID. This configuration file is read after the loadpoint directory and is used by users to save individualized configuration options.

3. **STARTUP DIRECTORY** The startup directory is the working directory for the current object. This is the last configuration file read by Pro/ENGINEER and any settings that are read from this file will override settings from previous configuration files.

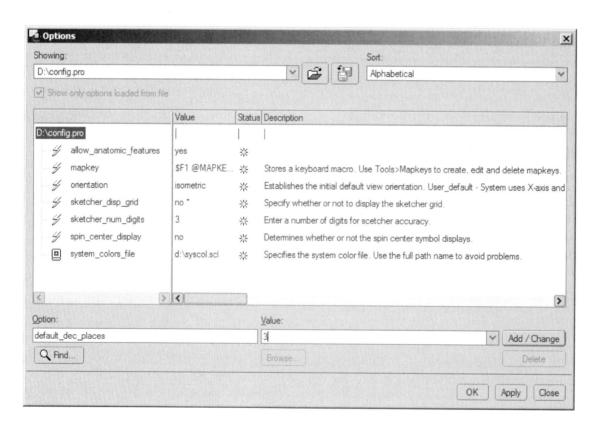

Figure 2–30   Configuration Options

MAPKEYS

Mapkeys are keyboard macros of frequently used command sequences. A possible use of a mapkey would be defining a macro for the command sequence to save a part file and then delete old version of the part. Without a defined mapkey, the following command sequence would be required to perform this task:

File >> Save >> Enter >> File >> Delete >> Old Versions >> Enter.

A Mapkey can be used to perform these steps in one keyboard selection.

The Mapkeys dialog box (Figure 2–31) is used to create and edit mapkeys. The following options are found under this dialog box:

- **New** The New option is used to define a new mapkey.
- **Modify** The Modify option is used to modify an existing mapkey.
- **Run** Run is used to execute an existing mapkey. Mapkeys can be run outside of the Mapkey dialog box.
- **Delete** The Delete option is used to delete an existing mapkey.
- **Save** The Save option is used to save the mapkey to the configuration file. Saving a mapkey will allow it to be used with other Pro/ENGINEER sessions.

DEFINING MAPKEYS

Mapkeys are created by recording keyboard command sequences. Pauses to allow for the selection of objects on the work screen are included automatically in the recording process. Perform the following steps to create a mapkey:

STEP 1: **Select TOOLS >> MAPKEYS on Pro/ENGINEER's Menu Bar.**

The Mapkeys dialog box will appear (Figure 2–31).

STEP 2: **Select NEW from the Mapkeys dialog box.**

The Record Mapkey dialog box will appear.

STEP 3: **Enter a key sequence, mapkey name, and mapkey description.**

The key sequence is the keyboard entry that will execute the mapkey. For function keys, enter a dollar sign in front of the function key sign (e.g., $F2).

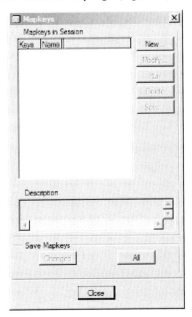

Figure 2–31 Mapkeys Dialog Box

STEP 4:   **Select RECORD and select command sequence from the keyboard and from Pro/ENGINEER's menu.**

After selecting Record, enter command sequence just as you would if you were performing the function. You can enter pauses to allow for the entry of sequences outside of the macro.

STEP 5:   **Select STOP to end the recording of a mapkey.**

STEP 6:   **Select OK on the Record Mapkey dialog box.**

STEP 7:   **Select SAVE on the Mapkey dialog box to permanently store the mapkey.**

Mapkeys are saved in the current configuration file (see Figure 2–30).

LAYERS

Layers are used to organize features and parts together to allow for the collective manipulation of all included items. Features and parts may be included on more than one layer. The Layer model tree on the navigator window is used to create and manipulate layers (Figure 2–32). The New Layer option is used to create new layers. Options are available for setting features and parts to selected layers or to remove items from a layer and to control how layers are displayed in a model.

CREATING A LAYER

Perform the following steps to create a new layer:

STEP 1:   **On the Navigator window, select SHOW >> LAYER TREE.**

The Navigator window allows for the display of model tree data or layer data.

STEP 2:   **On the Navigator window, select the LAYER >> NEW LAYER option.**

STEP 3:   **On the Layer Properties dialog box, enter a Name for the new layer (Figure 2–32).**

STEP 4:   **Select OK on the New Layer dialog box to create the new layer.**

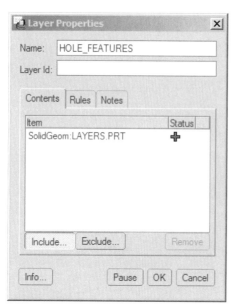

Table 2–3   Item Types Available to Include on a Layer

Item Type	Items Included
FEATURE	All features
AXIS	Datum axes and cosmetic threads
GEOM_FEAT	Geometric features
DATUM_PLANE	Datum planes
CSYS	Coordinate systems
DIM	Dimensions
GTOL	Geometric Tolerances
POINT	Datum points
NOTE	Drawing notes

Figure 2–32   Layers Dialog Box

Setting Items to a Layer

Perform the following steps to set an item or feature to a layer:

Step 1:   On the Navigator window, select SHOW >> LAYER TREE.

Step 2:   Select the Layer for defining then access the pop-up menu.

Step 3:   Select LAYER PROPERTIES on the pop-up menu.

Step 4:   On the Layers Properties dialog box, select the INCLUDE button.

Step 5:   Either on the work screen or the model tree, pick items to add to the layer.

Examples of items that can be added to a layer include features, curves, and quilts.

Step 6:   Select OK.

Default Layers

Default layers may be created for all new objects. Created items can be automatically placed on predetermined default layers. The following configuration file option is used to create a default layer and to automatically add an item type to the default layer:

Def_layer *item-type layername*

Def_layer is a configuration file option. This option may be used multiple times to correspond to as many default layers as might be needed in an object. The *item-type* value is used to specify a type of item to include on a default layer, while the *layername* value is the name of the layer that will be created. Table 2–3 shows a partial list of item name values that can be used as an *item-type*.

SELECTING FEATURES AND ENTITIES

Parts with many features, and assemblies with many parts, can make selecting entities, features, and objects difficult. Entities, features, and objects may be selected on the model tree or on the work screen. Common practices in Wildfire recommend the preselection of entities before performing an option. Pro/ENGINEER provides selection filters at the right end of the message area. Available filters include Smart, Geometry, Quilts, Datums, Annotations, and Features. The Smart option provides access to commonly picked entities within the context of a selected command. Pro/ENGINEER also allows for the querying of valid entities through a query list (see Figure 2–33). The Pick From List option is available with the right-mouse buttons pop-up menu.

Figure 2–33
Selections Dialog Box

SUMMARY

Pro/ENGINEER provides a variety of tools for interfacing with the modeling environment. Most of these tools are readily available on the toolbar. Other options can be found under the menu bar.

Pro/ENGINEER's data-based management system manipulates files differently than standard Windows applications. Options are available for saving and backing up object files. A unique feature of Pro/ENGINEER is that when it saves, a new version of the object file is created.

A variety of view manipulation options are available within Pro/ENGINEER. Objects can be dynamically zoomed, rotated, and panned. Views can be oriented and saved for later use. Additionally, the display of an object can be represented in one of four possible ways: wireframe, hidden line, no hidden, and shaded. Also, features and entities may be placed on layers and hidden from display.

Pro/ENGINEER's interface provides customization tools. The configuration file is a powerful tool for personalizing the work environment. Mapkeys can be created that provide a shortcut to commonly used menu pick sequences.

PRO/ENGINEER INTERFACE TUTORIAL

This tutorial will provide instruction for the establishment of a Pro/ENGINEER object. The part (*interface.prt*) shown in Figure 2–34 will serve as the part to be manipulated in this tutorial. This tutorial will cover:

- Manipulating an existing object file
- Viewing a Pro/ENGINEER object
- Setting up an object's units and materials
- Changing an object's tolerance display
- Creating a geometric tolerance
- Creating layers and setting items

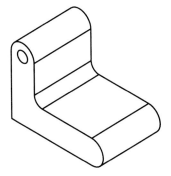

Figure 2–34
The Interface Part

OPENING AN OBJECT

This segment of the tutorial will cover the opening and saving of a Pro/ENGINEER part file. Highlighted within this segment will be the saving of a part as a different name by using the **Save As** option.

STEP 1: **Start Pro/ENGINEER**

STEP 2: **Use the FILE >> SET WORKING DIRECTORY option to change Pro/ENGINEER's working directory to a location where you have read/write privileges.**

From the File menu, select Working Directory then change the working directory to the directory where your part file will be saved.

STEP 3: **Use the FILE >> OPEN option to open the part *interface.***

Supplemental parts for this textbook can be found on the book's Web site at http://www.mhhe.com/kelley. Select the File >> Open option then manipulate the Look In box to find the location of this part file.

STEP 4: **Select FILE >> SAVE A COPY and save the part with the name *interface_part.***

If your working directory is different from the directory where the original *interface.prt* file is located, notice how the Save A Copy command is defaulting to save to the currently working directory. The Save A Copy command will allow you to change the directory and object name.

Notice, after performing this step, how the original *interface* part file is still active. The Save A Copy command will allow for a name change, but will keep the original object active.

STEP 5: **Use FILE >> ERASE >> CURRENT to erase the current object from memory.**

The Erase >> Current command will erase the *interface* part from session memory.

STEP 6: Use the FILE >> OPEN option to open the newly created part
interface_part.

Open the part file created in Step 4. This part will now be the active object in
session memory.

VIEWING THE OBJECT

This segment of the tutorial will introduce Pro/ENGINEER's viewing options. Covered
under this segment will be shading a model, dynamically rotating and zooming a model,
and setting the default orientation.

STEP 1: Select TOOLS >> ENVIRONMENT on the menu bar.

STEP 2: Ensure that the SPIN CENTER option is not checked then select OK.

Figure 2–35 Shading
a Model

STEP 3: Select the SHADING icon on the Toolbar.

As shown in Figure 2–35, select the Shade icon from the toolbar.

STEP 4: Dynamically rotate the model.

While selecting the mouse's middle button, dynamically rotate the object by
moving the cursor on the work screen. Notice how the center of spin for the
dynamic rotation is based on the location of the mouse when the middle
mouse button is selected. This is due to the Spin Center option not being
checked on the Environment dialog box (Step 2).

STEP 5: Dynamically zoom the model.

If your mouse is equipped with a scroll wheel, spinning the wheel will zoom
in and out on a model. Optionally, simultaneously selecting the keyboard's
Control key and the middle mouse button will dynamically zoom the object
by moving the cursor on the work screen. Moving the cursor from the bottom
of the screen toward the top will zoom out. Moving the cursor from the top of
the screen toward the bottom will zoom in on the model.

STEP 6: On the Environment dialog box (TOOLS >> ENVIRONMENT), set
Isometric as the object's default display setting (Figure 2–36).

Using the Environment dialog box found under the Tools menu option, select
Isometric as the default orientation then select OK.

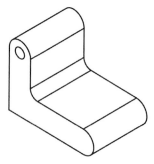

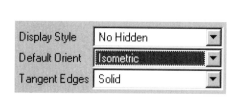

Figure 2–36 Model Display   **Figure 2–37** Isometric View of Model

Step 7:   Use VIEW >> ORIENTATION >> STANDARD ORIENTATION to change to the default orientation.

The default orientation is shown in Figure 2–37.

SETTING AN OBJECT'S UNITS

This segment of the tutorial will set the units for the part. A part's units can be set at any time during the modeling process. This segment will cover creating a new system of units and changing the units for your part.

Step 1:   Select EDIT >> SETUP on Pro/ENGINEER's Menu bar.

Step 2:   Select UNITS on the Menu Manager.

The Units option will display the Units Manager dialog box. This dialog box allows new systems of units to be created and/or selected.

Step 3:   Select the NEW option on the Units Manager dialog box.

As shown in Figure 2-38, select New from the dialog box. New will allow for the creation of a user defined System of Units.

Step 4:   Enter *UNITS* as the system name.

Step 5:   Enter the unit values shown in Figure 2–39.

Figure 2–39 displays the units to be set for this system. Enter these values as shown.

Step 6:   Select OK to close the Systems of Units Definition dialog box.

Step 7:   Select the newly create *UNITS* system on the Units Manager dialog box.

Step 8:   Select the SET option on the Units Manager dialog box.

Step 9:   Select INTERPRET EXISTING NUMBERS on the Warning dialog box then select OK

The *Interpret Existing Numbers (Same Dims)* option will interpret the existing unit numbers as the new unit numbers. As an example, a value of 2 in will be converted to a value of 2 mm.

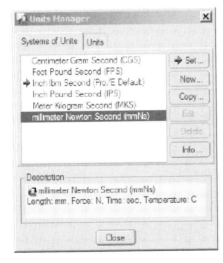

Figure 2–38   Units Manager Dialog Box

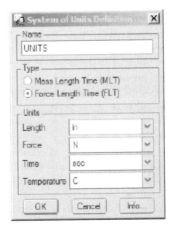

Figure 2–39   New System of Units

STEP 10:   **Close the Units Manager dialog box.**

STEP 11:   **Save your object.**

From the File menu, select Save. Enter the object to save.

ESTABLISHING LAYERS

This segment of the tutorial will establish layers. The model used in this tutorial is composed of datums and geometric features. You will create a layer for all geometric features and a layer for all datums.

STEP   1:   **On the Navigator window, select the SHOW >> LAYER TREE option (Figure 2–40).**

The Navigator displays the model tree and the layer tree in the left-most window.

STEP   2:   **Select LAYER >> NEW LAYER on the Navigator.**

After selecting the New Layer option, the Layer Properties dialog box will appear on the work screen (Figure 2–41).

STEP   3:   **On the New Layer dialog box, enter *GEOMETRIC_FEATURE* as a layer name (Figure 2–41).**

STEP   4:   **Select the INCLUDE option on the Layer Properties dialog box.**

STEP   5:   **Change the Selection Filter on the status bar to *Solid Geometry* (see Figure 2–42).**

The current model has multiple selectable entities (e.g., datum planes, coordinate system, axes). The filtering out of all elements except for solid geometry will help with the picking of entities for the current layer.

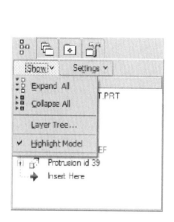

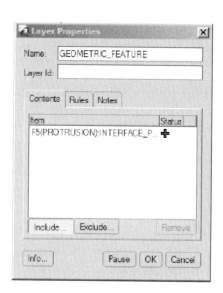

Figure 2–40   The Layers Menu   **Figure 2–41**   The New Layer Dialog Box

Figure 2–42   Selection Filter

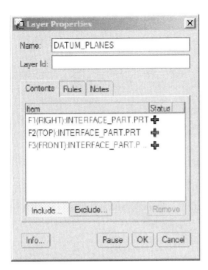

Figure 2–43 Datum Planes Layer Properties

Figure 2–44 Holes Layer Properties

Figure 2–45 Hole Selection

STEP 6: On the work screen, pick the geometry of the part.

STEP 7: Select OK on the Layer Properties dialog box.

STEP 8: Create a new layer that includes all the models datum planes (Figure 2–43).

Consider the following when performing this step:

- Name the layer *Datum_Planes*.

- Filter out all selectable entities except for Datum Planes.

- To include elements, on the navigator window, toggle back to the Model Tree display and pick the datum planes on the model tree.

STEP 9: Create a new layer that includes the protrusion's hole feature (Figure 2–44).

Consider the following when performing this step:

- Name the layer *Holes*.

- Filter out all selectable entities except for features (Feature).

- To include elements, access the right-mouse button's pop-up menu's Pick From List option then pick the Hole feature on the Selection dialog box (Figure 2–45).

POP-UP MENUS The Pick From List option allows for the selection of hard-to-reach entities. It is accessible through the right mouse button's pop-up menu. To access any mouse pop-up menu in Pro/ENGINEER, right-mouse select over the relevant area or option. The menu will appear after a slight delay.

STEP 10: With the Layer Tree visible in the navigator window, right-mouse select over the Datum_Planes layer to access the layer pop-up menu.

STEP 11: On the pop-up menu, select the BLANK LAYER option.

The Blank Layer option will hide selected layers.

STEP 12:   **Select the Refresh Screen icon on the toolbar.**

Your datum planes should be blanked.

STEP 13:   **Save your model.**

From the File menu, select the Save option. Enter the object to save.

PROBLEMS

1.   Create a material defining 1040 steel. Save this material to disk as a material file.

2.   Create a material defining aluminum. Save this material to disk as a material file.

3.   Create a system of units with the following configuration:

    •   Length = cm   •   Time = micro-sec

    •   Mass = kg   •   Temperature = K

4.   Create a configuration file with the following options and settings:

    •   *BELL* option set to NO

    •   *TOL_DISPLAY* option set to YES

    •   *TOL_MODE* option set to NOMINAL

    •   *SKETCHER_INTENT_MANAGER* option set to YES

    •   *SKETCHER_NUM_DIGITS* option set to 3

QUESTIONS AND DISCUSSION

1.   Describe the difference between the Backup command and the Save A Copy command? Give some examples of when each option would be appropriate.

2.   With the object filename *revolve.sec.2*, what does the number 2 represent? What does *sec* represent?

3.   What is a Pro/ENGINEER working directory?

4.   What file management option is used to close an object from Pro/ENGINEER's memory?

5.   Describe the process for creating a material file that can be utilized by other Pro/ENGINEER parts.

6.   How does Pro/ENGINEER's ANSI tolerance standard differ from its ISO tolerance standard?

7.   What is a Mapkey and how is it used in Pro/ENGINEER?

8.   What is the purpose of Pro/ENGINEER's configuration file?

3

CONSTRAINT-BASED SKETCHING

Introduction

Sketching is a fundamental skill within any parametric design application, including Pro/ENGINEER. Geometric features that are created as protrusions or cuts require the use of a sketch to define the features' section. Other features such as datum curves, trajectories for swept features, and sketched holes also require the use of sketching to define elements. A sketched Pro/ENGINEER section can be created within a feature creation option (e.g., Extrude, Revolve), or it can be created as a separate object, then used within later construction techniques. This chapter will cover the fundamental sketching techniques required to construct a Pro/ENGINEER feature. Upon finishing this chapter, you will be able to:

- Start a new Pro/ENGINEER object utilizing Sketch mode.
- Establish Pro/ENGINEER's sketcher environment.
- Sketch a section using Intent Manager.
- Apply dimensions to a Pro/ENGINEER sketch.
- Apply relations to section dimensions.
- Use geometric construction tools to create a sketch.
- Apply constraints to section entities.

DEFINITIONS

Constraint	An explicit relationship that exists between two sketched entities.
Entity	An element within the sketcher environment, such as a line, arc, or circle.
Parametric dimension	A dimension that is used as a parameter to define a feature.
Reference	An existing feature entity, such as a part edge or datum, used within a sketcher environment to construct a new feature. The referenced entity becomes a parent of the feature being constructed.
Section	A combination of sketched entities, dimensions, and constraints that defines a feature's basic geometry and intent.
Sketch	The entities of a section that define the basic shape of a feature.

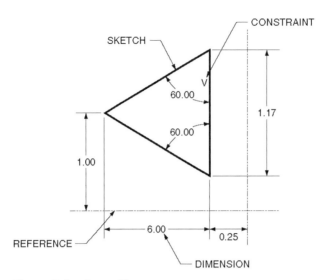

Figure 3–1   Section Elements

Sketcher environment	The environment within a feature creation process where sections are defined.
Sketch mode	A Pro/ENGINEER mode for the creation of sections. Sections may be created in Sketcher mode and placed in a sketcher environment at a later time.

FUNDAMENTALS OF SKETCHING

Sketched entities combined with dimensions, constraints, and references form a section (Figure 3–1). There are two categories of sections: those used to directly create a feature and those created in Sketch mode. Extrudes, Revolves, Sweeps, and Ribs are examples of features in Part mode that require a sketched section. In addition to Part mode, when creating a new Pro/ENGINEER object, Sketch is one of the modes that is available. Elements created in Sketch mode are saved with an *.sec file extension. When you are creating a feature in Part mode, selecting Save while sketching will save the section, not the part being created. A saved section can be inserted in a sketcher environment by using the Data From File option.

CAPTURING DESIGN INTENT

Pro/ENGINEER requires a fully defined section before completing a feature. The sketcher environment provides a variety of tools that will fully define a sketch and capture design intent. When sketching a section, take care to ensure that the intent of a design is met through the definition of the feature. The following are sketcher tools that can be used to capture a design's intent:

DIMENSIONS

Dimensions are the primary tool for capturing the intent of a design. Within a section, dimensions are used to describe the size and location of entities.

CONSTRAINTS

Constraints are used to define the relationship of section entities to other entities. As an example, a constraint might make two lines equal in length or it might confine two lines parallel to each other.

REFERENCES

When a feature is constructed in a sketcher environment, a section can reference existing features of a part or assembly. References can consist of part surfaces, datums, edges, or axes. An example of a reference would be aligning the end point of a sketched entity with the edge of an existing feature. Another example would be creating sketch entities from existing edges by utilizing the Use Edge option. Sections created in Sketch mode cannot utilize external feature references.

RELATIONS

Relationships can be established between two dimensions. Most algebraic and trigonometric equations can be utilized to form a mathematical relationship. Also, conditional statements can be applied to create a relationship.

SKETCHING ELEMENTS

Users of two-dimensional computer-aided design applications are accustomed to entering precise values for geometric elements. With low- and midrange CAD packages, if a line is 2 in in length, then it is drawn exactly 2 in in length. Many of the sketching tools found within Pro/ENGINEER resemble two-dimensional drafting options. Despite this, within Pro/ENGINEER, it is not important to precisely draw a section. Instead of creating exact elements, you can sketch geometry as you might sketch freehand. The following are guidelines that are important in sketching a section:

- In sketching a section, the shape of the sketch is important, not the size.
- In creating a section, the dimensioning scheme should match design intent.
- In creating a section, geometric constraints should match design intent.

After a section is sketched and dimensions and constraints are applied, it is unlikely that the section will be the correct size. The sketcher environment has dimension modification tools to solve this problem within Pro/ENGINEER, when a dimension is modified and the section is regenerated, if no conflicts exist, the sketch adjusts to the size of the dimension values.

SKETCH PLANE

When a section is constructed for the direct creation of a feature (e.g., by the Extrude command), the section has to be sketched on a plane. The only exception to this rule is when a geometric feature will be the first feature of a part; the sketch for this situation is not placed on a plane.

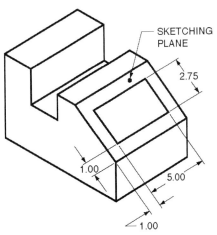

Figure 3–2   *Sketching on a Part Surface*

There are two types of planes that can be sketched on. The first type is a flat feature surface or plane (Figure 3–2). The second type is a datum plane. Datum planes can provide a suitable surface for sketching upon when one does not exist on a part. Datum planes can be completed on the fly while geometry features are being constructed. On-the-fly datum planes are not considered features and will not be displayed after the creation of the geometry under construction.

SECTION TOOLS

The sketcher environment is a customizable mode used to create Pro/ENGINEER sections. Initially, the sketcher environment is set to allow for two-dimensional sketching with grids on and Intent Manager activated. Each of these environmental options can be temporarily or permanently set to different values. As an example, a user can opt to sketch in three-dimensional mode by toggling off the Use 2D Sketcher option found under the Environment dialog box. Other section tools available include placing sections and toggling the display of vertices, constraints, and dimensions.

GRID OPTIONS

Grid parameters can be modified on the Sketcher Preferences dialog box (Figure 3–3). This dialog box is available by selecting Sketch >> Options on Pro/ENGINEER's menu bar while within a sketcher environment. The following options are available:

GRID DISPLAY

Grids can be toggled on or off by using the Grid On/Off toggle located on the Sketcher Preferences dialog box. Also, as shown in Figure 3–4, grids can be toggled

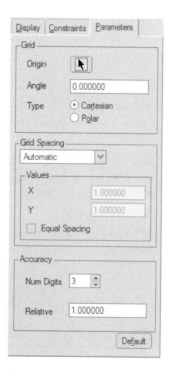

GRID DISPLAY

Figure 3–3 The Sketcher Preferences Dialog box

Figure 3–4 Grid Display Option

on or off using the Section Tools icon located on Pro/ENGINEER's main Toolbar menu.

GRID TYPE

The Parameters tab of the Sketcher Preferences dialog box allows for the selection of a Cartesian grid or a Polar grid. The Cartesian grid is the default.

GRID ORIGIN

The Origin option under the Parameters tab of the Sketcher Preferences dialog box allows the grid absolute origin to be moved. It can be set at the center or end of an entity, at a point and coordinate system, or at an entity vertex.

GRID PARAMETERS

The Grid Spacing option under the Parameters tab of the Sketcher Preferences dialog box allows for the modification of Cartesian X and Y direction grid spacing or Polar grid angle and distance spacing.

PLACING SECTIONS

Sections created within Part, Assembly, or Sketcher mode can be saved and used in other sketcher environments. When sketching a section, the File >> Save option is used to save the section. Perform the following steps to place a predefined section into a current sketcher environment:

STEP 1: In a Sketcher environment, from Pro/ENGINEER's menu bar, select SKETCH >> DATA FROM FILE >> FILE SYSTEM.

STEP 2: Select a section utilizing the OPEN dialog box.

The Open dialog box will default to the current working directory to allow for the selection of Pro/ENGINEER section files. You can move up or down on the directory tree to locate a section or you can select a new drive.

STEP 3: If necessary, adjust the handle points for moving, rotating, or scaling (see Figure 3–5) by right-mouse-button picking the desired handle, followed by dragging the handle to a new location.

STEP 4: Place the section by dragging the moving handle point to the desired location.

The selected section can be dynamically placed into the current sketcher environment. Place the section in a way that will allow for proper alignments when the section is regenerated.

STEP 5: If necessary, rotate the section by either dragging the rotation handle point or by adjusting the Rotate value in the Scale Rotate dialog box (Figure 3–5).

STEP 6: If necessary, enter a Scale value in the Scale Rotate dialog box.

STEP 7: When the section has been placed, select the Accept Changes checkmark on the Scale Rotate dialog box.

MODELING POINT For a new object, the display can be zoomed out to such an extent that the section cannot be readily recognizable. Use dynamic viewing (Control key and middle mouse button) to zoom in on the section being placed.

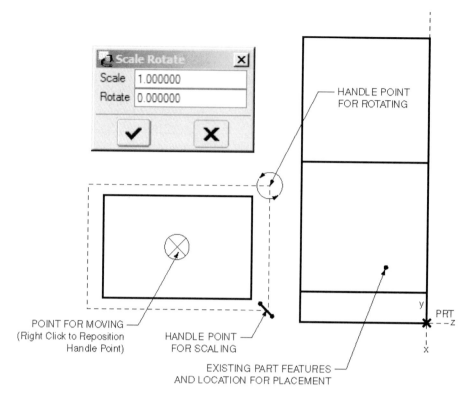

Figure 3–5   Placing a Section

Section Information

In a sketching environment, the Analysis menu provides options for obtaining information about entities within the current sketcher environment. The following are some of the available options:

- **Distance** The Distance option will measure the distance between two parallel lines, between two points, or between a point and a line.
- **Angle** The Angle option will measure the angle between two entities.
- **Entity** The Entity option will obtain information such as geometry type, endpoint tangencies, and endpoint coordinates on a selected entity. To obtain endpoint coordinate values, a coordinate system must exist.
- **Intersection Point** The Intersection Point option will locate the intersection point of two entities. For entities that do not physically intersect, an interpolated intersection point will be obtained.
- **Tangency Point** The Tangency Point option provides the location where two touching entities are tangent. For entities that do not touch, it will show the location on each entity that has the same slope.
- **Curvature** The Curvature option graphically displays the relative curvature of an entity, especially curved entities such as arcs and splines.

Constraints

Constraints are used within Pro/ENGINEER to capture design intent. A constraint is a defined relationship that exists between two geometric entities. An example would be constraining two lines parallel to each other. Figure 3–6 shows examples of several types of constraints.

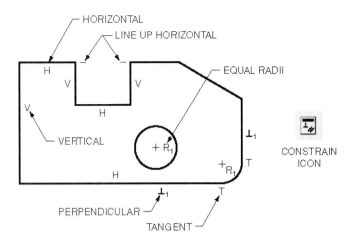

Figure 3-6 Constraint Examples

CONSTRAINTS WITH INTENT MANAGER

Pro/ENGINEER creates constraints on the fly while you are sketching. When an entity being sketched comes within a predefined tolerance value of meeting a constraint, such as being close to parallel to another entity, the entity is snapped to that constraint. Constraints can be applied manually using the Constraints icon on the sketcher toolbar or the Sketch >> Constrain option on the menu bar.

Constraints are considered weak or strong. Constraints created by Pro/ENGINEER during the sketching process are weak and can be overridden by a user-placed constraint or by a user-placed dimension. Constraints created through the use of the Constraints dialog box are strong and remain on the sketch unless explicitly deleted. Weak constraints can be converted to strong constraints by using the Edit >> Convert To >> Strong option.

CONSTRAINTS WITHOUT INTENT MANAGER

Intent Manager can be toggled off by unselecting the Sketch >> Intent Manager option. With Intent Manager deactivated, constraints are applied to sections after a successful re-generation. These constraints are based on the following assumptions:

- When arcs and circles are sketched with approximately the same radius, Pro/ENGINEER will assume that the radii are equal.
- When a line is sketched approximately horizontal or vertical, Pro/ENGINEER will assume that it is horizontal or vertical.
- When a line is sketched approximately parallel to another line, Pro/ENGINEER will assume that the two entities are parallel.
- When a line is sketched approximately perpendicular to another line, Pro/ENGINEER will assume that the two entities are parallel.
- When two entities are sketched approximately tangent to each other, Pro/ENGINEER will assume that the two entities are tangent.
- When lines are sketched approximately the same length, Pro/ENGINEER will assume that the lines are of equal length.
- When entities are sketched approximately symmetrical around a centerline, Pro/ENGINEER will assume that the entities are symmetrical.
- When a point entity (such as the endpoint of a line) is sketched near an arc, circle, or line, Pro/ENGINEER will assume that the point entity lies on the arc, circle, or line.
- When the centers or endpoints of arcs are sketched at approximately the same location, Pro/ENGINEER will assume that they have the same X and/or Y coordinate values.

Table 3–1 Constraint Options

Constraint Type	Symbol	Use
Same Points	⊕	The Same Points option is used to make two points coincident.
Horizontal	H	The Horizontal option is used to constrain a line horizontal.
Vertical	V	The Vertical option is used to constrain a line vertical.
Point On Entity	Q	The Point on Entity option is used to constrain a point on a selected entity.
Tangent	T₁	The Tangent option is used to constrain two entities tangent to each other.
Perpendicular	⊥	The Perpendicular option is used to constrain two entities perpendicular to each other.
Parallel	//₁	The Parallel option is used to constrain two entities parallel to each other.
Equal Radii	R₁	The Equal Radii option is used to constrain the radii of two arcs or curves equal to each other.
Equal Lengths	L₁	The Equal Lengths option is used to constrain two entities to have equal lengths.
Symmetric	→←	The Symmetric option is used to make entities symmetrical around a centerline.
Line Up Horizontal	--	The Line Up Horizontal option is used to line up two vertices horizontally.
Line Up Vertical	I	The Line Up Vertical option is used to line up two vertices vertically.
Collinear	—	The Collinear option is used to make two lines collinear.

With Intent Manager deactivated, there is no option for a user to apply a constraint after the regeneration of a section.

CONSTRAINT OPTIONS

Intent Manager allows for the dynamic application of constraints to sketched entities. Constraints applied through the Constraints Dialog box are considered strong and will override weak constraints and/or dimensions. Table 3–1 shows constraints that can be applied to sketcher elements.

SKETCHER DISPLAY OPTIONS

As shown in Figure 3–7, when the sketcher environment is entered, seven additional icons are added to the Toolbar menu. Five of these icons control the display of items exclusively within the sketcher environment. Each display option is also available under the Sketcher Preferences dialog box (Sketch >> Options). The following is a description of each icon option.

UNDO AND REDO

The Undo icon will undo the last executed sketching function (e.g., Line, Delete, Constrain), while the Redo icon will redo a function that was undone with the Undo option.

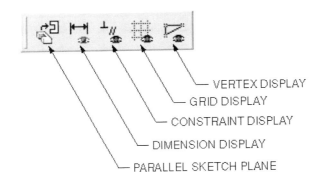

Figure 3–7   Sketcher Display Options

PARALLEL SKETCH PLANE

The default sketching environment is oriented parallel to the display screen. Since a sketcher environment can be dynamically rotated with the middle mouse button, the Parallel Sketch Plane icon will return the sketch plane to its default two-dimensional orientation.

DIMENSION DISPLAY

The Dimension Display icon controls the display of dimensions within the sketcher environment. This option allows dimensions to be toggled on or off.

CONSTRAINT DISPLAY

Constraint symbols are shown directly on the sketch. The Constraint Display icon controls the display of constraint symbols. Using this option, constraint symbols can be toggled on or off.

GRID DISPLAY

The Grid Display icon controls the display of grids within the sketcher environment. Often, grids can cause visualization problems while you are sketching a section.

VERTEX DISPLAY

Vertices are displayed at the end of sketcher entities. The Vertex Display icon controls the display of these vertices. Selecting this icon will turn off all vertices.

SKETCHING WITH INTENT MANAGER

Intent Manager is used by Pro/ENGINEER to facilitate the capturing of a part's design intent. It is activated by default. If desired, it can be deactivated with the Sketch >> Intent Manager option or it can be deactivated permanently by setting the configuration file option *sketcher_intent_manager* to a value of No. Intent Manager provides several sketching enhancements.

FULLY DEFINED SECTION

Pro/ENGINEER is a parametric modeling application. Parametric packages require sections to be fully dimensioned and constrained. Within conventional two-dimensional drafting standards, if two lines appear parallel or perpendicular, then they are assumed to be so. Within parametric modeling, a parallel or perpendicular constraint has to be applied. Intent Manager attempts to keep a sketch fully defined by applying dimensions and constraints during the sketching process. Additionally, Intent Manager will not allow a section to be overdefined.

CONSTRAINTS

Constraints are applied during the sketching process. Additionally, constraints can be applied with the Constraints dialog box and removed with the Delete key.

ALIGNMENTS AND REFERENCES

With Intent Manager deactivated, Pro/ENGINEER requires entities to be aligned to existing features using the Alignment option. Intent Manager requires that before the start of sketching, existing features are specified as references. Normally, the Specify Reference option is used for this purpose.

AUTOMATIC DIMENSIONING

Intent Manager applies dimensions automatically upon ending a sketch option. Dimensions initially created by Pro/ENGINEER are weak and can be overridden by user-placed dimensions and/or constraints.

Pro/ENGINEER does not know the intent for a design. Because of this, it does not know the correct dimensioning scheme. Since Pro/ENGINEER objects have stored parameters that can be accessed by other modules (e.g., Manufacturing and Assembly), proper placement of dimensions is important for capturing design intent. With the exception of simple geometric shapes, it is likely that dimensions will have to be changed on a sketch. When the Dimension option is used to change the dimensioning scheme, weak dimensions and constraints are subject to being overridden.

ORDER OF OPERATIONS

There is an established procedure for sketching a section with Intent Manager. For most sketching situations, the following steps should be followed:

STEP 1:   **Specify references.**

The position of a sketched section must be located with respect to existing part features. The Reference dialog box is used to identify references to use within the sketcher environment. Part edges, datum planes, vertices, and axes can be selected as references.

MODELING POINT   The References dialog box is used when you are creating a section within a feature creation option (e.g., Extrusions or Revolves). Since sketches created within Sketcher mode are stand-alone sections, there will be no references available and this step will be skipped. Additionally, when you are creating a Protrusion as the first feature of a part, references will not be available. When specifying references, pick references that correspond to design intent. Pro/ENGINEER will apply location dimensions from references that you specify.

STEP 2:   **Sketch the section.**

Sketch and/or construct the section using appropriate sketching tools.

STEP 3:   **Apply dimensions and constraints that match design intent.**

Intent Manager will apply dimensions and constraints automatically. These dimensions and constraints will be considered weak and can be overridden by the creation of new constraints or by adding new dimensions.

Step 4:   **Modify dimension values.**

Use the Modify option to individually change dimension values.

Step 5:   **Add dimension relationships (optional step).**

If necessary, use the Relations dialog box (Sketch >> Relations) to create dimension relationships.

Sketching Entities

Pro/ENGINEER's sketcher environment provides a variety of options for sketching two-dimensional entities. For experienced two-dimensional computer-aided design users, these tools will resemble basic geometry creation techniques.

Sketching Lines

Lines are created with the Line icon or with the Sketch >> Line option. Three entity types are available: Line, Centerline, and Two Tangents (Figure 3–8). Line is the default selection and is used to create feature entities. Centerline entities are used for construction techniques only (e.g., the axis of revolution for a revolved feature). The Two Tangent option creates a line entity tangent to two curved entities.

Sketching Arcs

Multiple arc creation tools are available within a sketcher environment. The following narrative and Figure 3–9 are a description of each:

- **Tangent Creation** The Tangent End option creates an arc that is tangent to the endpoint of a selected entity. The first endpoint of the arc or both endpoints can be selected as tangent points.
- **Concentric Arc** The Concentric Arc option places an arc concentric to an existing arc or circle.
- **Circular Fillet** The Circular Fillet option is used to create a fillet between two selected entities.
- **3 Point Arc** The 3 Point Arc option is used to create an arc by selecting the endpoints of the arc and then a third point on the arc.
- **Center/Ends Arc** The Center/Ends Arc option is used to create an arc by selecting the center point of the arc and the two endpoints of the arc.

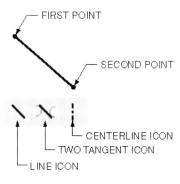

Figure 3–8   Line Construction

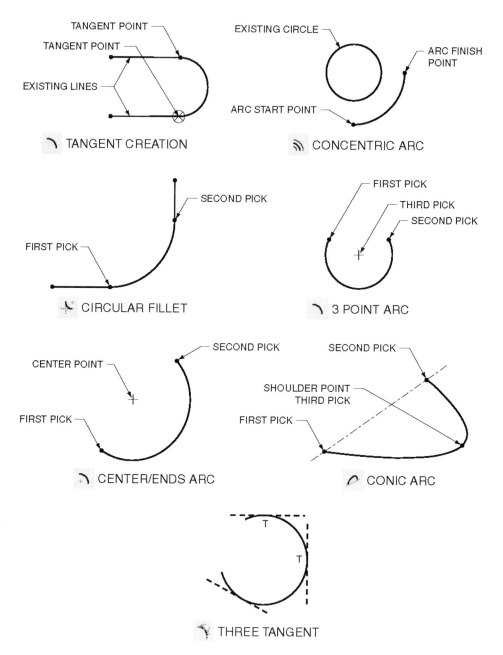

Figure 3–9   Arc Creation Options

- **Conic Arc** The Conic Arc option creates a cone-shaped entity through the selection of the conic's endpoints and the conic's shoulder point.
- **Three Tangent Arc** The Three Tangent Arc is used to create an arc tangent to three selected entities.

SKETCHING CIRCLES

Circles are created with available circle icons on the sketcher toolbar or though the Sketch menu on the menu bar. As shown in Figure 3–10, five sketching options are available.

- **Center/Point Circle** The Center/Point Circle option is used to create a circle by first selecting the circle's center point, then selecting a second point on the perimeter of the circle.

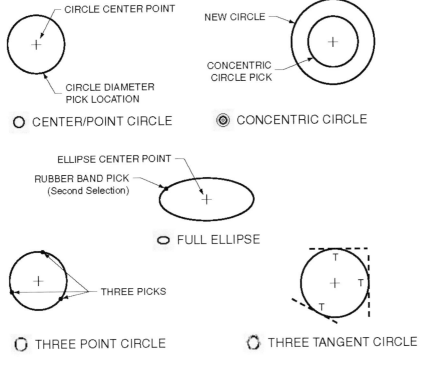

Figure 3–10 Circle Creation Options

- **Concentric Circle** The Concentric Circle option is used to construct a circle with a center point common to a second circle. The referenced circle can be a sketched entity or an entity referenced from an existing part feature. When you select an existing feature, this feature will become a section reference.

- **Full Ellipse** The Full Ellipse option creates an ellipse by first selecting the ellipse's center point, then selecting a point on the perimeter of the ellipse.

- **Three Point Circle** The Three Point Circle option creates a circle through the location of three points that lie on the perimeter of the circle.

- **Three Tangent Circle** The Three Tangent Circle option is used to create a circle tangent to three selected entities.

SKETCHING A RECTANGLE

A rectangle can be created by using the Rectangle icon or the Sketch >> Rectangle option. As shown in Figure 3–11, a rectangle is created by, first, selecting one vertex of the rectangle and, second, by selecting the opposite vertex.

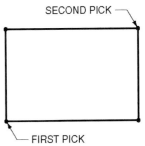

Figure 3–11
Creating a Rectangle

SPLINES

A spline is a variable-radius curve that passes through multiple interpolation points (see Figure 3–12). A spline is created by using either the Spline icon on the sketcher toolbar or the Sketch >> Spline option. To draw a spline, on the workscreen, pick control point locations that lie on the spline. A spline can be constructed tangent to sketched geometry or existing part features. If required, Intent Manager creates dimensions to define spline endpoints; the Dimension option can be used to manually create dimensions to additional control points. To modify a spline's interpolation points or control points, preselect the spline then access the pop-up menu's Modify option.

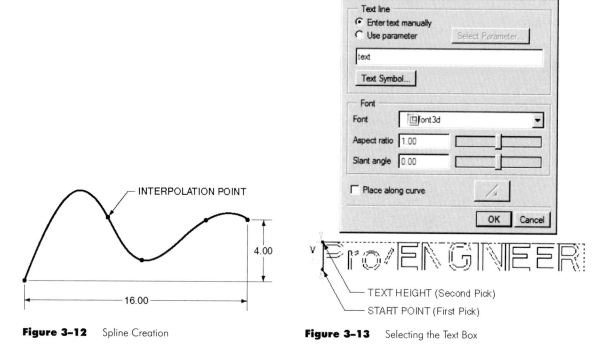

Figure 3–12   Spline Creation

Figure 3–13   Selecting the Text Box

SKETCHED TEXT

Text can be used in solid extruded Protrusion and Cut features, and with Cosmetic features. Sketched text is created by using the Sketch >> Text option on the menu bar or the Text icon on the sketcher toolbar. Perform the following steps to create text:

STEP 1:   Select the TEXT icon on the sketcher toolbar.

STEP 2:   On the workscreen, pick the start point for the string of text (Figure 3–13).

STEP 3:   On the workscreen, pick a point on the screen that will define the height of your text font.

This selection point will also define the orientation of your string of text.

STEP 4:   Enter text string.

STEP 5:   If necessary, on the Text dialog box, adjust the text font, aspect ratio, and slant angle (see Figure 3–13).

POINTS

Points created within a sketcher environment are extruded as axes. Points are created using the Sketch >> Point option on Pro/ENGINEER's menu bar.

ELLIPTICAL FILLET

An elliptical fillet is a sketch entity created between two selected entities. As shown in Figure 3–14, the fillet is created in an elliptical shape. To create an elliptical fillet, either use the Elliptical Fillet icon or the Sketch >> Fillet >> Elliptical option.

CONSTRUCTION ENTITIES

Normal sketch entities such as lines, arcs, circles, and splines can be converted to construction entities through the Edit >> Toggle Construction option (Figure 3–15) or the pop-up menu's Construction option. This is an extremely useful technique for creating complex

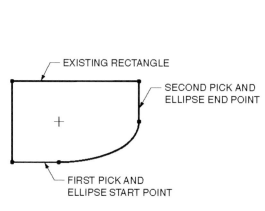

Figure 3–14 An Elliptical Fillet

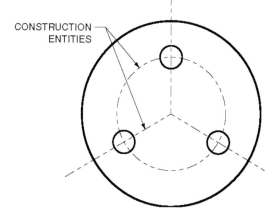

Figure 3–15 Construction Entities

shapes and features. Construction entities are used as references only and will not protrude with feature creation processes.

SKETCHING WITHOUT INTENT MANAGER

Intent Manager was first introduced with Release 20 of Pro/ENGINEER. The following are some of the differences between sketching without Intent Manager and sketching with Intent Manager:

- Without Intent Manager, constraints are applied after regeneration and are based on assumptions applied by Pro/ENGINEER.

- Without Intent Manager, dimensions have to be added manually by the user. The Automatic Dimensioning option is available, though, for fully defining a sketch.

- Without Intent Manager, a section does not remain fully defined during section modification.

- Without Intent Manager, section references are created from existing part features by using the Alignment option or by dimensioning to the edge of an existing feature.

- Without Intent Manager, mouse functions are available for sketching lines, circles, and arcs.

ALIGNMENTS

Pro/ENGINEER uses alignments to reference existing part geometry. As shown in Figure 3–16, the design intent for the endpoints of sketched entities might require them to lie on the edge of an existing feature. Without Intent Manager, the Alignment option is used to create section references. There are two ways that Alignment can be used to create a reference:

- A reference can be created by picking the part geometry, such as an edge, with the right mouse button, then entering the element as a known entity with the middle mouse button.

- Picking the part geometry, then picking the sketched entity can create a reference.

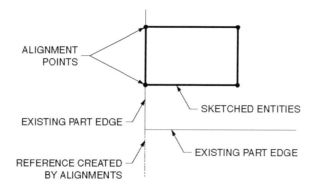

Figure 3-16 Aligning an Entity

ORDER OF OPERATIONS

Just as in sketching with Intent Manager activated, sketching without Intent Manager requires that an established procedure be followed. The following is the recommended order of operations for sketching without Intent Manager.

STEP 1:   **Sketch section.**

Sketch the section with tools found under the Sketcher menu. Additionally, geometric construction tools such as Use Edge and Offset Edge are available under the Geom Tools menu.

STEP 2:   **Align section entities to existing geometry (optional step).**

Use the Alignment option to align sketched entities to existing feature edges. This is an optional step and may not be needed for all sections. As an example, it can be used when a sketched cut does not intersect an existing feature edge.

STEP 3:   **Dimension the section, using appropriate dimensioning techniques.**

Use the Dimension option to dimension the sketch.

STEP 4:   **REGENERATE the section.**

Use the Regenerate option to regenerate the section. In the message area, the valued words *Section Regenerated Successfully* will identify when a section is fully defined (Figure 3–17). If a section is not fully regenerated, add additional dimensions and/or alignments.

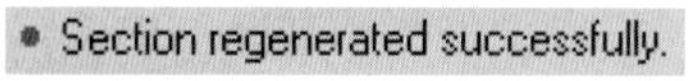

> ● Section regenerated successfully.

Figure 3-17 The Valued Message

MODELING POINT   If the automatic dimension option is used to fully define a section, ensure that the section's dimensioning scheme meets design intent before modifying any dimension values. This will help to ensure that the model's dimensions meet design requirements.

STEP 5:   **MODIFY dimension values.**

Use the Modify option to change any dimension values.

STEP 6:   **REGENERATE the section.**

Use the Regenerate option to update the section to the new dimension values.

STEP 7:   **Add a dimensional RELATION to the section (Optional Step).**

If necessary, create relationships between dimensions.

STEP 8:   **Regenerate the section.**

DIMENSIONING

Pro/ENGINEER provides a variety of dimensioning types for defining a sketch. Typical computer-aided design dimensions such as linear, radial, and angular are available. Additionally, dimensioning options such as perimeter and ordinate exist. All dimensioning types are available under the sketcher environment's Dimension option.

LINEAR DIMENSIONS

Linear dimensions are used to indicate the length of line segments and to measure the distance between two entities, such as two parallel lines or the distance from the center of a circle to a line. The following procedures describe how to dimension entities within the sketcher environment. Refer to Figure 3–18 for a graphical representation.

LINE SEGMENT

To dimension a line segment, select Dimension, then with the left mouse button pick the line to dimension. Next, with the middle mouse button, pick the location on the workscreen for the dimension placement.

PARALLEL LINES

To dimension the distance between two parallel lines, select Dimension, then with the left mouse button, pick the parallel line segments. Next, with the middle mouse button, pick the location on the workscreen for the dimension placement.

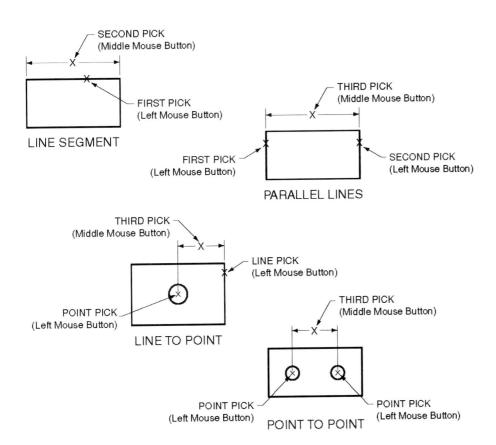

Figure 3–18 Dimensioning Options

LINE TO POINT

To dimension the distance between a line segment and a point, select Dimension, then with the left mouse button, pick the line and the point. Next, with the middle mouse button, pick the location on the workscreen for the dimension placement.

POINT TO POINT

To dimension the distance between two points, select Dimension, then with the left mouse button, pick the two points. Any dimension to a line entity will assume that the dimension will be oriented perpendicular to the line. With a point-to-point dimension, a menu provides the option for creating a horizontal, vertical, or slanted dimension. After selecting the dimension orientation, with the middle mouse button, pick the location on the workscreen for the dimension placement.

RADIAL DIMENSIONS

Arcs and circles are considered radial entities. The Dimension option is available for creating radius and diameter dimensions. Refer to Figure 3–19 for a graphical representation on dimensioning radial entities.

RADIUS DIMENSIONS

A radius is the distance from the center of an arc or circle to the perimeter of the entity. To dimension a radial entity as a radius, select Dimension, then with the left mouse button, pick the radial entity. Next, with the middle mouse button, pick the location on the workscreen for the dimension placement.

DIAMETER DIMENSIONS

A diameter is the maximum distance across a circle. To dimension a radial entity as a diameter, select Dimension, then with the left mouse button, pick the radial entity two times. Next, with the middle mouse button, pick the location on the workscreen for the dimension placement.

ANGULAR DIMENSIONS

The angle formed between two line segments can be dimensioned within the sketcher environment. Additionally, the angle formed by the endpoints of an arc can be dimensioned. Refer to Figure 3–20 for a graphical representation on dimensioning angular entities.

LINE SEGMENTS

To dimension the angle between two line segments, select Dimension, then with the left mouse button, pick the two line segments. Next, with the left mouse button, pick the location on the workscreen for the dimension placement. The dimension placement location determines whether an acute or obtuse angle is formed.

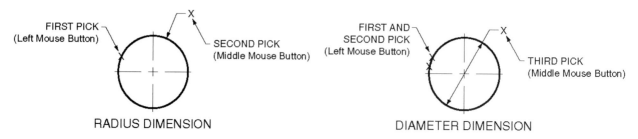

Figure 3–19   Radial Dimensioning Options

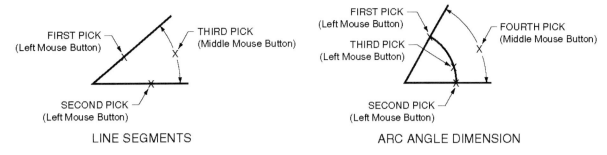

Figure 3–20   Angular Dimensioning Options

ARC ANGLE DIMENSION

To dimension the angle formed by the endpoints of an arc, select Dimension, then with the left mouse button, pick the endpoints of the arc. Next, select the arc being dimensioned. Finally, pick the location on the workscreen for the dimension placement.

PERIMETER DIMENSIONS

The Perimeter option under the dimension menu measures the perimeter of a sketched loop or chain of entities. Since a perimeter's dimension value can be modified, it requires the selection of a variable dimension. Because of a perimeter dimension's unique characteristic, this variable dimension is used as the dimension to vary when the perimeter dimension's value is modified. The variable dimension cannot be modified.

BASELINE DIMENSIONS

Pro/ENGINEER provides an option for creating baseline dimensions (Figure 3–21). Before a baseline dimension can be created, a baseline is required. The Dimension option uses this baseline as a reference for creating the remaining baseline dimension. Perform the following steps to create a baseline dimension.

STEP 1:   **Select SKETCH >> DIMENSION >> BASELINE from Pro/ENGINEER's menu bar.**

Before an baseline dimension can be created, Pro/ENGINEER has to know the baseline from which to measure each dimension.

STEP 2:   **Pick the baseline entity (Figure 3–21).**

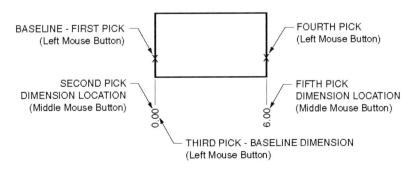

Figure 3–21   Baseline Dimensioning

STEP 3:   Select the location of the baseline dimension with the middle mouse button.

STEP 4:   ⟷   Select the Dimension icon on the sketcher toolbar (or select Sketch >> Dimension >> Normal).

STEP 5:   Pick the baseline dimension's numeric value.

STEP 6:   Pick the entity to dimension (fourth pick in Figure 3–21), then select the placement location for the dimension.

REFERENCE DIMENSIONS

Reference dimensions are used to show the size or location of an entity, but their values cannot be modified. Reference dimensions do not play a role in the definition of a feature. They can be used, however, for defining dimensional relationships. Reference dimensions are created in a similar manner to linear, angular, and radial dimensions. When a dimension is placed that creates an overconstrained situation, Pro/ENGINEER will recognize this and require you (through the Resolve Sketch dialog box) to either undo the creation, delete the dimension, or convert the dimension to a reference (see Figure 3–22).

MODIFYING DIMENSIONS

When you are sketching a feature in Pro/ENGINEER, the first priority is to create a section that meets design intent through the dimensioning scheme and through the proper utilization of constraints. Once a section is sketched and all dimensions and constraints applied, dimension values can be modified to their correct value.

 INDIVIDUAL MODIFICATION

An individual dimension can be modified by double picking the dimension's value with the pick icon. The section will be regenerated automatically upon entering the new value.

 MULTIPLE MODIFICATION

The Modify Dimensions dialog box is accessible through the Modify Dimensions icon and can be used to modify one or more dimension values (Figure 3–23). There

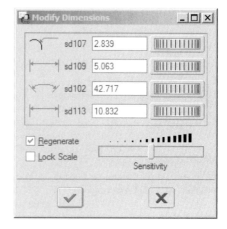

Figure 3–22   Resolve Dimension Dialog Box

Figure 3–23   Modify Dimension Dialog Box

are two techniques for selecting dimensions. With the first technique, the user can preselect dimensions with the Pick icon by using the Shift key or by drawing a pick box around all necessary dimension values. The user can then individually modify each dimension value through the dialog box.

 SCALE MODIFICATION

Often, it is difficult to modify the values of a complicated section. A complicated section can lead to dimensions that conflict. A conflict in dimensions will lead to a failed regeneration. A solution to this problem is using the Lock Scale option on the Modify Dimensions dialog box (Figure 3–23). The Lock Scale option allows for the modification of one dimension. After the new dimension value is entered, the remaining dimensions are scaled the same factor.

SKETCHER RELATIONS

The Relation option is a powerful tool for capturing design intent. Within the sketcher environment, this option creates relationships between dimension values. As an example, if the intent for a design is to have the width of a prism two times the length, a mathematical equation can be entered through the Relation option that makes the width value always twice the length.

When a feature is created, dimensions are used to define size and location. Each dimension has a value associated with it and a dimension symbol. Figure 3–24 shows an example of a section with normal dimension values provided in one view and respective dimension symbols provided in the second. The dimension symbol is used within a relation equation to establish a dimensional relationship.

Notice from Figure 3–24 that the format of a dimension symbol while in the sketcher environment is sd*. The *s* represents the sketcher environment (it is not used in part mode) and the *d* represents a dimension type. Pro/ENGINEER sequentially provides the dimension symbol number.

There are two types of relationships that can be created: equality and comparison. An equality relationship requires an algebraic equation. The following are examples of equality relationships:

$$sd1 = sd2$$
$$sd3 = sd2 + sd1$$
$$sd4 = sd1*(sd2 + sd3)$$
$$sd5 = sd4/(sd1 + sd2 + sd3)$$
$$sd5 = sd4*SQRTsd2$$

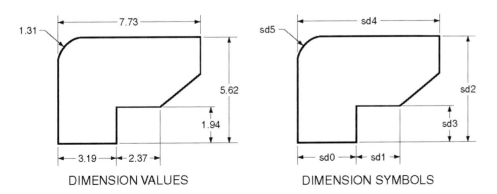

DIMENSION VALUES   DIMENSION SYMBOLS

Figure 3–24 Dimension Symbols

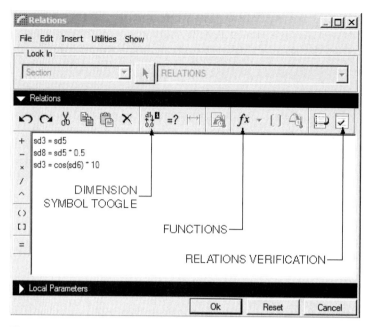

Figure 3–25   •   Relations Dialog Box

A comparison relationship can take the form of an equation or a conditional statement. The following are examples of comparison relationships:

$$sd1 < sd2$$
$$(sd2 + sd1) < (sd3 + sd4)$$
$$IF\ (sd1 + sd3) > (sd3 + sd4)$$

The Relations dialog box (Sketch >> Relations) is used to create a relationship between dimensions and parameters. When this dialog box is open, the section's dimensions are converted from the value format to the symbol format (see Figure 3–25). This allows for the selection of appropriate dimension symbols. Perform the following steps to add a relation to a section:

STEP 1:   **Determine dimensional relationships that will satisfy design intent.**

Relations are used to intelligently incorporate design intent into a model. Relations should be used only when appropriate.

STEP 2:   **Select TOOLS >> RELATIONS on the menu bar.**

The Relations dialog box is used to create dimensional relationships.

> **MODELING POINT**   The Relations dialog box provides options in addition to adding basic dimensional relationships. A variety of conditional statements and mathematical operations can be utilized within a relations statement. One useful tool is the Insert Functions dialog box. Assessable through the Functions icon (Figure 3–25), this dialog has a defined list of available mathematical options from which to choose (trig functions, exponents, square, etc).

STEP 3:   **Enter a relations equation in the Relations dialog box.**

Add a valid relational equation, then enter the value. Dimension symbols can be entered through the keyboard or by selecting on the work screen (Figure 3–25).

STEP 4:   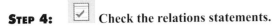 **Check the relations statements.**

The verify option ensures that no conflicts exist between your entered relations statements.

STEP 5:   **Select OK on the dialog box.**

GEOMETRIC TOOLS

A variety of geometric construction tools are available for manipulating sketched entities. These options are available as icons on the sketcher toolbar.

 **DYNAMIC TRIM**

The Dynamic Trim option will trim a selected entity up to its nearest vertex point or points (Figure 3–26). Pick the segment of the entity to delete.

 TRIM

The Trim option trims two selected entities at their intersection point. As shown in Figure 3–27, this option deletes the selected entities on the opposite side of the intersection point from where each entity was selected. Figure 3–27 also demonstrates the Bound option. This option requires the selection of a bounding entity. Selected entities are trimmed at the boundary entity or extended to the bounding entity.

 DIVIDE

The Divide option will divide an entity into two segments. The entity is divided at the selection point.

 MIRROR

The Mirror option will mirror selected entities over a picked centerline (Figure 3–28). The first step in the mirror process is to pick the entities to mirror. Multiple entities can be picked through the shift key and pick icon combination. Follow entity selection by executing the Mirror option, then picking the centerline to mirror about.

 SCALE AND ROTATE

The Scale and Rotate option is used to scale and/or rotate the entire section or selected entities within a section. Use the Edit >> Select >> All option to ensure the selection of all entities in the section.

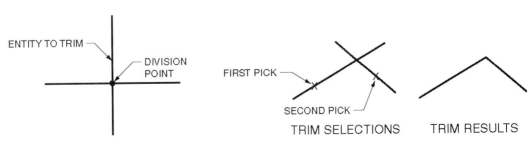

Figure 3–26   The Dynamic Trim Option

Figure 3–27   Trim Options

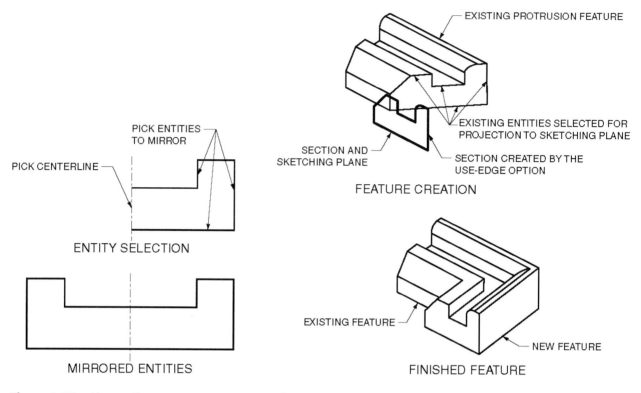

Figure 3–28   Mirroring Entities   **Figure 3–29**   Using Existing Feature Edges as Sketch Geometry

USE EDGE

The Use Edge option creates sketcher geometry from existing feature edges. Selected feature edges are projected onto the sketching plane as sketcher entities. As shown in Figure 3–29, the selected edges do not have to lie on or parallel to the sketching plane. Once entities are projected onto the sketching plane, they can be trimmed, divided, and filleted.

OFFSET EDGE

The Offset Edge option creates sketcher geometry offset from existing feature edges. This option is similar to the Use Edge option except an offset value is required. The projected edges will be offset the specified distance.

SUMMARY

Many of Pro/ENGINEER's feature construction tools require the sketching of a section. Sections are composed of geometric entities, dimensions, constraints, and references. Sections can be sketched with or without Intent Manager. Intent Manager applies constraints and dimensions during the sketch construction process. Without Intent Manager, constraints and dimensions are applied during regeneration.

SKETCHER TUTORIAL 1

Tutorials in this chapter will explore sketching with Intent Manager. Figure 3–30 shows the section to be sketched. When creating a section with Intent Manager activated, adhere to the following order of operations:

1. Sketch section. (*Note:* In creating a section within Part or Assembly mode, the first step in the order of operations is to specify references. Since in Sketch mode no existing features exist to reference, this step is skipped.)

2. Apply dimensions that match design intent.

3. Modify dimension values. (*Note:* Do not modify dimension values until the dimensioning scheme matches design intent.)

4. Regenerate.

5. Apply relations (if required).

6. Regenerate (if necessary).

This tutorial will cover the following topics:

- Starting a new object file in Sketch mode
- Sketching with Intent Manager
- Creating entities with the Line, Arc, and Circle options
- Saving a section

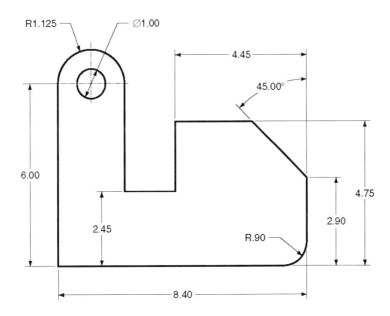

Figure 3–30 Section Sketch

CREATING A NEW OBJECT IN SKETCH MODE

This segment of the tutorial will start a section file in Sketch mode.

STEP 1: **Start Pro/ENGINEER.**

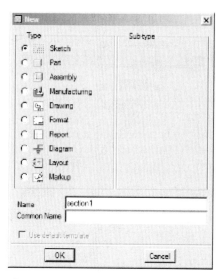

Figure 3–31   New Dialog Box

STEP 2: **Select FILE >> SET WORKING DIRECTORY, then select an appropriate working directory.**

The working directory is the default directory where Pro/ENGINEER will save model files and where Pro/ENGINEER will look when the Open command is selected. You should select a directory where you have read and write privileges.

STEP 3: **Select FILE >> NEW (or select the New icon).**

STEP 4: **Select SKETCH as the model type, then enter *section1* as the object's name (Figure 3–31).**

STEP 5: **Select the OK option to create the section file.**

SKETCH ENTITIES

Pro/ENGINEER provides multiple tools for sketching geometric shapes (line, circle, arc, etc.). The most used of these options are available on the sketcher toolbar, while other options are available under the Sketch option on the menu. While you are sketching, Pro/ENGINEER will dynamically apply geometric constraints that could affect the capturing of model design intent. Examples of such constraints include Parallel, Horizontal, Vertical, and Equal Distance. When working this tutorial, ensure that all required geometric constraints are captured dynamically.

STEP 1: **From the Start Point shown in Figure 3–32, use the Line option to sketch the entities shown.**

The left mouse button of your mouse is used to pick line endpoints and the middle mouse button is used to cancel the line command.

When sketching entities, you should focus on sketching the approximate shape; do not worry about the size of the entities being sketched. Appropriate dimensions will be modified in a later step. The start point and the endpoint of the sketch should lie approximately horizontal from each other.

Note: Disregard any dimensions that will be created when you cancel the line command.

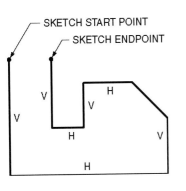

Figure 3–32   Sketching the Section

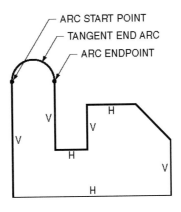

Figure 3–33   Creating a Tangent End Arc

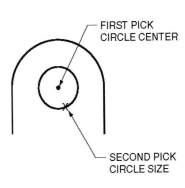

Figure 3–34   Creating a Center/Point Circle

MODELING POINT Lines that appear vertical or horizontal should be sketched accordingly. Intent Manager will apply constraints that will snap each line horizontal or vertical. On Figure 3–32, the lines labeled H and V represent horizontally and vertically constrained lines respectively.

After you terminate the Line option, Pro/ENGINEER's Intent Manager will fully define the sketch with appropriate dimensions. The dimensioning scheme for your sketch will probably not match design intent.

STEP 2:   Use the ARC option to create the arc shown in Figure 3–33.

STEP 3:   Use the CIRCLE option to create the circle shown in Figure 3–34.

As shown in Figure 3–34, select the center point of the circle, making it coincident with the center vertex of the existing arc. Next, define the size of the circle by dragging the perimeter of the entity. The diameter value will be modified in a later step.

STEP 4:   Use the CIRCULAR FILLET option to create the fillet shown in Figure 3–35.

Use the Circular Fillet option to create the fillet. This option requires the selection of two entities to fillet between. The fillet's radius will be defined in a later step.

STEP 5:   Use the DIMENSION option to create the dimensioning scheme shown in Figure 3–36.

Use the Dimension option to place dimensions according to Figure 3–36. When you are defining the dimensioning scheme for a section, it is helpful to disregard the current weak dimensions on the sketch. Weak dimension and weak constraints will be overridden through the placement of strong dimensions. Use the following dimensioning techniques when placing your dimensions:

- The left mouse button is used to select entities to dimension and the middle mouse button is used to place the dimension.

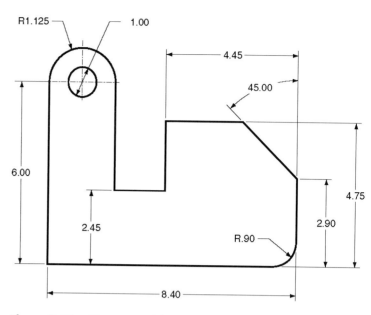

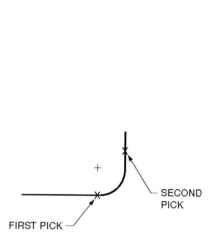

Figure 3–35   Creating a Filleted Arc

Figure 3–36   Dimensioning Scheme

- To dimension the distance between parallel lines, pick each line with the left mouse button, then place the dimension with the middle mouse button.

- Radius dimensions are created with a single selection of the entity, while diameter dimensions are created with two selections of the entity.

STEP 6:   Use the **MODIFY** option to modify the dimension values to match Figure 3–36. (Modify smaller dimensions first.)

Use the Modify Dialog box to change dimension values. After selecting the Modify icon, you can select all available dimensions. It is advisable to start with smaller dimensions first, followed by larger dimensions last. This helps to avoid unusual regeneration problems. Use the Undo option to undo any modification errors.

STEP 7:   SAVE the section.

From the File menu select the Save option. Your object will be saved as a section file (*.sec).

STEP 8:   Select the Continue icon to exit sketch mode.

Selecting the continue icon will close the current sketch window. The object will remain in session memory.

SKETCHER TUTORIAL 2

This tutorial will create the section shown in Figure 3–37. Within this tutorial, the following options will be used:

- Trim
- Delete
- Mirror
- Divide
- Constraints
- Centerline
- Rectangle

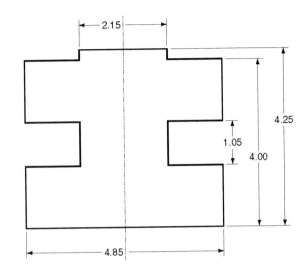

Figure 3–37   Finished Sketch

CREATING A NEW SECTION

This segment of the tutorial will describe the creation of a new Sketch mode object file.

STEP 1:   Start Pro/ENGINEER.

STEP 2:   Select an appropriate working directory.

Use the File >> Set Working Directory option to select a working directory.

STEP 3:   Create a new section file.

Using the File >> New option, select Sketch as the type and enter *Section2* as the object name.

CREATING THE SKETCH

This segment of the tutorial will describe the process for creating the sketch shown in Figure 3–37.

STEP   1:   Use the CENTERLINE icon to construct a vertical centerline (Figure 3–38).

The Centerline icon is located behind the Line icon on the sketcher toolbar. Since centerlines do not extrude into features, they are useful as construction lines. Additionally, many geometric tools, such as mirrored entities and

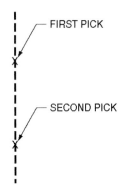

Figure 3–38
A Vertical Centerline

revolved features, require a centerline. It is important to note that your centerline may not actually look like a centerline on the workscreen.

STEP 2: ⬜ **Use the RECTANGLE option to create the rectangular entity shown in Figure 3–39.**

As shown in Figure 3–39, the Rectangle option requires the selection of opposite corners of the rectangle. At this time, do not worry about entering a precise size for the rectangle. When sketching, be sure to align the left edge of the rectangle with the centerline.

STEP 3: **Create a second rectangle as shown in Figure 3–40.**

Sketch the second rectangle the same size as the first rectangle. When sketching, use Intent Manager to ensure that the length of the horizontal side of the new rectangle is the same length as the first rectangle. If necessary, dynamically zoom out from the sketch (Control key and middle mouse button).

STEP 4: ╲ **Sketch the line shown in Figure 3–41.**

Use the Line option to sketch the Vertical line between the points shown in Figure 3–41. A midpoint geometric constraint is available for dynamically snapping to the midpoint of a line. Don't apply this constraint to this entity.

STEP 5:  **Use the DYNAMIC TRIM option to trim the entities as shown in Figure 3–42.**

The Dynamic Trim option will trim an entity up to its nearest vertex point(s). Select the segment of each entity that you want to delete.

STEP 6: **↖ Use the Pick icon to select the entities shown in Figure 3–43, then use the keyboard's delete key to delete each entity.**

Be careful not to delete the vertical centerline.

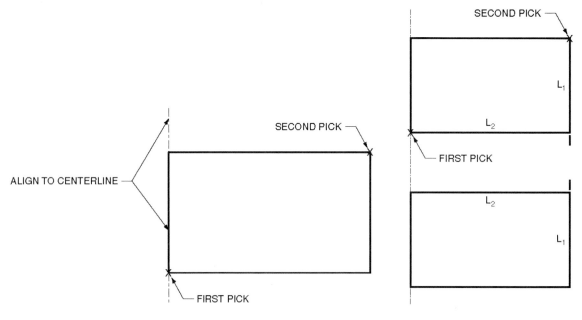

Figure 3–39   Creating a Rectangle

Figure 3–40   The Second Rectangle

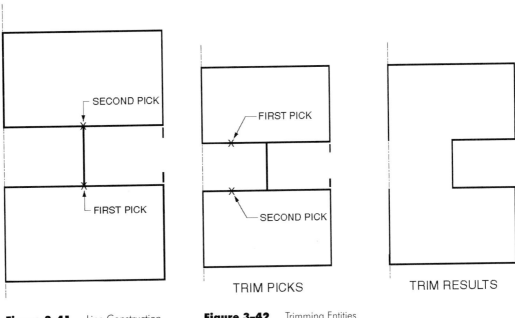

Figure 3–41   Line Construction   **Figure 3–42**   Trimming Entities

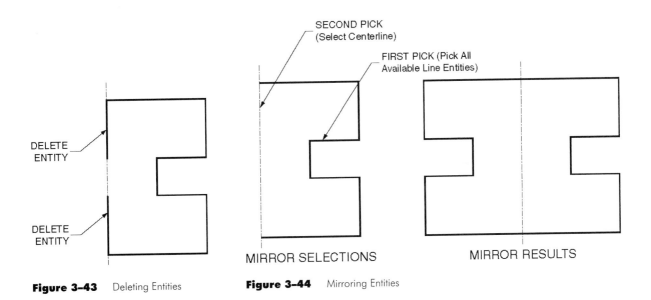

Figure 3–43   Deleting Entities   **Figure 3–44**   Mirroring Entities

STEP 7: Using the PICK option, drag a selection box around all available entities within your sketch.

The next step of this tutorial will mirror your entities about this centerline. To use the Mirror option, entities to mirror must be preselected.

STEP 8: Select the MIRROR option then pick the centerline to mirror the entities about (Figure 3–44).

All line entities must be preselected before you can select the Mirror option. If your results do not match Figure 3–44, undo the last command, then start over at Step 7.

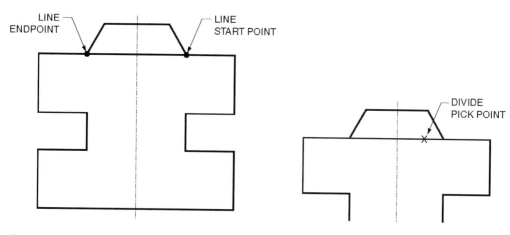

Figure 3–45   Sketching a Line

Figure 3–46   Divide Selection

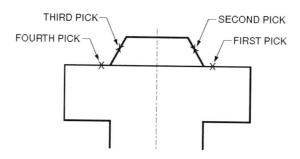

Figure 3–47   Trim Selection

STEP 9: Sketch the line entities shown in Figure 3–45.

STEP 10: Use the DIVIDE option to break the single entity at the point shown in Figure 3–46.

The Divide option will break an entity at the point selected. Its icon is located behind the Dynamic Trim option.

STEP 11: Use the TRIM option to trim the four lines as shown in Figure 3–47. (Ensure you select Trim, not Dynamic Trim.)

STEP 12: Add perpendicular constraints to the entities shown in Figure 3–48.

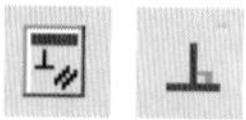

To add a constraint, select the Constraints icon on the sketcher toolbar, then select the Perpendicular constraint icon. Select the entities at the points shown in Figure 3–48.

STEP 13: Add LINE UP VERTICAL constraints to the vertices shown in Figure 3–49.

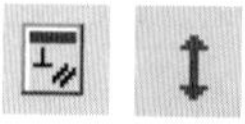

The Line Up Vertical constraint option will line up entity endpoints in a vertical orientation. Select each vertex as shown in the figure.

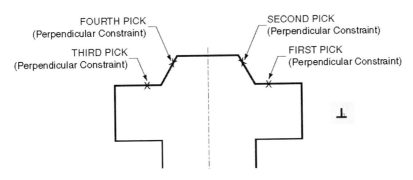

Figure 3–48   Perpendicular Constraints

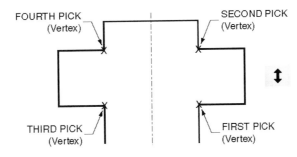

Figure 3–49   Line Up Vertical Constraints

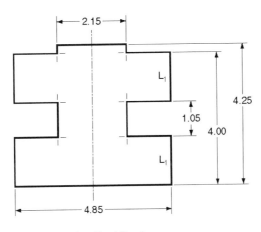

Figure 3–50   Final Sketch

STEP 14:   Use the **DIMENSION** option to create the dimensioning scheme shown in Figure 3–50.

STEP 15:   Use the **MODIFY** option to change dimension values to match Figure 3–50.

The Modify Dimension dialog box is used to change dimension values. You can also double selection each dimension with the Pick icon.

STEP 16:   Save your section.

STEP 17:   Select the Continue icon to exit the sketcher environment.

PROBLEMS

Use Pro/ENGINEER's Sketch mode to create the following sections. The constraints and dimensions shown match design intent.

1. Problem 1

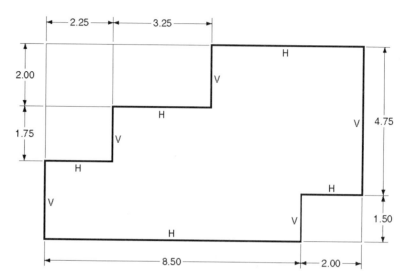

Figure 3–51   Problem 1

2. Problem 2

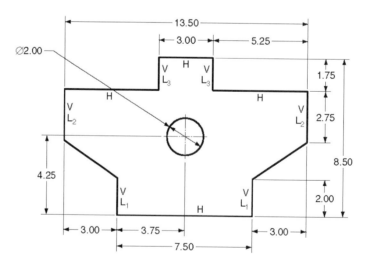

Figure 3–52   Problem 2

3. From Problem 2, add a dimensional relationship that will make the circle centered horizontally within the base geometry. Within your relationship, your hole's location should be controlled by the 13.50 dimension.

4.   Problem 4

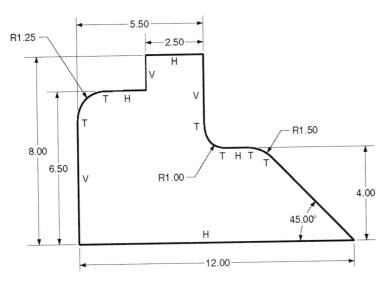

Figure 3-53   Problem 4

QUESTIONS AND DISCUSSION

1.   Describe ways of capturing within a parametric model the intent of a design.
2.   What makes a suitable sketching surface within Pro/ENGINEER?
3.   How does the application of geometric constraints differ when Intent Manager is activated when compared to when Intent Manager is not activated?
4.   List possible constraints that can be applied within a Pro/ENGINEER section.
5.   Describe the order of operations for sketching without Intent Manager.
6.   Describe the order of operations for sketching with Intent Manager.
7.   Write a valid relations equation for making dimension *sd1* equal to twice dimension *sd2*.
8.   Within a sketcher environment, what is a feature reference?

4

EXTRUDING, MODIFYING, AND REDEFINING FEATURES

Introduction

This chapter introduces extruded features, a concept associated with basic modeling fundamentals. Within Pro/ENGINEER, extruded cuts and protrusions can be created. Additionally, this chapter will introduce the Edit Definition command, feature modification techniques, and datum plane construction. Upon finishing this chapter, you will be able to:

- Model solid features as extruded protrusions.
- Remove material from features using extruded cuts.
- Modify feature dimension values using the Modify command.
- Modify feature definitions using the Edit Definition command.
- Create datum planes.

DEFINITIONS

Base feature	The first geometric feature created in a part. It is the parent feature for all other features.
Child feature	A feature whose definition is partially or completely referenced to other part features. A feature that is referenced by a child feature becomes a parent of this feature.
Cut	A negative space feature created from a sketched section.
Definition	A parameter of a part. An example of a definition of a hole feature would be the depth of the hole.
Protrusion	A positive space feature created from a sketched section.
Negative space feature	A feature created by removing material from a model. Examples of negative space features include holes, cuts, and slots.
Parent feature	A feature referenced by another feature.
Positive space feature	A feature created by adding material to a model. Examples of positive space features include protrusions, shafts, and ribs.

FEATURE-BASED MODELING

Parametric design packages are often referred to as feature-based modelers. A feature is a subcomponent of a part that has its own parameters, references, and geometry (Figure

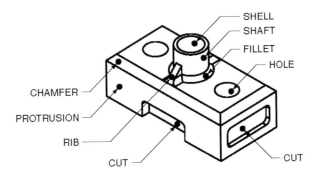

Figure 4–1   Features of a Part

4–1). Geometry is the graphic description of a feature. Geometry may be sketch-defined or predefined. Sketch-defined features consist of sketched sections that are protruded or cut to form either positive or negative space. Predefined geometry has a common section such as a hole, round, or chamfer. Parameters are the dimensional values and definitions that define a feature. A hole may have a diameter of 1.00 in and can be extruded completely through all existing features. The diameter is a parameter, as is the through-all definition. Parametric modeling packages allow users to modify parameters after the feature has been modeled. This is one of the unique properties that separate parametric modelers from Boolean-based modelers. References are ways that features are related to other features in a part or assembly. Examples of references include axes, sketch planes, placement planes, reference planes, and reference edges. The surface of one feature may serve as the sketch plane for a second feature. The edges of the first feature may also serve as reference lines for parameters defining the second feature. In both examples, the first feature is a parent of the second feature.

PARENT-CHILD RELATIONSHIPS

Parametric models are composed of features that have established relationships. Features are built upon other features in a way that resembles a family tree, hence the phrase *parent-child relationship*. Actually, a history tree of the relationships between features in a Pro/ENGINEER model resembles a web. The first feature created in a part is the center of the web and is the parent feature for all features. Child features branch off from the base feature and themselves become parent features. Unlike a typical family tree, a child feature may have several parent features.

Parent-child relationships can be established between features implicitly or explicitly. Implicit relationships can be established through the adding of a numeric equation using the Relations option. An example of this would be making two dimensions of equal value. In this process, one dimension governs the value of another. The feature with the governing dimension is the parent feature of the feature with the governed dimension. Care should be taken when modifying a feature that has a dimension that governs another. If a parent feature is selected for deletion, Pro/ENGINEER will provide an error message requesting an action to be accomplished to satisfy the void relationship. The user has the option of deleting, modifying, redefining, or rerouting the relationship.

Explicit relationships are created when one feature is used to construct another. An example would be selecting a plane of one feature as the sketch plane for a second feature. The new feature will become a child of the feature that is being sketched upon. Another similar example of an explicit relationship would be using existing feature edges within the sketcher environment to create a new feature. When you specify references while sketching, these selected references will create a relationship between the feature being sketched and the existing feature being referenced. The new sketched feature becomes a child of any referenced feature.

THE FIRST FEATURE

Pro/ENGINEER parts are composed of features. Determining what will be the first feature, or base feature, of a part is an important decision. Because of the Parent-Child Relationships that exist between features, the first feature created will likely become the parent feature for all the features of a part. There are three possible first features of a part. The following is a discussion of each one.

DATUM PLANES

Datum planes are the recommended first features created for a new part. Datum planes can be used for future feature creation techniques, they are valuable within Assembly mode, and they make good reference features. Pro/ENGINEER has an option for creating a set of default datum planes. These three datum planes are created orthogonal to each other and are listed as the first three features on the model tree. By default, these default datum planes are named DTM1, DTM2, and DTM3.

PROTRUSION

A Protrusion is a positive space geometric feature. Examples of protrusions include extruded, revolved, and swept features. When a protrusion is created as the first feature, this feature will become the parent feature for all subsequent features in the part.

MODELING POINT   Pro/ENGINEER's part and assembly modes utilize template files as starting points for object creation. The default template file for part mode comes standard with a set of default datum planes, created layers, a default coordinate system, and units set to inch-pounds-seconds. The three default datum planes within this template file have been renamed *Front, Top,* and *Right.* Using this default template file guarantees that datum planes will be the first feature created in a part. The configuration file option *template_solidpart* can be utilized to set a user-defined default part file.

USER-DEFINED FEATURE

A user-defined feature is a feature that has been created by a user and saved to disk for use in later modeling applications. Several features can be combined in the creation process to form one grouped user-defined feature. When a part is created, a user-defined feature can be retrieved as the first feature.

STEPS FOR CREATING A NEW PART

Pro/ENGINEER follows an established procedure for starting a new part. This procedure is valid for most solid and thin protrusions. Listed below is a recommended sequence for starting a new part. There are three assumptions to keep in mind when following these steps. The following describes these assumptions:

- It is assumed that Pro/ENGINEER's default template file will be used. As mentioned in the preceding Modeling Point, a part template is an existing file that has preexisting parameters, settings, and features. During the creation of a new part file, the default template can be changed by deselecting the Use Default Template option on the New dialog box. Additional template files come preexisting with Pro/ENGINEER to include an *Empty* template file.

- It is assumed that the first features created will be Pro/ENGINEER's default datum planes. This assumption affects two steps in the process. First, one of the steps in this sequence is the creation of the default datum planes. Second, establishing the sketcher environment would not be necessary if the default datum planes are not available.

- It is assumed that a user wants to establish model parameters such as part material and units. A part can be created without entering the Set Up menu. As noted in the first assumption, Pro/ENGINEER's default template comes with units already established.

Perform the following generic steps to create a new part.

STEP 1: **Establish the correct Working Directory.**

You can change Pro/ENGINEER's default working directory at anytime during the modeling process. When searching for a file, Pro/ENGINEER, by default, first looks in the working directory. Additionally, when saving a new part, Pro/ENGINEER saves the object in the current working directory. The Working Directory option can be found under the File menu.

INSTRUCTIONAL NOTE   For more information on Steps 1 through 3, see appropriate sections in Chapter 2.

STEP 2: ☐ **Create a NEW object with Pro/ENGINEER's default template.**

The File >> New option or the New icon on the menu bar will allow you to create a new object file. The New dialog box gives options for selecting a Pro/ENGINEER mode, with Part mode being the default. A file name can be entered, or the default name can be selected. File names must be limited to 32 characters (usually alphanumeric) or less and have no spaces.

STEP 3: **Set up the Model.**

Several options are available for setting up a part. Two common selections are part material and modeling units. Both part material and units can be set later in the modeling process if desired. You can access setup options with the *Edit >> Setup* command.

STEP 4: **Select a suitable sketch plane.**

Datum plane and planar surfaces make suitable sketching surfaces. Datums are the preferred planes. Pro/ENGINEER's default datum planes provide three possible choices.

STEP 5: **Sketch the feature's section.**

Extrusions, revolves, blends, sweeps and a variety of advanced feature creation tools require at least one sketched section.

STEP 6: **Select the Feature Creation Method.**

Common feature creation tools available include the exrude, revolve, and blend tools. The sweep tool and it's variations, such as the variable section sweep, require addition sections or geometric entities.

STEP 7: **Finish the feature.**

Pro/ENGINEER manages the creation of features through a dashboard interface. This intuitive user interface allows for the defining of feature

Figure 4–2   Extrude Dashboard

parameters in a logical sequence. Figure 4–2 shows the dashboard of options available under the Extrude command.

> **MODELING POINT**   Often, when a mistake is made in the modeling process, it is tempting to cancel the creation process and start over. When you make a mistake, or when you skip a modeling step, continue with the modeling process and modify the definition later under the appropriate dashboard option.

STEP 8:   **Perform file management requirements.**

It is recommended that you save an object file after finishing a feature. Other file management options available include Backup and Save As. If a part is complete, purging old part files can be accomplished by using the File >> Delete >> Old Versions option.

FEATURE CREATION TOOLS

Extruded protrusions, cuts, and surfaces are forms of features created under the extrude dashboard. The primary difference between the protrusion option and the cut option is that a protrusion is a positive space feature and a cut is a negative space feature. When you protrude a feature, you are actually creating a solid object. With the cut option, an extruded feature removes material from existing features.

Extruded protrusions are commonly used geometric features, but other feature tools are also available. The following is a partial list of solid feature creation tools.

EXTRUDE

The Extrude tool sweeps a sketched section along a straight trajectory. The user draws the section in the sketcher environment and then provides an extrude depth. The section is protruded the depth entered by the user.

REVOLVE

The Revolve tool sweeps a section around a centerline. The user sketches a profile of the revolved feature and a centerline to revolve about. The user then inputs the degrees of revolution.

SWEEP

The Sweep command protrudes a section along a user-sketched trajectory. The user sketches both the trajectory and the section.

BLEND

The Blend command joins two or more sketched sections. The trajectory may be straight or revolved.

Shown in Figure 4–3 is an illustration of how one section can be used to create an Extrude, Revolve, Sweep, or Blend feature.

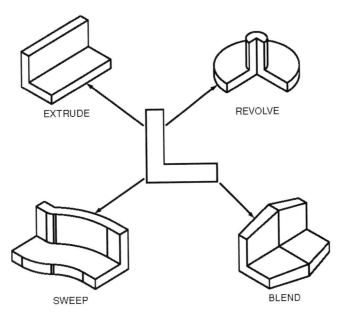

Figure 4-3 Variations in Menu Option Features

INSTRUCTION POINT Starting with Release 2000i, by default, the Shaft, Slot, Neck, and Flange commands have been removed from Pro/ENGINEER's Insert menu. The configuration file option *allow_anatomic_features* set to YES will return these commands to the menu. See the "Configuration File" section in Chapter 2 for information on modifying this file.

SOLID, SURFACE, AND THIN FEATURES

When you are creating an extrusion, Pro/ENGINEER provides the option of choosing either a solid feature, a thin feature, or a surface feature (Figure 4–4). Solid features are objects that are completely enclosed with material. Thin features are often confused with surfaces. In Pro/ENGINEER, surfaces are quilts with no defined thickness, whereas thin features are actually solids with a user-defined thickness. As shown in Figure 4–4, when the section is extruded as a solid, the section's feature is completely enclosed with material. When the section is extruded as a thin feature, the walls of the section are protruded with the provided wall thickness only.

Thin features can be used with all forms of the Extrude, Revolve, Sweep, and Blend commands. An example of an extruded thin cut is shown in Figure 4–5. The Thin option may be used with the protrusion option for the base or secondary feature of a part or with the cut option for secondary features.

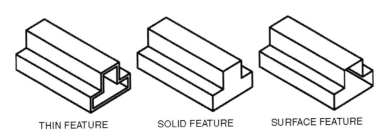

Figure 4-4 Thin versus Solid Features

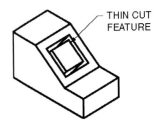

Figure 4-5 A Thin Cut Feature

EXTRUDED FEATURES

The following section will explore options available within extruded protrusions and cuts.

EXTRUDE DIRECTION

When you are sketching on a plane, Pro/ENGINEER, by default, specifies an extrude direction. When you are sketching on a datum plane, this direction is in the positive direction. When you are sketching on an existing feature, a Protrusion, as shown in Figure 4–6, will be extruded away from the feature. Since the objective of a cut is to remove material, a cut will be extruded toward the feature.

The Extrude option gives the user the option of flipping the direction of extrusion. It also gives the option of specifying a symmetric extrusion (Figure 4–7). The Symmetric selection under the Extrude dashboard's Options menu protrudes a section outward from the sketch plane equally in both directions. If the extrude depth is input to be 1.00 in, the total extrusion will be 1.00 in, not 1.00 in in both directions. A typical symmetric extrusion will divide the specified depth and extrude equally on both sides of the sketching plane. An option does exist for extruding a feature in both directions from the sketching plane. This two-sided blind depth technique allows the user to input unequal extrusion distances on both sides of the sketch plane.

> **MODELING POINT**   Features are composed of definitions that are established by the user. A definition is not permanently set. It can be changed before finishing a feature, or with the Edit Definition command. If a definition such as the extrude direction, depth option, or material removal side is incorrectly set, do not cancel or delete the feature. Redefine it later.

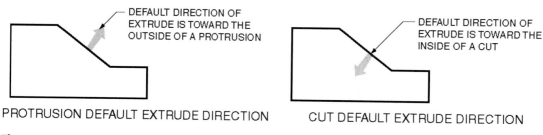

DEFAULT DIRECTION OF EXTRUDE IS TOWARD THE OUTSIDE OF A PROTRUSION

DEFAULT DIRECTION OF EXTRUDE IS TOWARD THE INSIDE OF A CUT

PROTRUSION DEFAULT EXTRUDE DIRECTION

CUT DEFAULT EXTRUDE DIRECTION

Figure 4–6   Default Extrude Directions

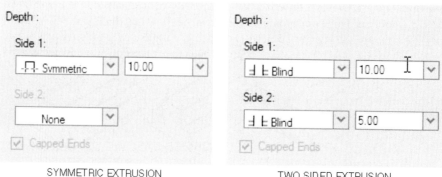

SYMMETRIC EXTRUSION

TWO SIDED EXTRUSION

Figure 4–7   Extrusion Direction Options

DEPTH OPTIONS

For extruded features, an important parameter is the distance of extrusion. Pro/ENGINEER provides eight basic ways to specify an extrusion's depth. Four common depth options are shown in Figure 4–8. The depth for an extrusion is entered for a feature after exiting the sketching environment.

VARIABLE (BLIND)

Blind is the simplest and most basic of the depth options. The Blind option allows a user to input a specified depth value. It is the most common option for extruded base features.

TWO-SIDED BLIND

A two-sided blind is used for two-sided extrusions only. This depth option will allow the user to enter separate extrude depths for both sides of the sketch plane.

TO NEXT

The To Next option extrudes a feature to the next part surface. Part geometry must exist prior to using this option.

THROUGH ALL

Through All is one of the most common depth options for cut features. It extrudes a feature through the entirety of a part. The design intent of many material removal features (such as a Cut or Hole) is to cut completely through a part. Entering a variable depth that will extrude through the part may not be adequate if the part thickness changes. The Through All option adjusts for changing part dimensions.

TO SELECTED

The To Selected option extrudes a feature until a user-selected surface, point, vertex, or curve. The extrusion will stop at the selected entity.

THROUGH UNTIL

The Through Until option extrudes a feature up to a selected surface or plane.

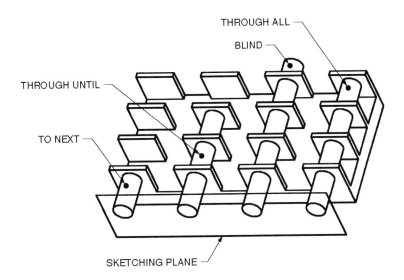

Figure 4–8   Depth Options Available for Extruded Features

OPEN AND CLOSED SECTIONS IN EXTRUSIONS

Extruded sections can be sketched opened or closed. With the obvious exception of a base feature, many sections for an extruded protrusion or cut will suffice with an open section. The following are guidelines to follow when considering an open or closed section:

- Sections may not branch and they can have only one loop. As shown in Figure 4–9, in sketching a section aligned with the edges of an existing feature, often it is not necessary to sketch over the existing geometry. Aligning the required sketch with the existing geometry will usually create a successful section. If Pro/ENGINEER is not sure which side of the section to protrude or cut material, it will require the user to select a side.

- Thin feature sections can be open or closed.

- For thin features, sections can be open when not aligned with existing geometry.

- Multiple closed sections may be included in a sketch. As shown in Figure 4–10, when a section is included within another, the inside section creates negative space.

MATERIAL SIDE

There are two definitions associated with Material Side. The most common is the Material Removal Side. Within the Extrude option, the Material Removal Side definition is relevant only to the Cut command. The Material Removal Side definition is used to specify what side of a section that material will be removed from. By default, the removal side is toward the inside of a section. The user has the option of flipping the direction. Material Side definitions can be used with protrusions when a sketch is an open section. Often, Pro/ENGINEER

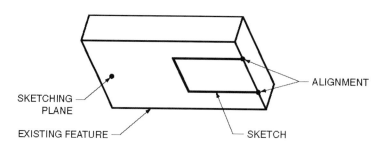

Figure 4–9   Aligning a Sketch with Existing Part Geometry

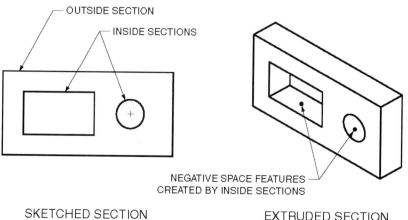

Figure 4–10   Creating Negative Space with Included Sections

cannot determine which side of the sketch should be extruded. When this situation occurs, the user has to input the material side.

CREATING EXTRUDED FEATURES

The Extrude tool is one of the most used feature creation options. Options are available on the extrude dashboard (Figure 4–11) for creating protrusions, cuts, and surface features. An extrusion is a section that is protruded along a straight line. It is a common option for creating primary and secondary features and is one of the fundamental skills needed to model in Pro/ENGINEER.

The following steps will outline the process for creating an extruded protrusion (solid). Extruded cuts and surfaces are created with similar steps and options.

STEP 1:   Select the Sketch Tool.

The Sketch Tool option is located on the Datum toolbar.

STEP 2:   On the work screen, pick a suitable sketching plane.

Notice on your screen and in Figure 4–12 how the Sketch dialog box is used to manage the sketching plane and the sketching plane orientation.

STEP 3:   Orient the sketching plane.

Notice in Figure 4–12 how Pro/ENGINEER will provide a default sketch plane orientation. In this example, you are sketching on datum plane Front and orientating datum plane Right toward the right of the work screen. You can see this setup in the Sketch dialog box's Reference and Orientation options. You can pick any planar reference to include part surface and you can also orient a plane in toward the Top, Bottom, Left, or Right.

STEP 4:   Select the SKETCH option on the Sketch dialog box.

STEP 5:   Sketch the Section.

Sketch the section according to design intent. The following is the order of operations for sketching a section:

1.   Specify references.
2.   Sketch section.
3.   Place dimensions according to design intent.
4.   Modify dimensional values.
5.   Add relations.

Note: See Chapter 3 for more information on sketching.

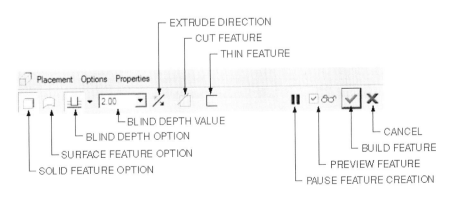

Figure 4–11   Extrude Dashboard

Figure 4–12   Sketch Dialog Box   **Figure 4–13**   Additional Depth
                                                              Options

STEP 6:   **Select the Continue icon when the section is complete.**

STEP 7:   **Select the Extrude icon on Pro/ENGINEER's Base Feature toolbar.**

The Extrude toolbar is used to create extruded protrusions, cuts, and surfaces. It utilizes a dashboard to define parameters for extruded features (Figure 4–11). On dashboards, parameters are defined left to right, with the more commonly used elements available with minimum menu manipulation.

STEP 8:   **Select the appropriate option for creating a solid or a surface feature (see Figure 4–11).**

Remember that you are working from the left to the right on the dashboard.

STEP 9:   **Select a Depth option and define depth references.**

Depth options available include Variable, Through Until, To Next, and Through All. The reference you define is based on the option selected. As an example, if you select Variable, you will have to enter a blind depth. If you select Through Until, you will have to select a feature to serve as the bounding reference.

The Options selection on the dashboard provides additional depth options (Figure 4–13). For symmetrical extrusions, a user can enter a two-sided blind extrusion.

MODELING POINT   Often, design intent can be met better by extruding both directions from the sketching plane. Additionally, having a datum located in the middle of the part can enhance future modeling techniques, such as the Copy >> Mirror option.

STEP 10:   **Select a material creation direction.**

This option determines the direction that the feature will extrude from the sketching plane.

STEP 11: 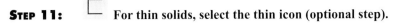 **For thin solids, select the thin icon (optional step).**

By default, the thin option is not selected. Unlike the solid option, thin features protrude or cut with a user-defined wall thickness. This wall thickness is defined after exiting the sketching environment.

STEP 12: **Preview the feature.**

At this point, you can still modify any definitions established for the feature. Definition parameters are shown along the dashboard.

STEP 13: **Build the feature.**

Select the Build Feature option on the dashboard to finish the feature.

DATUM PLANES

Datum planes are used as references to construct features. Datum planes are considered to be features, but they are not considered to be model geometry. When a datum plane is created, the datum plane will show as a feature on the model tree. Datum planes can be created and used as a sketch plane where no suitable one currently exists. As an example, a datum plane can be constructed tangent to a cylinder. This will provide a sketching environment that can be used to construct an extruded feature through the cylinder (Figure 4–14). When a feature is sketched on a datum plane, the datum plane is considered a parent feature.

A datum plane continues to infinity in all directions, with one side yellow and the opposite side red. Protrusions and orientations occur initially toward the yellow side of the datum plane. By default, datum planes are named in sequential order starting with DTM1. As a note, template files may have datum planes that have been renamed (i.e., Front, Top, and Right). Use the Pop-up menu's Rename option to change the name of a feature.

Many times, a datum plane would be useful, but cluttering the model with additional features could be detrimental. One solution to this problem would be to make a datum plane that is available only for the creation of a feature. To do this, datum planes can be created on the fly within the feature creation process. Datum planes created on the fly are internal to the feature that is undergoing creation. These datum planes are grouped under their parent features on the model tree and become invisible after the feature is created.

PRO/ENGINEER'S DEFAULT DATUM PLANES
Datum planes are commonly used as the first feature of a part or assembly. Pro/ENGINEER provides three orthogonal datum planes within its template files, referred to as the default

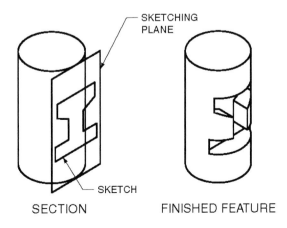

Figure 4–14 Datum Plane Tangent to Cylinder

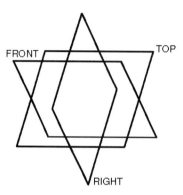

Figure 4–15   Default Datum
Planes

datum planes (Figure 4–15). When utilizing an empty template file, selecting the datum plane icon will automatically create Pro/ENGINEER's default datum planes.

CREATING DATUM PLANES

A datum plane can be created at any point in the modeling process. One of the primary uses of a datum plane is as a sketching surface. Datum planes can be used as a mirror reference plane within the Copy command, or they can be used as references in sketching a feature. Datum planes are powerful features for aligning and mating parts within Assembly mode. Additionally, datum planes can be combined with other features by using the Group option and then patterned.

Creating datum planes is a vital skill needed by all Pro/ENGINEER users. Several constraint options exist for datum plane definition. Some constraint options are stand-alone, while some are not. Paired constraint options are not stand-alone and require two or more constraint definitions during the datum plane construction process. As an example, the Angle option requires the selection of an existing plane from which to reference the angle, then a Through constraint option to pass the datum plane through an axis or edge. Stand-alone constraint options require one option only.

STAND-ALONE CONSTRAINT OPTIONS

THROUGH PLANE

The Through Plane constraint option creates a datum plane that passes through an existing part plane. Figure 4–16 shows the Datum Plane dialog box employing this option.

OFFSET PLANE

The Offset Plane constraint option creates a datum plane that is offset from an existing plane. The user selects the plane from which to offset. Two offset options are available:

• **Point** Select a point to pass the datum plane through. The plane will be created through the point and parallel to the existing part plane.

• **Enter Value** Enter an offset distance value. When selecting this option, the user enters an offset value in the Datum Plane dialog box (Figure 4–16). An arrow in the graphics screen shows the default direction of the offset. To offset in the opposite direction, enter a negative value. The plane will be created offset from the reference plane at the value entered.

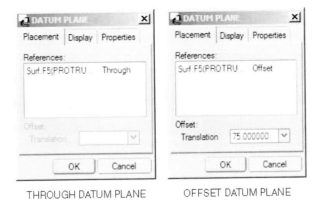

THROUGH DATUM PLANE   OFFSET DATUM PLANE

Figure 4–16 *Stand-Alone Datum Creation*

OFFSET/COORD SYSTEM

The Offset Coordinate System constraint option creates a datum plane offset from the coordinate origin and normal to a selected coordinate axis. A coordinate system has to exist prior to the use of this option.

BLEND SECTION

The Blend Section constraint option creates a datum plane through a section used to create a feature.

PAIRED CONSTRAINT OPTIONS

Some datum plane constraint options will not fully define the location of a datum plane. As an example, placing a datum plane through an axis still leaves a rotational degree of freedom. This degree of freedom has to be constrained with a paired constraint. To create a paired constraint, use the keyboards control key to pick the necessary constraining references.

THROUGH AXIS (ALSO EDGE OR CURVE)

This option is similar to Through Plane, except this option places a plane through an axis, edge, or curve. An axis, edge, or curve selected with the through option will not fully constraint a datum plane; hence, a second constraint option is required.

THROUGH POINT/VERTEX

This option places a datum plane through a point or vertex. Similar to the Through Axis option, this option will not fully define a datum plane and needs an additional constraint option.

NORMAL TO AXIS (ALSO EDGE OR CURVE)

This option places a datum plane perpendicular to an axis, edge, or curve. As with the Through Axis option, an additional constraint is needed.

TANGENT TO CYLINDER

This constraint option places a datum plane tangent to a hole or cylindrical surface. This is an extremely useful option since it allows a feature to be constructed on the surface of a cylinder. This constraint option is often paired with the Normal >> Plane option or the Angle to Plane option.

ANGLE TO PLANE

This option places a datum at an angle to an existing plane. It is often paired with the Through option, or the Tangent to Cylinder option.

ON-THE-FLY DATUM PLANES

Datums created with one of the options on the Datum toolbar are considered part features. Datums, especially datum planes, can clutter a part and make viewing and selecting features difficult. Datums intended for use as a modeling tool for individual features can be created on the fly during feature modeling. On-the-fly datum planes are available only for the feature in which they are intended. They are not considered part features and will be grouped with their parent feature on the model tree. Additionally, when the feature creation modeling process is finished, the datum will disappear from the workscreen.

MODIFYING FEATURES

What separates parametric modeling packages, such as Pro/ENGINEER, from Boolean-based modeling packages are their feature modification capabilities. Features created within Pro/ENGINEER are composed of parameters. Examples of parameters include parametric dimensions, extrude depth, and material side. Parameters such as these are established during feature construction. These feature parameters, and other feature definitions such as a section's sketch and sketch plane, can be modified later in the part modeling process.

DIMENSION MODIFICATION

Parametric dimensions are dimensions that are used to define a feature. They can be modified at any time. Modifying a dimension value is the most common dimension modification function, but other modification tools do exist. The number of decimal places in a dimension can be modified along with the tolerance format.

> **DIMENSION POP-UP MENU** Dimension values, tolerances, display style, and other cosmetics can be modified through the Dimension Properties dialog box. This dialog box is accessible through the Properties option on the dimension pop-up menu. The pop-up menu is available by right-mouse-button picking a dimension. Control-key picking more than one dimension will allow for the modification of multiple dimensions simultaneously. Two methods are used to show feature dimensions for modification: using the Edit option on the model tree's pop-up menu and double picking features on the workscreen.

Most dimension modification options are available within the Dimension Properties dialog box (Figure 4–17). The Properties option on the dimension pop-up menu will open this dialog. The following are dimension modification techniques that are available within Pro/ENGINEER.

MODIFYING A DIMENSION VALUE

Dimension tolerances can be displayed in a variety of modes. To modify the tolerance mode for an individual dimension, access the Dimension Properties dialog box by picking the dimension on the work screen, then access the pop-up menu's Properties option. As shown in Figure 4–17, an option exists on the Dimension Properties dialog box that will allow for the changing of a tolerance mode.

TOLERANCE MODE — DECIMAL PLACE
MODIFICATION   MODIFICATION

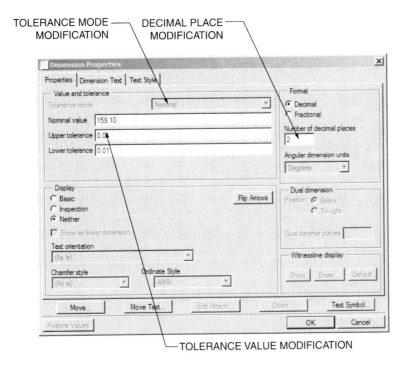

└─ TOLERANCE VALUE MODIFICATION

Figure 4–17   The Modify Dimension Dialog Box

MODELING POINT   To display tolerances for all objects created within Pro/ENGINEER, the configuration file option *tolerance_mode* must be set (to a value such as Limit) and the option *tolerance_display* must be set to Yes. To display tolerances for individual objects, select Tolerance Display on the Environment dialog box.

MODIFYING A TOLERANCE MODE

Dimension tolerances can be displayed in a variety of modes. To modify the tolerance mode for an individual dimension, access the Dimension Properties dialog box by picking the dimension on the work screen, then access the pop-up menu's Properties option. As shown in Figure 4–17, an option exists on the Dimension Properties dialog box that will allow for the changing of a tolerance mode.

MODIFYING TOLERANCE VALUES

Modifying dimensions directly on the workscreen will allow for the modification of tolerance values. As an example, if a dimension is set to Limits as the tolerance mode, either the upper or the lower dimension value can be changed. A problem with this approach is that the nominal value of the limit dimension cannot be modified. Another approach is to modify the nominal value and/or the tolerance values with the Dimension Properties dialog box (Figure 4–17).

DIMENSION DECIMAL PLACES

Initially, dimension decimal places are set to 2. This value can be changed permanently with the configuration file option *default_dec_places*. To change

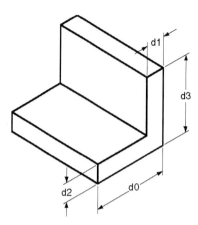

Figure 4–18   Dimension Symbols

decimal places for individual dimensions, use the Dimension Properties dialog box's Number of Decimal Places option (Figure 4–17).

TOLERANCE FORMAT

A tolerance's format mode can be changed by using the Tolerance mode option found on the Dimension Properties dialog box. Available formats include nominal, limits, plus-minus symmetric, and plus-minus.

ADDING TEXT AROUND A DIMENSION

Text can be added around a dimension value by using the Dimension Text tab on the Dimension Properties dialog box. A dimension's value is shown in a dimension note with the symbol [@D]. This symbol must remain in the dimension note. A pallet of commonly used symbols is available for inserting into notes.

CHANGING DIMENSION SYMBOLS

Dimensions within Pro/ENGINEER can be displayed in two ways. The first way is by showing the actual dimension or tolerance value. The second way is by showing the dimension symbol. As shown in Figure 4–18, when a dimension is created, it is provided with a dimension symbol. For the first dimension created on a part, the default symbol is [d0]. This number increases in sequential order for every new dimension. To change a dimension's symbol to make it more descriptive, utilize the Name option on the Dimension Properties Dimension Text tab.

REDEFINING FEATURES

Features are composed of parameters. Parameters can be modified and changed through the Edit Definition command or through dimension modification. Many different varieties of parameters exist. The following is a partial list:

- A feature's section
- The sketch plane for a feature
- The depth option for a feature
- The material removal side
- The extrude direction of a feature
- The trajectory of a swept feature
- The value of a dimension

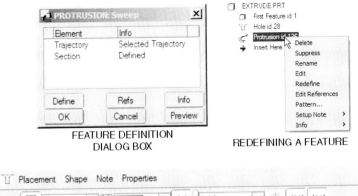

FEATURE DEFINITION
DIALOG BOX

REDEFINING A FEATURE

DASHBOARD DEFINITIONS

Figure 4–19   Feature Definitions

Figure 4–19 shows two examples of how Pro/ENGINEER organizes feature definitions. The first example, a feature definition dialog box, was the common interface for accessing feature definitions before the release of Wildfire. With this option, elements of a feature can be selected on the dialog and redefined through the Define menu option. Wildfire introduced a new approach to manipulating feature definitions. Through a dashboard, feature definitions are readily modifiable without the cascading menus normally associated with feature creation in Pro/ENGINEER.

MODELING POINT   Not all of Pro/ENGINEER's feature creation options have been converted to a dashboard interface. Users of Pro/ENGINEER can expect advanced features such as sweeps, blends, and variable-section sweeps to make this transition in later releases.

SUMMARY

The power of Pro/ENGINEER is its ability to capture product design intent through the intelligent incorporation of feature parameters and definitions. Properly modeled, parts are robust representations of products that can be utilized throughout the design process. The Extrude command is a basic modeling tool for building parts and capturing design intent. Suboptions exist for constructing solids, surfaces, and thin solids.

EXTRUDE TUTORIAL EXERCISE

This tutorial provides step-by-step instruction on how to model the part shown in Figure 4–20. This tutorial is the beginning guide for part modeling in Pro/ENGINEER and will reveal many of the modeling commands and modification techniques that are basic to part construction in Pro/ENGINEER.

This tutorial will cover:

- Starting a new model.
- Setting default datum planes.
- Creating an extruded protrusion.
- Creating an extruded cut.
- Dimension modification.
- Redefining a feature.
- Saving a part.

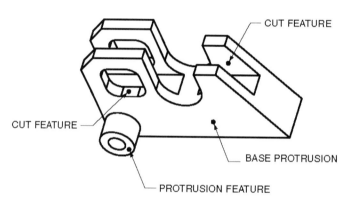

Figure 4–20   Finished Part

STARTING A NEW MODEL

This segment of the tutorial will explore the starting of a new part file.

STEP 1:   **Start Pro/ENGINEER.**

STEP 2:   **Use the FILE >> SET WORKING DIRECTION option to establish a working directory for your part file.**

Pro/ENGINEER utilizes a working directory to help manage files. When a new file is saved, it is saved in the current working directory, unless a new directory is specified. When the Open command is executed, the default directory is the current working directory.

STEP 3:   **Use the FILE >> NEW option to create a new part file.**

From the File menu, select the New option. Part is one of the modes of Pro/ENGINEER, and is the default mode. Enter a name for the new part. Part names must be less than 31 characters and cannot include spaces.

Figure 4–21   New Dialog Box   **Figure 4–22**   Sketch Dialog Box

Notice on your New dialog box how the Use Default Template option is checked (see Figure 4–21). This setting will use a specific part file as a seed file for the new object. Pro/ENGINEER's initial default template file for a new part includes a set of three default datum planes and the default coordinate system. The part file's default units is set at Inch lbm Second. This book assumes you will use this template file. The default template file for a part can be changed with the configuration file option *template_solidpart.*

STEP 4: **Enter a name for the new part file, then select OK.**

STEP 5: **Select EDIT >> SET UP >> UNITS option on the main bar.**

STEP 6: **On the Units Manager dialog box, select *Inch Pound Second (IPS)*, then select the SET icon.**

STEP 7: **On the Warning dialog box, select the CONVERT EXISTING NUMBERS option, then select OK.**

STEP 8: **Close the Units Manager dialog box.**

STEP 9: **Select DONE on the Part Setup menu.**

SKETCHING THE SECTION

Extruded protrusion features require a sketched, closed section. This segment of the tutorial will provided instruction on the sketching of the section for the base extruded feature.

STEP 1: **Select the Sketch Tool on the datum toolbar.**

Most features require some form of sketching. Because of this, sketching is a fundamental skill in Pro/ENGINEER. After selecting the sketch option, notice how Pro/ENGINEER reveals the Sketch dialog box (Figure 4–22). This dialog is used to define and orient the sketching environment.

STEP 2: **On the work screen, select Datum Plane FRONT as the Sketching Plane.**

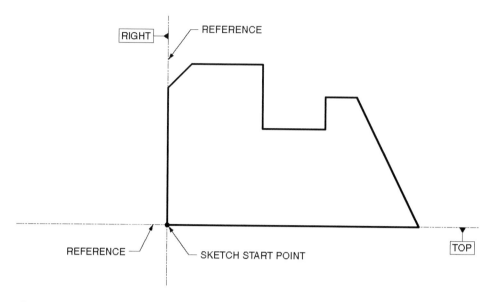

Figure 4–23   First Sketch

This extruded section will be sketched on datum plane FRONT. You can select the label associated with this datum (FRONT), you can select any portion of the boundary of the datum, or you can select the datum on the model tree.

STEP 3: **To orient the sketching environment, select Right as the orientation on the Sketch dialog box (see Figure 4–22) then on the work screen pick datum plane Right.**

You orient your sketching environment by selecting a direction of orientation (e.g., Top, Right, Left, and Bottom) followed by a plane to face in the direction of orientation. In this example, selecting Right from the Sketch dialog box followed by picking datum plane Right will orient datum plane Right toward the right of the work screen.

STEP 4: **Select the SKETCH option on the Section dialog box.**

STEP 5: Use the LINE icon to sketch the section shown in Figure 4–23.

In Figure 4–23, the dimensioning scheme defining the size of the sketch has been purposely hidden. When sketching, you should not worry about the size of your sketch. What is important is the sketching of geometry that matches the shape of your design intent. Start your sketch at the intersection of datum planes RIGHT and TOP, sketching in a counterclockwise direction. The left mouse button is used to pick entity locations, and the middle mouse button is used to cancel a command. Align the bottom edge of the feature with datum plane TOP, and align the vertical edge of the sketch with datum plane RIGHT.

STEP 6: **Dynamically zoom to the region shown in Figure 4–24.**

It is good sketching strategy to zoom in on the area being sketched. Use the following dynamic viewing capabilities to zoom to the shown region:

- **Dynamic Zoom** Scroll wheel on mouse or a combination of the Control key and the middle mouse button.
- **Dynamic Pan** Shift key and the middle mouse button.

-  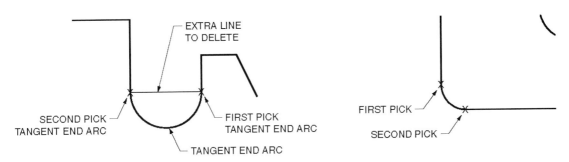 **Parallel Screen** Use this tool to orient the sketching plane parallel to the screen..

STEP 7:   Use the ARC icon option to add a tangent arc to the sketch (see Figure 4–24).

STEP 8:   Delete the extra line shown in Figure 4–24 by first selecting the entity with the Pick icon, then selecting the Delete key on your keyboard.

EXTRA LINE
TO DELETE

SECOND PICK
TANGENT END ARC

FIRST PICK
TANGENT END ARC

TANGENT END ARC

FIRST PICK

SECOND PICK

Figure 4–24   Tangent Arc

Figure 4–25   Filleted Arc

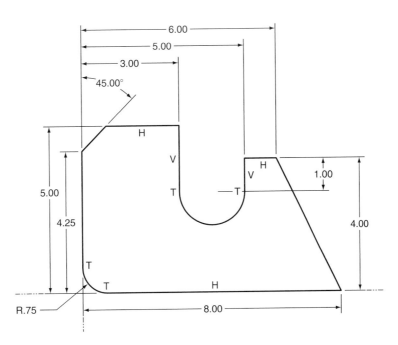

Figure 4–26   Dimensioning Scheme

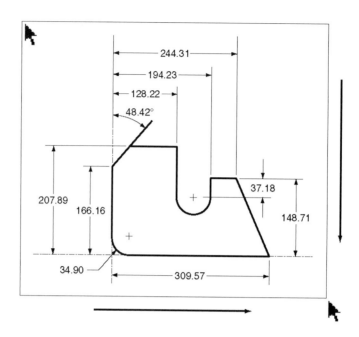

Figure 4–27   Select Dimension Using the Pick Icon

STEP 9:   Use the CIRCULAR FILLET icon to add a filleted arc to the sketch.

Add an additional arc as shown in Figure 4–25. The Circular Fillet option will require you to select two nonparallel entities. A fillet will be created between the selected entities.

STEP 10:   Place dimensions according to design intent.

Use the Dimension icon to match the dimensioning scheme shown in Figure 4–26. Placement of dimensions on a part should match design intent. With Intent Manager activated (Sketch >> Intent Manager), dimensions and constraints are provided automatically that fully define the section. Pro/ENGINEER does not know what dimensioning scheme will match design intent, though. Because of this, it is usually necessary to change some dimension placements.

Note: Use the left mouse button to locate dimension extension lines and use the middle-mouse button to place dimension line.

MODELING POINT   If possible, a good rule of thumb to follow is to avoid modifying the section's dimension values until your dimension placement scheme matches design intent.

STEP 11:   Using the Pick icon, drag a box around the entirety of the sketch making sure to include all dimensions (see Figure 4–27).

Your next task is to modify the sketch's dimension values. By preselecting all available dimensions, you will be able to simultaneously modify each dimension.

MODIFY THE LARGEST
VALUE TO EQUAL 8.00

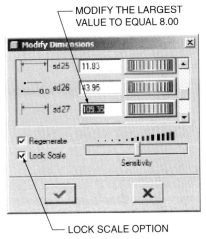

LOCK SCALE OPTION

Figure 4-28   Select Dimension Using the Pick Icon

STEP 12: Select the Modify icon then check the Lock Scale option (Figure 4-29).

The Lock Scale option will allow you to modify one dimension, with the remaining dimensions scaling the same factor.

STEP 13: Modify the largest dimension value to equal 8.00, select ENTER on your keyboard, and then select the CHECK icon on the dialog box.

At this point, it does not matter which dimension you modify to 8 units, though it is best to select the largest dimension. Your objective is to scale the model down to a workable size. The next step of this tutorial will require you to modify each dimension individually.

STEP 14: Using the Pick icon, double select the arc's radius dimension value then modify its value to equal 0.750.

Double picking a dimension value will allow it to be modified through Pro/ENGINEER's textbox. Remember to select Enter on your keyboard after modifying the value. The sketch will be modified automatically. Often, it is best to modify smaller dimension values first. If your sketch has unexpected results, select the Undo icon on the toolbar.

STEP 15: Starting with the smallest dimension values first, modify the remaining dimensions to match Figure 4-26.

Use the Undo icon on the toolbar to undo unexpected results.

MODELING POINT   Sketched protrusion sections have to be completely enclosed and with no intersections. If a section has intersecting entities or open geometry, an open-loop error will occur.

STEP 16: Select the Continue icon to exit the sketching environment.

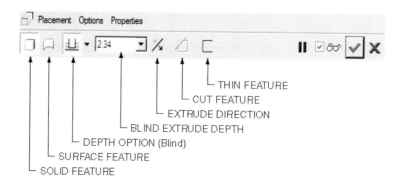

Figure 4–29   Dimension Scale Modification

CREATING AN EXTRUDED PROTRUSION

Extruded protrusion features are arguably the most common positive space feature found in a Pro/ENGINEER solid model. The extrude option is utilized to create extruded solids, surfaces, and cuts. This segment of the tutorial will create an extruded solid.

STEP 1:   Select the EXTRUDE icon on the basic feature toolbar.

An Extruded Protrusion is the most common type of geometry feature first created in a solid part. Other commands available include Revolve, Sweep, Blend, Helical Sweep, and Swept/Blend. You will be creating a solid feature in this exercise.

> **MODELING POINT**   The Extrude command utilizes a dashboard to define required and optional feature elements (Figure 4–29). Most features can be created by working left to right on the dashboard. For an extruded protrusion, the feature elements shown in Figure 4–29 are required elements.

STEP 2:   Select the BLIND depth option on the dashboard.

STEP 3:   Enter 2.00 as the variable depth.

STEP 4:   Dynamically view your part.

Dynamically view your part using the following options:

- Dynamic Rotation   Middle mouse button
- Dynamic Pan   Shift key and middle mouse button
- Dynamic Zoom   Mouse scroll wheel

STEP 5:   Select the Preview feature icon and observe model.

STEP 6:   Select the Build Feature icon to finish the feature.

Selecting the Build Feature option will create the feature. Observe the Model Tree to see the new feature (Figure 4–30).

STEP 7:   Save your part file.

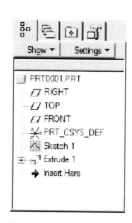

Figure 4–30
New Feature Added to Model Tree

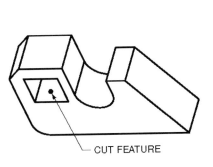

TOP ORIENTATION
(Pick Surface)

SKETCHING PLANE
(Select Surface)

CUT FEATURE

Figure 4–31   Cut Feature   **Figure 4–32**   Orienting the Sketch

CREATING AN EXTRUDED CUT

This segment of the tutorial will create the extruded Cut feature shown in Figure 4–31. Additionally, the Edit Definition option will be used to modify the sketch.

STEP 1: **Select the Sketch Tool option icon on the Datum toolbar.**

STEP 2: **Pick the front of the part (Figure 4–32) as the sketching plane.**

You will be sketching on the front of the part. Any planar surface can be used as a sketch surface.

MODELING POINT   If you select the first protruded feature as your sketching plane, this new Cut feature will become a child feature of the protruded feature. Any changes to a parent feature can affect its children.

STEP 3: **To orient the sketching environment, select Top as the orientation on the Section dialog box then on the work screen pick the top of the part (Figure 4–32).**

To orient the sketching environment, from the Sketch View menu, select Top, then pick the top of the part. When selecting a surface for orientation, the selected surface becomes a parent feature of the feature under construction. Try to select an orienting surface that is already a parent feature.

STEP 4: **Select the SKETCH option on the Sketch dialog box.**

MODELING POINT   Don't forget the dynamic viewing capabilities of Pro/ENGINEER. Dynamic viewing can be utilized during the execution of most Pro/ENGINEER commands.

STEP 5: **Select SKETCH >> REFERENCE on the menu bar then utilize the References dialog box, pick the two references shown in Figure 4–33 then close the dialog box.**

The design intent for this part requires this cut feature to be located from the top and left edges of the first part feature. As shown in Figure 4–33, select these two edges as references. Note that these two references may already be picked.

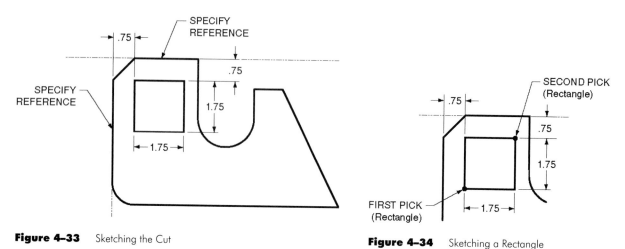

Figure 4–33   Sketching the Cut

Figure 4–34   Sketching a Rectangle

STEP 6:   Select **NO HIDDEN** as the display for your model.

You will find it easier to sketch with No Hidden or Hidden selected as the model's display mode.

STEP 7:   Use the Rectangle icon to sketch the section.

With the Rectangle option, you select diagonal corners of the rectangular feature.

STEP 8:   Apply the correct dimensioning scheme.

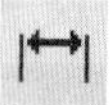

Use the Dimension icon to apply the dimensioning scheme that matches the design intent. In this example, the dimensioning scheme shown in Figure 4–34 matches the intent of the design.

STEP 9:   MODIFY dimension values.

Use the Modify icon to change the dimension values to match Figure 4–34. Each dimension can be preselected with the Pick icon and the Control key before you select the Modify option. Another valid option would be to modify each dimension individually by double selecting it with the pick option.

STEP 10:   On the menu bar, use the **TOOLS >> RELATIONS** option to add a dimensional Relationship.

The design intent for the part requires that this cut feature remain square. Use the Add option on the Relations menu to add a dimensional relationship that will make the horizontal dimension equal to the vertical dimension. As shown in Figure 4–35, the dimension symbol for the vertical dimension is *sd2* while the dimension symbol for the horizontal dimension is *sd3*. Your dimension symbols may be different. For the dimension symbols shown in Figure 4–35, enter the equation sd3 = sd2 in the text editor region of the dialog box. Again, your dimension symbols may be different from those in the figure.

STEP 11:   Add the dimension relationship shown in Figure 4–35.

STEP 12:   On the Relations dialog box, select the Verify Relations icon to verify the accuracy of your entered relationship.

While this relationship is easy to enter, more complex equations can be formed. The Verify Relations option is used to ensure that you have entered a proper dimensional relationship.

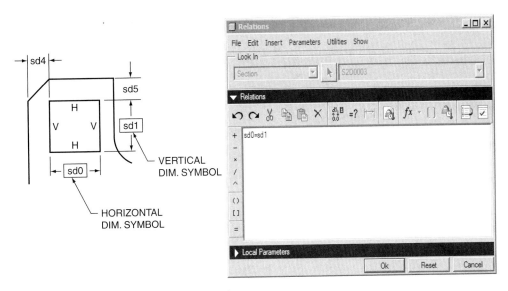

Figure 4–35 Adding Dimensional Relationships

STEP 13: If your relations verified correctly, select OK to exit the Relations dialog box.

STEP 14: Select the Continue icon to exit the sketcher environment.

STEP 15: Select the Extrude option on the toolbar.

Cut is a suboption on the Extrude dashboard.

STEP 16: Select the CUT option on the extrude dashboard.

STEP 17: Dynamically rotate your model.

Before selecting an extrude depth, dynamically rotate your object using the middle mouse button. Observe the current feature creation direction. Your cut should extrude into the part. If it doesn't, you will need to flip the direction of feature creation.

CAUTION In the following step you will need to change the direction of feature creation. Notice on the toolbar that there are two options with the same icon. The first icon is the feature creation direction, while the second icon is the material removal side.

STEP 18: If your cut does not extrude into the part, change the direction of extrusion.

Make sure that you pick the icon for changing the feature creation direction.

STEP 19: Select THROUGH ALL depth option.

The Through All depth option will create a cut whose depth always extrudes completely through any previously created features.

STEP 20: Preview your feature, then deselect the preview option.

Select the Preview icon on the dashboard. Dynamically rotate your part to observe changes.

STEP 21:  **Select the Build Feature icon to finish the feature.**

Selecting the Build Feature option will create the feature. Observe the Model Tree to see the new feature (Figure 4–30).

STEP 22: ⬛ **Save your part file.**

> **MODELING POINT** Previewing a feature is an important step during part modeling. Any conflicts that might exist between the new feature and existing features will normally be revealed during the preview process. If an error does occur, definitions associated with the feature can be redefined by using the Define option on the Feature definition dialog box.

EDITING THE FEATURE'S DEFINITION

In this section of the tutorial, you will modify the sketch of the cut feature.

STEP 1: **On the model tree, pick the sketch assocated with the last extruded feature.**

You can expand the second extruded feature on the model tree to see its associated sketch.

STEP 2: **Select the EDIT DEFINITION option on the mouse pop-up menu.**

STEP 3: 📝 **Modify the dimension value as shown in Figure 4–36.**

As shown in the figure, use the Modify option to change the 0.75 dimension value to equal 0.70.

Note: You can also double-pick dimension values to modify dimensions.

STEP 4: ✔ **Select the Continue icon to exit the sketching environment.**

STEP 5: **Dynamically rotate your part to observe the new feature.**

STEP 6: **Pick the last extruded feature on the model tree and select the EDIT DEFINTION option.**

Notice how you can edition the definition of a feature. In example, you could edit this feature's defintion to include depth, material removal side, extrude direction, etc.

STEP 7: ✔ **Select the Build Feature icon on the dashboard to finish the feature.**

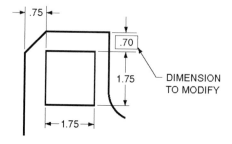

Figure 4–36 Modifying a Dimension

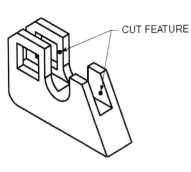

Figure 4–37   Cut Feature

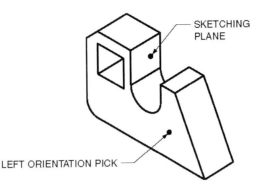

Figure 4–38   Sketching Plane and Left Orientation

Step 8:   Save the Part.

Creating a Second Extruded Cut

This segment of the tutorial will create the Cut feature shown in Figure 4–37.

Step 1:   Select the SKETCH tool on the dashboard.

Step 2:   Pick the sketching plane shown in Figure 4–38.

Step 3:   Observe the direction of cut.

Notice the arrow on your part protruding from the sketching plane. Since you will be cutting in both directions, this arrow represents the first direction of cut. This arrow is also important for the orientation of the sketcher environment. Within a sketcher environment, this arrow shows the direction that your sketch plane will be facing.

Step 4:   On the Sketch dialog box, select LEFT as the orientation then pick the plane shown in Figure 4–38.

By selecting Left and by selecting the front of the part, you will be orienting the front of the part toward the left of the sketcher environment.

Step 5:   Select the SKETCH option on the Sketch dialog box.

Step 6:   On Pro/ENGINEER's toolbar, turn off the display of datum planes.

Step 7:   Select SKETCH >> REFERENCES on the menu.

Step 8:   Use the References dialog box to specify the two references shown in Figure 4–39

If necessary, you can also use the References dialog box to delete unwanted references. Since regeneration errors are normally caused by conflicted references, it is usually advisable to delete references that are not needed.

Step 9:   Use the Line icon options to sketch the three lines representing the section.

Start the first line by using Line option. Make sure that the start point and endpoint of the sketch are aligned with existing geometry.

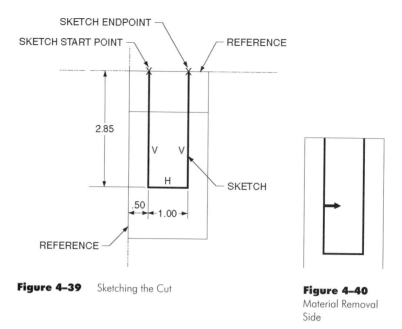

Figure 4–39   Sketching the Cut

Figure 4–40
Material Removal
Side

STEP 10:   Dimension geometry according to design intent.

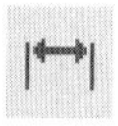

Figure 4–39 portrays design intent for the feature. Use the Dimension icon to dimension the feature.

STEP 11:   Modify dimension values.

Use the Modify option to modify your dimension values, or double-pick each dimension with the Pick icon.

STEP 12:   Select the Continue icon to exit the sketcher environment.

If you do not have a fully defined section, Pro/ENGINEER will give you an error message stating this fact. If this occurs, try to realign your elements or try adding dimensions that will fully define the sketch.

STEP 13:   Select the EXTRUDE tool.

STEP 14:   Dynamically rotate your model.

Dynamically rotate your model to observe the direction of cut.

STEP 15:   Working left to right on your dashboard, select the SOLID option.

STEP 16:   Select THROUGH ALL as the depth option.

STEP 17:   Select the OPTIONS menu item on the dashboard (Figure 4–41).

STEP 18:   Select THROUGH ALL as the depth option for both sides of the cut (see Figure 4–41).

STEP 19:   Deselect the Options menu item.

STEP 20:   Select the CUT option on the extrude dashboard.

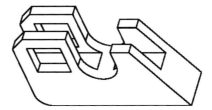

Figure 4–41   Two Sided
Extrusion

Figure 4–42   Finished Cut

STEP 21:   **Observe the direction of material removal.**

Pro/ENGINEER tries to determine the side of the section to remove material. As shown in Figure 4–40, an arrow points in the direction of material removal. If this is not the correct side, you use the Change Material Side icon to change the material removal side.

STEP 22:   **Select the Build Feature icon to finish the part.**

Your part should look like Figure 4–42.

STEP 23:   **SAVE your part.**

CREATING AN EXTRUDED PROTRUSION

This segment of the tutorial will create the extruded protrusion feature shown in Figure 4–43. This section highlights the Use Edge option, which turns existing part features and edges into sketch geometry.

STEP 1:   Select the Sketch option on the dashboard then pick the front of the part as the sketching plane (Figure 4–44).

STEP 2:   On the Sketch dialog box, select TOP as the orientation then pick the top of the part (Figure 4–44).

The Top option will orient the selected surface toward the top of the sketcher environment. After selecting the orienting surface, Pro/ENGINEER will launch the sketching environment.

STEP 3:   Select SKETCH on the Section dialog box.

STEP 4:   On the toolbar, turn off the display of datum planes.

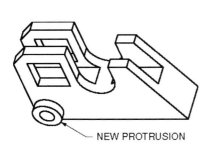

NEW PROTRUSION

Figure 4–43   Extruded Feature

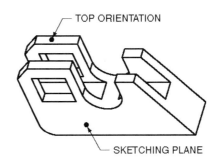

TOP ORIENTATION

SKETCHING PLANE

Figure 4–44   Sketching Plane and
Orientation

STEP 5:   On the toolbar, select No Hidden as the model's display mode.

STEP 6:   Utilizing the USE EDGE icon, pick the edge shown in Figure 4–45. (Make sure you select the edge only once.)

During this portion of the tutorial, you will pick an existing edge to project as an entity in this feature's section. The Use Edge option will turn existing feature edges into entities of the current sketch. When you select existing feature edges, the feature from which the edge is obtained becomes a parent feature of the feature under construction.

Pick the edge only once. Since the newly formed sketch entity will lie coincident with its parent edge, it may not be easily identifiable. Selecting a second time will create a second entity. This will cause an error when you attempt to exit the sketcher environment.

STEP 7:   Use the ARC icon to create the tangent end arc shown in Figure 4–45.

You will create an arc that joins the projected arc created in the previous step. In combination with the arc from the previous step, this newly created arc will form a circle. As shown in Figure 4–45, select at the first tangent point, then the second tangent point. When moving the cursor from Point 1 to Point 2, ensure that the arc is created toward the inside of the part.

STEP 8:   Use the Concentric circle icon to create the circle shown in Figure 4–46.

The Concentric Circle icon can be found under the normal Circle icon on the sketch toolbar. Perform the two picks shown in the figure then select the middle mouse button to end the concentric circle creation process.

STEP 9:   Modify the circle's diameter dimension value.

Modify the circle's dimension value to equal 0.75.

STEP 10:   Select the Continue icon to exit the sketcher environment.

After selecting continue, if you get the error message *Cannot have mixture of open and closed sections,* you probably picked the edge more than once while executing the Use Edge option. If necessary, use the Pick option and the Delete key to delete any extra entities.

STEP 11:   Select the Extrude tool.

This section of the tutorial will create an extrusion on the front surface of the part.

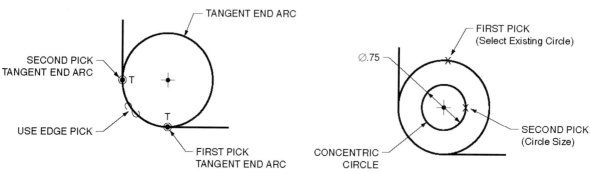

Figure 4–45   Tangent End Arc   **Figure 4–46**   Concentric Circle

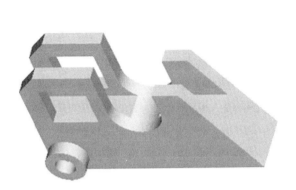

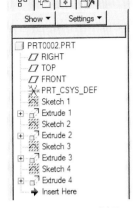

Figure 4–47   Preview of Model   **Figure 4–48**   Model Tree

STEP 12: **Dynamically rotate the model to observe the direction of extrusion.**

Dynamically rotate the part to get a better view of the extrusion direction.

STEP 13: **Select BLIND as the depth option.**

The Blind option will allow you to enter an extrude depth.

STEP 14: **Enter 0.500 as the variable extrude depth.**

On the extrude dashboard, enter 0.500 as the extrude depth.

STEP 15: **Shade your model and dynamically rotate to observe the protrusion.**

Your model should appear as shown in Figure 4–47.

STEP 16: **Select the Build Feature option on the dashboard.**

STEP 17: **Observe the model tree.**

The model tree for the part is shown in Figure 4–48. This part contains four extrusions and each extrude's respective associated sketch. In order of geometry creation:these extrusions consist of a protrusion, a cut, a second cut, and a second protrusion.

STEP 18: **Save your part.**

DIMENSION MODIFICATION

This segment of the tutorial will modify the dimensions used to define the second extruded cut. Covered will be options for modifying a dimension's value and tolerance format.

STEP 1: **On the model tree, right-mouse pick the third extrusion (the second cut feature) to access the pop-up menu (Figure 4–49).**

STEP 2: **On the pop-up menu, select the EDIT option.**

This step of the exercise requires you to select the second cut feature. You can select this feature directly on the work screen or you can select the

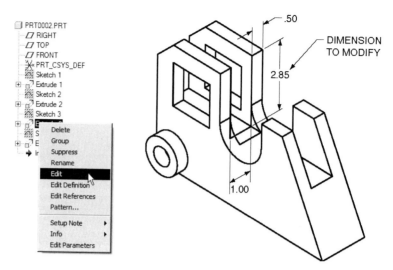

Figure 4–49
Model Tree Pop-Up Menu

Figure 4–50   Dimension Modification

feature from the model tree. When you select the feature, parametric dimensions used to construct the feature will appear (Figure 4–50).

> **MODELING POINT**   Often, it is difficult to select a feature directly on the workscreen. When it is necessary to select a feature or element directly on the workscreen, consider using the Activate Query List option from the pop-up menu. The query list will give you an option to toggle through available features. You can also use the filter option at the lower right corner of your screen to filter out all selectable entities except dimensions.

STEP   3:   Double-pick the 2.85 dimension and modify its value to equal 2.90.

After picking the dimension, Pro/ENGINEER requires you to enter the new dimension value in the textbox.

STEP   4:   Select the Regenerate icon on the toolbar.

After a dimension value has been modified, Pro/ENGINEER requires a regeneration of the part. Select the Regenerate option.

STEP   5:   On the Menu Bar, select the TOOLS >> ENVIRONMENT option.

STEP   6:   On the Environment dialog box, check the DIMENSION TOLERANCES option, then select OK.

The Environment dialog box is located under the Tools menu. Check the Dimension Tolerances setting. This will display dimensions with tolerances.

STEP   7:   Using the model tree's pop-up menu and the EDIT command, edit the second cut feature (the third extrusion as shown in Figure 4–51).

Right-mouse-button select the second cut feature and select the Edit option on the pop-up menu.

STEP   8:   Use the mouse's scroll wheel to dynamically zoom in on the cut feature's dimensions.

It will be easier to pick dimensions on the workscreen if you are zoomed in on the feature.

STEP   9:   Using the Control key for multiple item selection, pick the three dimensions associated with the second cut feature (Figure 4–51).

In Pro/ENGINEER, as in a standard Windows application, the Control key will allow for multiple item selection. Be sure to select only the text. While

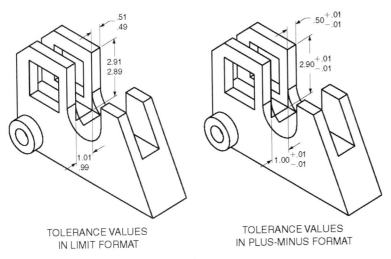

TOLERANCE VALUES
IN LIMIT FORMAT

TOLERANCE VALUES
IN PLUS-MINUS FORMAT

Figure 4–51   Dimension Tolerance Values

selecting dimension values, you might inadvertently pick a geometric
feature on the workscreen. This will cause dimensions associated with this
feature to be displayed also.

MODELING POINT   When you have to pick difficult to select items, such as the dimensions in Step 9,
consider changing the selection filter. The Filter box is located at the bottom of the screen on the status bar. By default, the Smart Filter allows you to pick "high-level" items such as features and datums. Other filter options include
dimensions, quilts, and annotations.

STEP 10:   **With the dimensions from the previous step highlighted, access the
mouse button's pop-up menu and select the Properties option.**

In Pro/ENGINEER, pop-up menus are accessible when the right mouse
button is selected anywhere within the work screen window.

STEP 11:   **On the Dimension Properties dialog box, select PLUS-MINUS as the
Tolerance mode (Figure 4–52).**

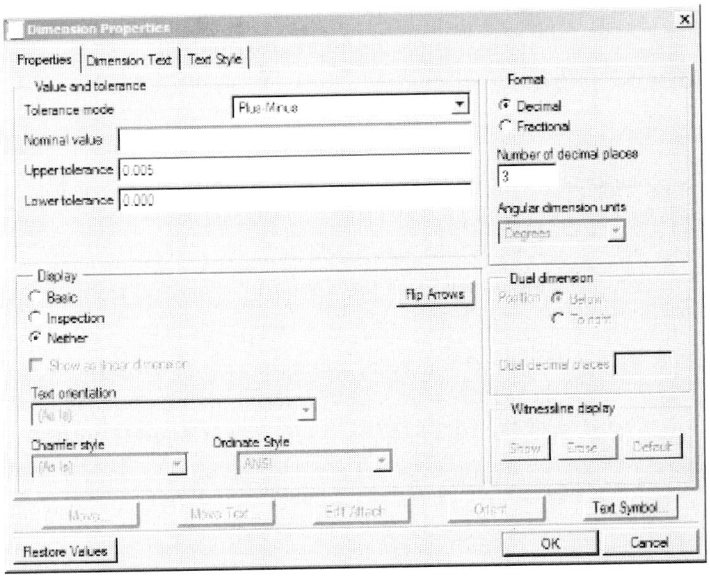

Figure 4–52   The Dimension Properties Dialog Box

STEP 12: **On the Dimension Properties dialog box, enter 3 as the number of decimal places.**

As shown in Figure 4–52, enter 3 as the number of decimal places.

STEP 13: **Enter Upper and Lower Tolerance values.**

As shown in Figure 4–52, enter 0.005 as the upper tolerance value and 0.000 as the lower tolerance value.

MODELING POINT Other options are available on the Dimension Properties dialog box. The Dimension Text tab is used for entering additional text, notes, and symbols around an existing dimension value. The Text Style tab is used to modify the font of a text.

STEP 14: **Select the OK option on the dialog box.**
STEP 15: **Save your model.**

REDEFINING A FEATURE'S DEPTH

This segment of the tutorial will redefine the extrusion depth of the last extruded feature. Within this redefinition, you will change the depth from a value of 0.500 to a value of 1.00. The Edit Definition command is one of Pro/ENGINEER's most useful and powerful options. It is used to redefine attributes and parameters associated with features. Feature parameters such as Feature Creation Direction, Material Removal Side, and Depth value can be modified.

STEP 1: **On the model tree, right-mouse select the last extrude feature and use the pop-up menu to access the EDIT DEFINITION command (Figure 4–53).**

As shown on the model tree, this tutorial has required the creation of two protrusion features and two cut features. This section of the tutorial will require you to modify the last protrusion feature.

STEP 2: **On the dashboard, enter 1.00 as the new Depth value.**
STEP 3: **Dynamically rotate your model to observe feature.**

STEP 4: **Select the Build Feature option.**
STEP 5: **Save your model.**

Figure 4–53 Redefining a Feature

REDEFINING A FEATURE'S SKETCH

This section of the tutorial will use the Edit Definition option to add filleted corners to the first cut feature.

STEP 1: **On the model tree, select the sketch definition associated with the first cut feature (the second extrude) with your right mouse button then select the EDIT DEFINITION command (Figure 4–54).**

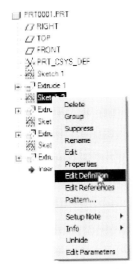

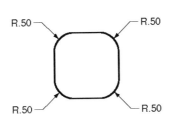

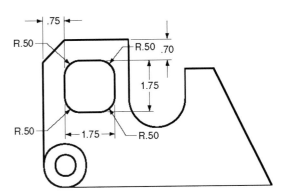

Figure 4–54   Redefining a
Section

Figure 4–55   New Section

Figure 4–56   Final Section

You can redefine a sketch directly on the model tree as performed in this
section of the tutorial or you can access the sketch through the feature's
dashboard by redefining the feature on the model tree.

STEP 2:   Select No Hidden as the model's display mode and turn off
the display of datum planes.

It is easier to sketch with the model not shaded and the display of datum
planes turned off.

STEP 3:   Using the Circular Fillet option, construct the filleted arcs shown
in Figure 4–55.

The Circular Fillet option requires you to pick the two entities bordering the
arc. The first pick will define the initial radius of the fillet.

STEP 4:   Modify the radius dimensions of each fillet to equal 0.500.

When selecting a dimension value to modify, Pro/ENGINEER requires that
you select the text of the dimension. Your section should appear as shown in
Figure 4–56.

STEP 5:   On the Menu bar, select the TOOLS >> RELATIONS option.

MODELING POINT   The original sketch for this feature was created with the Rectangle option. The default rectangle scheme dimensions the length of each line of the rectangle. A previous step in this tutorial required
the establishment of a dimensional relationship between the two defining dimensions. The Circular Fillet option will
create new dimensions with new dimension symbols which in turn will create invalid dimension symbols in your relations equation. Use the Relations dialog box to adjust the symbols defining the dimensional relationship.

STEP 6:   Using the dimension symbols from your sketch, modify the relations
equation to create a valid dimensional relationship.

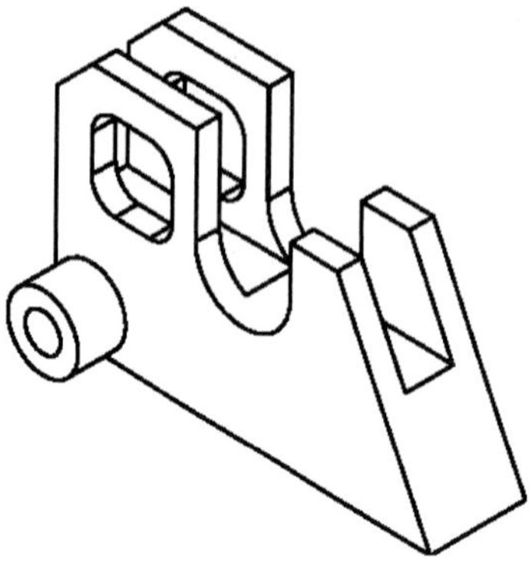

Figure 4–57 Completed Part

STEP 7:  **Select the Verify Relations icon on the Relations dialog box.**

If you don't get a positive verification, reenter your equation with proper dimension symbols.

STEP 8: **If you receive a valid verification, select OK to confirm the verification then select OK to exit the Relations dialog box.**

STEP 9:  **Select the Continue icon to exit the sketching environment.**

STEP 10: **Shade and dynamically rotate your part.**

Your model should look similar to Figure 4–57.

STEP 11: **Save your part.**

STEP 12: **Use the FILE >> DELETE >> OLD VERSIONS option to purge old versions of the part.**

Every save of the part creates a new version. Use the File >> Delete >> Old Versions option to delete every version except for the last saved.

DATUM TUTORIAL

This tutorial will provide step-by-step instruction on how to create various datums within Pro/ENGINEER. The part shown in Figures 4–58 and 4–59 will be the base for this tutorial. Creating this part will be the first segment of this tutorial.

This tutorial will cover:

- Creating a part
- Creating datum planes
- Creating datum points
- Creating datum axes
- Creating a coordinate system

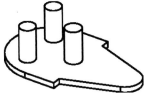

Figure 4–58   Base Part

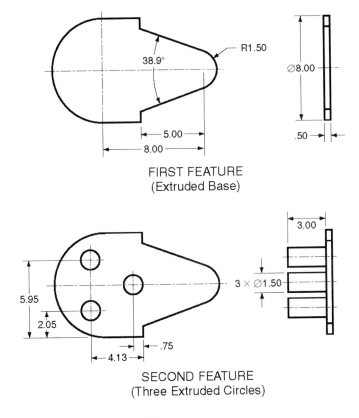

FIRST FEATURE
(Extruded Base)

SECOND FEATURE
(Three Extruded Circles)

Figure 4–59   Drawing of Part

CREATING THE PART

This first step in this tutorial is to model the part shown in Figure 4–58. This part (*datum1.prt*) is also available for download from the book's Web page. Because of the nature of this tutorial, do not use Pro/ENGINEER's default datum planes. This requirement can be accomplished by deselecting the Use Default Template option on Pro/ENGINEER's New dialog box, followed by selecting Empty as the template file.

Create this part with two extruded protrusions. Figure 4–59 shows details for each feature. Consider the following when creating this part:

- Name the part file *datum1*.
- Do not use Pro/ENGINEER's default datum planes.
- Sketch the flat base feature first as an extruded protrusion (Figure 4–59). Use the Centerline icon (located behind the Line icon) to create the two shown centerlines.
- Use the Edit >> Toggle Construction option to create construction entities as needed.
- Remember to utilize available entity constraints.
- Create the three shaft features as one extruded protrusion, utilizing the top of the first feature as the sketching plane.
- When creating the part, use the dimensioning schemes shown in Figure 4–59. These schemes match the design intent for the part.

CREATING DATUM PLANES

This section of the tutorial will explore Pro/ENGINEER's datum plane creation capabilities. Do not start this section of the tutorial until you have created the part shown in Figures 4–58 and 4–59. The following datum planes will be created:

- As shown in Figure 4–60, a datum plane will be placed through datum axes A_2 and A_3. This datum plane will be used to construct the remaining datum planes.
- A datum plane will be offset from the first datum plane.
- A datum plane will be made tangent to two selected surfaces.
- A datum plane will be constructed through an edge and at an angle to the first datum plane.

STEP 1:   Select the Datum Plane icon on the datum toolbar.

STEP 2:   **Change your selection filter to AXIS (Figure 4–61).**

The Selection Filter option is available in the lower right-hand corner of your workscreen. Filtering out all elements except for axes will ensure that you only pick axes in the next step.

STEP 3:   **Holding down the Control key, pick the first two axes shown in Figure 4–60.**

The Control key will allow for the multiple selection of entities. For this datum, your plane will pass through the two axes.

STEP 4:   **On the Datum Plane dialog box, ensure that both selected axes have THROUGH as their constraint option (Figure 4–62).**

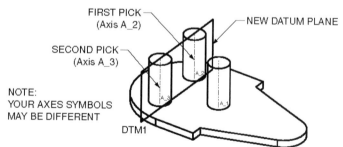

FIRST PICK (Axis A_2)

NEW DATUM PLANE

SECOND PICK (Axis A_3)

NOTE: YOUR AXES SYMBOLS MAY BE DIFFERENT

DTM1

Axis

Figure 4–60   Through Axes Datum Plane

Figure 4–61   Selection Filter

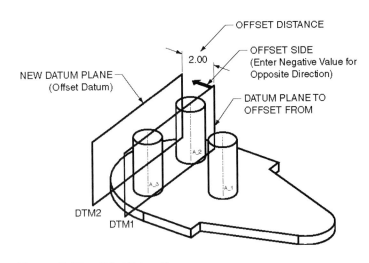

Figure 4–62 Datum Plane Creation   **Figure 4–63** Offset Datum Plane

The Through constraint option allows datum planes to be constructed through an axis, edge, curve, point, vertex, cylinder, or plane. With the exception of Through/Plane datum planes, all Through constraint options require a second paired constraint option. In this example, the second constraint option will place the datum through the second axis.

STEP 5: **Select OK on the Datum Plane dialog box.**

Your datum should appear as shown in Figure 4–60.

STEP 6: **Select the Datum Plane icon on the Datum toolbar.**

The next several steps of this tutorial will require you to create a datum plane offset from an existing plane (Figure 4–63).

STEP 7: **Change your selection filter to DATUM PLANE.**

STEP 8: **On the workscreen, pick the previously created datum plane (Figure 4–63).**

You will offset from the first datum plane.

STEP 9: **On the Datum Plane dialog box, ensure that OFFSET is selected as the constraint option.**

The Offset option will create a datum plane offset from an existing plane. The existing plane may be any planar surface.

STEP 10: **On the workscreen, observe the direction that your datum plane will offset.**

STEP 11: **On the Datum Plane dialog box, enter a translation of 2.00 in the direction shown in Figure 4–63.**

Enter a negative value if necessary to place the required datum plane.

STEP 12: **Select OK on the Datum Plane dialog box.**

The datum plane should appear similar to Figure 4–63. If the offset side is incorrect, modify the offset to a value of °2.00.

MODELING POINT Don't forget the point of this tutorial. The creation of datum features, especially datum planes, is an important skill for part modeling. Datum planes can be used as sketch planes and for other feature construction purposes.

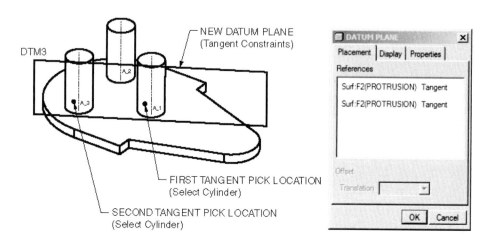

Figure 4-64   Tangent Datum Plane

STEP 13:   **Deselect any current entity by pick any empty space on your screen.**

STEP 14:   Select the Datum Plane icon on the Datum toolbar.

The next several steps of this tutorial will require you to create a datum plane tangent to two cylinders (Figure 4–64).

STEP 15:   **Change your selection filter to ALL.**

STEP 16:   **Using your control-key, pick the two extruded cylinders shown in Figure 4–64.**

Your new datum plane will be tangent to the two selected surfaces.

STEP 17:   **On the Datum Plane dialog box, ensure that TANGENT is selected for both constraint options (Figure 4–64).**

The Tangent constraint option will create a datum plane tangent to a cylindrical surface.

STEP 18:   **Select OK on the Datum Plane dialog box.**

Your model should appear as shown in Figure 4–64.

STEP 19:   **Deselect any current entity by pick any empty space on your screen.**

STEP 20:   Select the Datum Plane icon on the Datum toolbar.

The next several steps of this tutorial will require you to create a datum plane through an edge and at an angle to an existing plane (Figure 4–65).

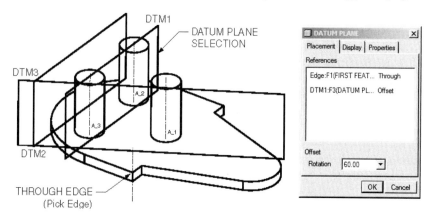

Figure 4-65   Through Edge Datum Plane

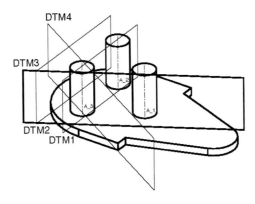

Figure 4–66   Finished Datum Planes

STEP 21:   Change your selection filter to EDGE.

STEP 22:   On the workscreen, pick the edge shown in Figure 4–65.

Your datum plane will pass through this edge.

STEP 23:   On the Datum Plane dialog box, ensure that THROUGH is selected as for the first constraint option (Figure 4–65).

STEP 24:   Change your selection filter to DATUM PLANE.

STEP 25:   Using the Control key, pick datum plane DTM1 as shown in Figure 4–65.

The Control key will allow you to use multiple constraint options in the Datum Plane dialog box. In this example, you will create a datum plane through an edge and at an angle to datum plane DTM1.

STEP 26:   Ensure that OFFSET is selected for the second constraint option (Figure 4–65).

STEP 27:   On the Datum Plane dialog box, enter 60.00 as the angle of rotation.

Note that negative values can be entered. If your datum plane does not resemble Figure 4–66, enter °60.00 as the value.

STEP 28:   Select OK on the Datum Plane dialog box.

STEP 29:   Save your model.

CREATING A DATUM AXIS

This section of the tutorial will explore the creation of a datum axis. The Datum axis shown in Figure 4–67 will be created.

STEP   1:   Select the Datum Axis icon on the Datum toolbar.

STEP   2:   Change the selection filter to SURFACE.

STEP   3:   Pick the curved surface shown in Figure 4–67.

The Datum Axis will be created through the center of the selected surface.

STEP   4:   Ensure that THROUGH is selected as the constraint option.

The Through Cylinder option will place a datum axis through the center of a cylinder or circular surface.

STEP   5:   Select OK on the Datum Axis dialog box.

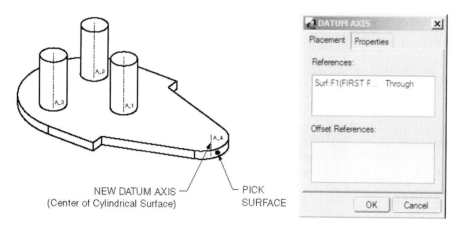

Figure 4–67   Datum Axis

CREATING A COORDINATE SYSTEM

This section of the tutorial will create the Coordinate System shown in Figure 4–68.

STEP 1:   Select the Coordinate System icon on the Datum toolbar.

Coordinate systems are considered by Pro/ENGINEER to be a type of datum feature.

STEP 2:   Turn off the display of datum planes.

Use the datum plane display option on the toolbar to turn off the display of datum planes.

STEP 3:   Change your selection filter to SURFACE.

STEP 4:   Using the Control key, pick in order the three surfaces shown in Figure 4–69.

STEP 5:   On the Coordinate System dialog box, change to the Orientation tab.

STEP 6:   Under the Orientation tab, change the axis orientations to match Figure 4–70.

You should manipulate the orientation options so that your datum axes match Figure 4–70. Notice in the figure how the To Determine options have been set to Z and X respectively.

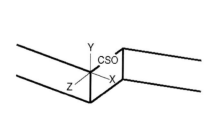

Figure 4–68   Coordinate System

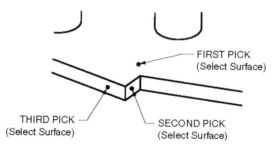

Figure 4–69   Plane Selection

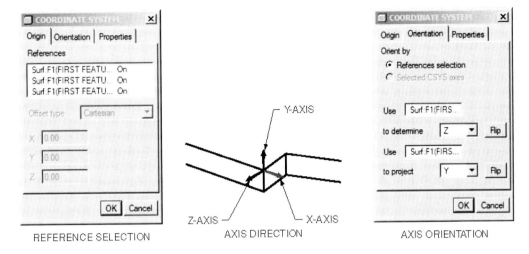

Figure 4–70 Axis Definitions

STEP 7: **When your axes match Figure 4–70, select OK on the dialog box to create the coordinate system.**

STEP 8: **Save your part then delete old versions.**

From the File menu, select the Save option then use the File >> Delete >> Old Versions option to delete old versions of the part file.

PROBLEMS

1. Use Pro/ENGINEER to model the parts shown in Figures 4–71 through 4–75.

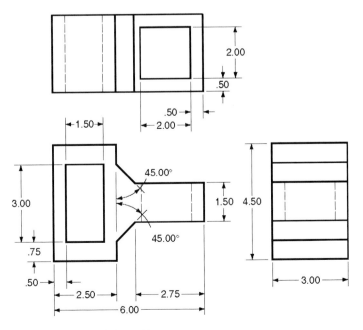

Figure 4–71 Problem 1 (a)

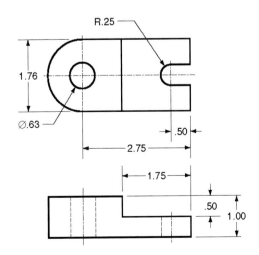

Figure 4–72 Problem 1 (b)

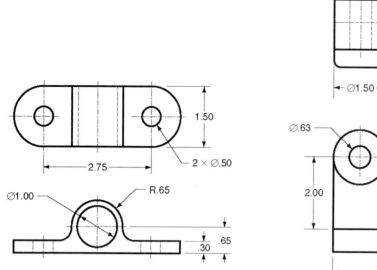

Figure 4–73   Problem 1 (c)

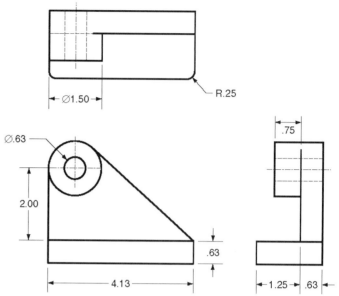

Figure 4–74   Problem 1 (d)

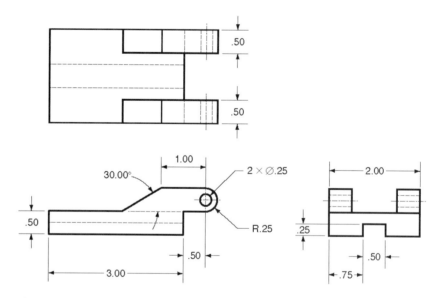

Figure 4–75   Problem 1 (e)

2. Use Pro/ENGINEER to model the part shown in Figure 4–76. Use millimeter-newton-second as the unit system for the part.

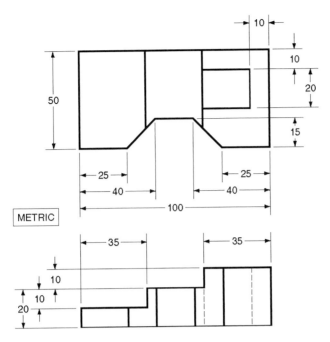

Figure 4–76   Problem 2

QUESTIONS AND DISCUSSION

1. Describe different forms of features available under the Extrude command.

2. What is the difference between a solid feature, a thin feature, and a surface feature?

3. How does a symmetric extrusion differ from a one-sided extrusion?

4. Compare the Through-Until depth option with the Through-Next depth option.

5. Describe methods available within Part mode to modify a dimension's value.

6. How does a cosmetic feature differ from a geometric feature?

7. What option within Pro/ENGINEER is available for changing the depth definition for an extruded feature?

8. What can be utilized as a sketching plane within Pro/ENGINEER?

5

FEATURE CONSTRUCTION TOOLS

Introduction

Pro/ENGINEER provides many feature construction tools that do not rely upon a sketcher environment. Examples of these include holes, rounds, and chamfers. This chapter will cover the basics of these and other common feature construction commands. Upon finishing this chapter, you will be able to:

- Construct a straight linear hole.
- Construct a straight coaxial hole.
- Construct a standard hole.
- Create fillets and rounds on a part.
- Create a chamfer on a part.
- Tweak features using the draft option.
- Create a part shell.
- Create a cosmetic thread.
- Create a linear pattern.

DEFINITIONS

Draft surface	A part surface that is angled to allow for ease of removal of the part from a mold or cavity.
Cosmetic feature	A part feature that allows for cosmetic details that do not require complicated regenerations. An example of a cosmetic feature would be a cosmetic thread, or a company logo sketched on a surface.
Sketched hole	A hole, such as a counterbored or countersunk hole, that has a sketched profile.
Standard hole	A hole with a size and attributes based on recognized thread standards.

HOLE FEATURES

The Hole command is used to create either straight holes, holes with varying profiles, or holes defined by standard fastener tables. A hole is considered a negative space feature, and straight holes can be created as a cut feature. Why use the Hole command instead of Cut?

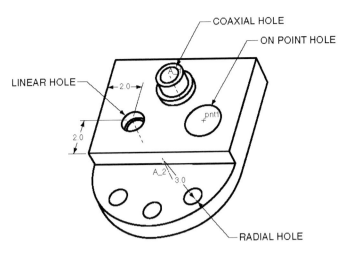

Figure 5-1   Hole Placement Options

The Hole command has a predefined section that does not require sketching. Additionally, the Hole command's algorithms are not as complicated as a cut feature's, thus requiring less computer power to regenerate.

HOLE PLACEMENT OPTIONS

There are four hole placement options available (Figure 5–1). The following is a description of each:

LINEAR

The Linear option locates a hole from two feature planes or edges. Planes must be perpendicular to the hole's placement plane.

DIAMETER AND RADIAL

The Radial and Diameter options will locate a hole at a distance from an axis and at an angle to a reference plane. It is used often for radial patterns such as might be found on a bolt-circle pattern. This option is covered in detail in chapter six.

COAXIAL

The Coaxial option locates the center of a hole coincident with an existing axis. The user is required to provide the axis, placement surface, and hole diameter.

ON POINT

The On Point option locates the center of a hole on a datum point. The user is required to provide the point, placement surface, and hole diameter.

HOLE TYPES

Pro/ENGINEER provides three options for defining the profile of a hole: Simple, Standard, and Sketched. The Straight option produces a hole that has a rectangular profile throughout the length of the hole. The bottom of the hole can be flat or pointed to represent a drilled hole. The Sketched option requires the user to sketch the profile of the hole within a sketcher environment. Figure 5–2 is an example of a sketched profile of a counterbored hole. The Sketched Hole option is covered in chapter six.

A standard hole is defined through a fastener table. An example would be a 1.00 unified national course thread. It has 8 threads per inch with a tap drill of 7/8 in. As shown in

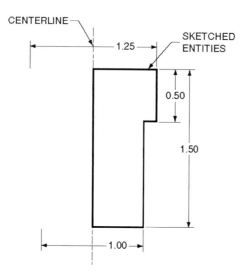

Figure 5–2   A Sketched Hole

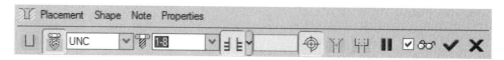

Figure 5–3   Standard Hole Options

Figure 5–3, a standard hole is specified along with a tapped screw size equal to *1-8*. This will produce an interior thread minor diameter equal to .875 (or 7/8 in). Pro/ENGINEER derives this information from three default text files: *ISO.hol, UNC.hol.,* and *UNF.hol.* Additional user-defined hole charts can be created and specified with the configuration file option *hole_parameter_file_path.*

The following options are available for standard holes:

- **Tapped Hole** The Tapped Hole option is selected when a fastener's threads are engaged within the hole. Setting this option will provide hole parameters for an internal thread (i.e., minor diameter and thread length). This option is only available with through-all standard holes.

- **Clearance** This is the defined parameter when Tapped Hole is not selected. The Clearance option is selected when a fastener's threads are not engaged within the hole. Setting this option provides a clearance fit for the fastener.

- **Include Thread Surface** This option is available only when the Tapped Hole option is selected. It provides a cosmetic representation of the thread. This option is available under the Shape menu option on the hole dashboard.

- **Thread Length** This option provides the length of the threads within the standard hole and is available only with tapped holes. This option can be found under the Shape menu option.

- **Counterbore** As the name implies, the Counterbore option creates a counterbored hole. The user must enter the counterbore's diameter and depth.

- **Countersink** Like the Counterbore option, the Add Countersink option creates a countersunk hole. The user must enter parameters for the hole.

MODELING POINT Standard hole parameters can be viewed and modified under the Properties option on the hole dashboard (Figure 5–3). Property values are received from a hole's fastener table. The Note option on the dashboard allows for the viewing and modification of a hole's standard note.

HOLE DEPTH OPTIONS

Like features created in the Extrude option, straight and standard holes have several available depth options. As with extruded features, the selection of a hole's depth option is important for incorporating design intent. As an example, it is common for holes to cut completely through a part. If this is the intent of a design, then the Through All option should be used. The following are available depth options.

- **Variable** The Variable option cuts a hole to a user-defined depth. The constructed hole has a flat bottom.

- **Through Next** This option will cut the hole to the next part surface.

- **Through All** This option will cut a hole completely through a part.

- **Through To Surface** This option cuts a hole to a user-selected surface. The constructed hole has a flat bottom.

- **To Reference** This option cuts a hole to a user-selected point, vertex, curve, or surface. The reference has to exist before this option can be executed. The hole will have a flat bottom after construction.

- **Symmetric** This option creates a two-sided hole with equal depths on both sides of the placement plane.

CREATING A STRAIGHT LINEAR HOLE

Perform the following steps to create a Straight-Linear Hole:

STEP 1: Select the Hole icon on the toolbar.

STEP 2: Pick the hole's placement plane.

STEP 3: Select the Straight Hole icon on the hole dashboard.

STEP 4: Select the hole's placement references.

There are two ways to select a hole's placement references:

- Drag the hole's placement references on the workscreen to an appropriate edge or plane.
- Use the Placement menu option on the dashboard to pick placement references (Figure 5–4). Use the Control key to pick multiple references.

INSTRUCTIONAL NOTE Planar placement references such as part surfaces and datum planes have to be normal to the hole's placement plane.

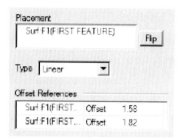

Figure 5–4   Hole Placement References

STEP 5:   **Enter distance values for the hole's placement references.**

Depending upon the referencing method selected in Step 4, you can enter dimension values directly on the screen or you can enter them under the Placement (Figure 5–4).

STEP 6:   **Select a depth parameter (e.g., Through All, Variable), then perform the operation appropriate to the option.**

INSTRUCTIONAL NOTE   If you select a depth parameter such as Variable or Through Until, you will have to perform an additional step. As an example, if you select the To Reference option, you will have to pick a point, curve, or surface to which to cut the hole.

STEP 7:   **Enter a Diameter value for the hole.**

STEP 8:   **Select the Build Feature Checkmark.**

CREATING A STRAIGHT COAXIAL HOLE

The Coaxial option locates the center of a hole coincident with an existing axis. Perform the following steps to create a straight coaxial hole:

STEP 1:   Select the Hole icon.

STEP 2:   **Select the Axis reference on the model.**

An axis is the primary reference for a coaxial hole. If necessary, change your selection filter to Axis.

STEP 3:   **Select the Placement menu option on the dashboard.**

STEP 4:   **Using the control-key, pick the hole's Placement Plane.**

Using the control key will allow you to pick the secondary reference without undefining the previously picked primary reference. For coaxial holes, the placement plane is the secondary reference.

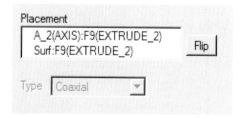

Figure 5–5   Coaxial Hole Placement

STEP 5:   Enter a Diameter value for the hole.

STEP 6:   Select a depth parameter (e.g., Through All, Variable), then perform the operation appropriate to the option.

INSTRUCTIONAL NOTE   If you select a depth parameter such as Variable or Through Until, you will have to perform an additional step. As an example, if you select the To Reference option, you will have to pick a point, curve, or surface to which to cut the hole.

STEP 7:   Select the Build Feature Checkmark.

ROUNDS

Pro/ENGINEER utilizes the Round command to create both fillets and rounds. Despite its apparent simplicity, the Round command can be one of Pro/ENGINEER's most difficult tools to master. Rounds constructed on complex features often result in failures. Several modeling techniques should be followed to help avoid round conflicts:

- Create rounds toward the end of the modeling process.
- Create smaller-radius rounds before larger.
- Avoid using round geometry as references for the creation of features.
- If a surface is to be drafted, draft the surface first, then create any necessary rounds.

ROUND-RADIUS OPTIONS

Several different Round options are available within Pro/ENGINEER (Figure 5–6):

- **Constant**   Constant rounds are created with a constant radius.
- **Variable**   Variable rounds are created with a variable radius. Radius values are defined from the end of chained segments.
- **Through Curve**   The Through Curve option defines a round's radius on the basis of a selected datum curve.
- **Full Round**   The Full Round option creates a round in place of a selected surface.

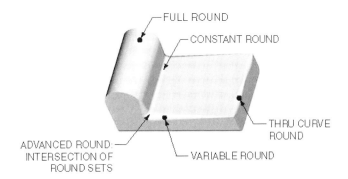

Figure 5–6   Round Options

ROUND REFERENCE OPTIONS

Rounds are normally created on the edge of a feature. Pro/ENGINEER provides the Edge Chain option to perform this task. Other reference selection options are available to allow for flexibility in the round creation process:

- **Edge Chain** This option allows for the selection of edges to place rounds. Suboptions exist for selecting edges one at a time or for selecting tangent edges.
- **Surf-Surf** This option allows for the selection of two surfaces to place a round. The round will be formed between the two surfaces.
- **Edge Pair** This option leads to the Full Round radius option. With this option, the surface between two selected edges will be replaced with a round.

ROUND SETS

A round set can hold a variety of parameters to include radius values, shape elements, and transitions. In Wildfire, round sets are manipulated with the Control key. To start the first round set, continue holding down the Control key while selecting references. To start a new round set, release the Control key, then pick the first reference in the new round set. To continue adding references to the new round set, use the Control key. The Sets menu option on the dashboard allows for the manipulation of round set radius values and shape elements.

SHAPE ELEMENTS AND TRANSITIONS

The default shape of a round in Pro/ENGINEER is a circular arc round with a rolling ball shape. A circular arc profile can be changed to a conic arc profile and the rolling ball shape can be changed to a normal-to-spine shape. Three transitions are available (Figure 5–7): Corner Sweep, Patch, and Round Corner. The corner sweep option sweeps the round with the radius value of the large radius in the available round sets. The patch option will place a patch element at the intersection of joining rounds.

CREATING A SIMPLE ROUND

Perform the following steps to create a simple round:

STEP 1:   Select the Round option on the toolbar.

STEP 2: Use the Control key to pick edges to round.

Selecting edges with the Control key will allow for all rounds to be included in one round set. Releasing the Control key and picking a new edge will define a new round set.

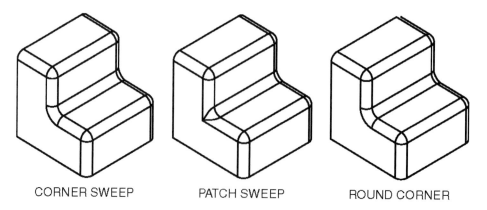

CORNER SWEEP   PATCH SWEEP   ROUND CORNER

Figure 5–7   Round Transitions

Step 3:   Enter a radius value for your round set.

Step 4:   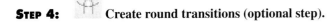Create round transitions (optional step).

Step 5:   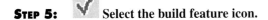Select the build feature icon.

Chamfer

Pro/ENGINEER provides a construction option for creating edge and corner chamfers (Figure 5–8). An Edge Chamfer creates a beveled surface along a selected edge. The Corner Chamfer option creates a beveled surface at the intersection of three edges.

Chamfers are created in similar way to rounds. More specifically, as with rounds, multiple chamfer sets can define difference round attributes. Perform the following steps to create an edge chamfer:

Step 1:   Select the Chamfer option.

The Chamfer command creates a beveled surface at a selected solid edge. *Note:* the Corner chamfer option is available under the Insert >> Chamfer menu option.

Step 2:   Use the Control key to pick edges to chamfer.

Selecting edges with the Control key will allow for all chamfers to be included in one chamfer set. Releasing the Control key and picking a new edge will define a new chamfer set.

Step 3:   Select an edge chamfer dimensioning scheme.

The following dimensioning schemes are available:

- **45 x D** This option creates a chamfer with a 45 degree angle and with a user-specified distance.

- **d x d** This option creates a chamfer at a user-specified distance from an edge.

- **d1 x d2** This option creates a chamfer at two specified distances from an edge.

- **Ang x d** This option creates a chamfer at a user-specified distance from an edge and at a user-defined angle.

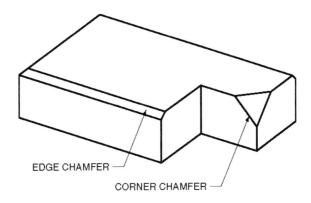

EDGE CHAMFER

CORNER CHAMFER

Figure 5–8   Chamfer Types

Figure 5–9   Chamfer Dashboard

STEP 4:   **Enter dimension values appropriate to the scheme selected in Step 3 (see Figure 5–9).**

STEP 5:   Create chamfer transitions (optional step).

STEP 6:   Select the build feature icon.

DRAFT

Cast and molded parts often require a drafted surface for ease of removal from the mold. Pro/ENGINEER provides multiple options for creating drafted surfaces. The maximum angle that may be created is ±30 degrees.

Selected surfaces are drafted by pivoting around a neutral plane or curve. Planes may be surface planes or datum planes, while curves may be datum curves or edges. Additionally, a surface can be split at the neutral plane or curve.

NEUTRAL PLANE DRAFTS

With a neutral plane draft, picked surfaces are pivoted around a selected draft hinge. The hinge can be a part surface or a datum. Three split options are available (Figure 5–10):

NO SPLIT

The No Split option creates a draft without a split in the drafted surface (Figure 5–10). The user selects the draft hinge and the draft surface, then enters a draft angle. The angle can be positive or negative.

SPLIT AT DRAFT HINGE

The Split at Draft Hinge option creates a draft with the drafted surface split at the selected draft hinge (Figure 5–10). The portion of the surface selected for drafting will be the surface to which the draft angle is applied. The user selects the neutral plane and the draft surface then enters a draft angle.

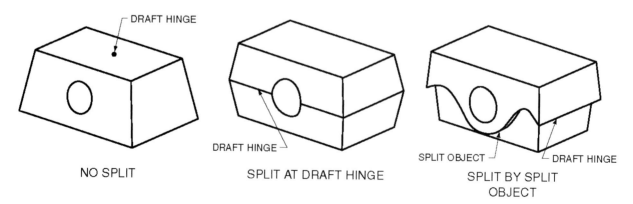

Figure 5–10   Neutral Plane Drafts

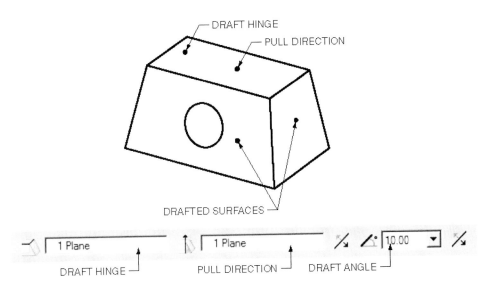

Figure 5-11 Neutral Plane No Split Draft

SPLIT BY SPLIT OBJECT

The Split by Split Object option creates a drafted surface out of a user-sketched section (Figure 5–10). The sketched section is pivoted around the neutral curve. The user selects the neutral plane and a surface to sketch upon. A sketcher environment provides the user with the requirement of sketching the surface to be drafted.

CREATING A NEUTRAL PLANE NO SPLIT DRAFT

Perform the following steps to create a neutral plane draft. The part (*draft1.prt*) shown in Figure 5–11 is available on the book's Web page.

STEP 1: Select the Draft tool on the engineering features toolbar.

STEP 2: Using the Control key, pick surfaces to draft.

STEP 3: Select the Draft Hinges reference box (see Figure 5–11).

Draft options available include Neutral Plane for pivoting around a plane or surface and Neutral Curve for pivoting around a datum curve or edge.

STEP 4: On the workscreen, select the surface to serve as the draft hinge (or neutral plane).

There are two important points to remember when selecting a draft hinge. First, the selected surface must be normal to the surfaces being drafted. Second, the draft hinge defines the pivot point for rotating the drafting surfaces. The draft hinge surface will keep its form and size.

STEP 5: Select the Pull Direction reference box (optional step).

The pull direction plane has to lie perpendicular to the draft surfaces. Since this surface is often the same as the draft hinge, Pro/ENGINEER automatically will make the pull direction plane the same as the draft hinge.

STEP 6: Select the Pull Direction plane.

STEP 7: Enter a draft angle.

STEP 8: Select the Build Feature option on the dashboard.

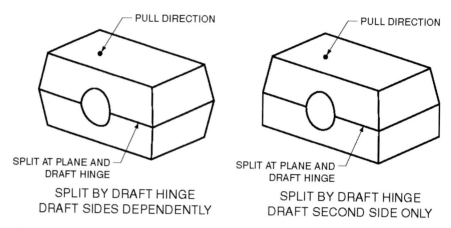

PULL DIRECTION   PULL DIRECTION

SPLIT AT PLANE AND   SPLIT AT PLANE AND
DRAFT HINGE   DRAFT HINGE

SPLIT BY DRAFT HINGE   SPLIT BY DRAFT HINGE
DRAFT SIDES DEPENDENTLY   DRAFT SECOND SIDE ONLY

Figure 5-12   Neutral Plane Split Draft

CREATING A NEUTRAL PLANE SPLIT DRAFT

A split draft occurs when surfaces on either or both sides of the neutral plane pivot separately. The surfaces can draft with the same angle (dependent option) or they can have their own angle definitions (independent option).

Perform the following steps to create a neutral plane draft with a split surface (see Figure 5–12). The part for this guide (*draft2.prt*) is available on the book's Web page.

STEP   1:   Select the Draft tool on the engineering features toolbar.

STEP   2:   Using the Control key, pick surfaces to draft.

STEP   3:   Select the Draft Hinges reference box.

Draft options available include Neutral Plane for pivoting around a plane or surface and Neutral Curve for pivoting around a datum curve or edge.

STEP   4:   On the workscreen, select the surface to serve as the draft hinge (or neutral plane).

There are two important points to remember when selecting a draft hinge. First, the selected surface must be normal to the surfaces being drafted. Second, the draft hinge defines the pivot point for rotating the drafting surfaces. The draft hinge surface will keep its form and size.

STEP   5:   Select the Pull Direction reference box (optional step).

The pull direction plane has to lie perpendicular to the draft surfaces. Since this surface is often the same as the draft hinge, Pro/ENGINEER automatically will make the pull direction plane the same as the draft hinge.

STEP   6:   Select the Pull Direction plane.

STEP   7:   Open the Split menu option on the dashboard (Figure 5–13).

STEP   8:   Select the SPLIT BY DRAFT HINGE option on the References menu.

STEP   9:   Select a Side Option on the References menu.

The following side options are available:

• **Draft Sides Independently** Surfaces on each side of the draft hinge have independent draft angles.

• **Draft Sides Dependently** Surfaces on each side of the draft hinge have the same draft angle.

• **Draft First Side Only** The first side surfaces are drafted.

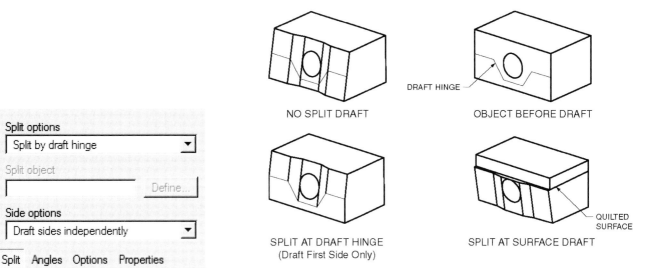

NO SPLIT DRAFT

DRAFT HINGE

OBJECT BEFORE DRAFT

SPLIT AT DRAFT HINGE
(Draft First Side Only)

QUILTED
SURFACE

SPLIT AT SURFACE DRAFT

Split options

Split by draft hinge ▼

Split object

[   ] Define...

Side options

Draft sides independently ▼

Split Angles Options Properties

Figure 5-13   Draft References Menu

Figure 5-14   Neutral Curve Drafts

- **Draft Second Side Only** The second side surfaces are drafted (Figure 5–12).

STEP 10:   **Enter a draft angle.**

STEP 11:   ✓ **Select the Build Feature option on the dashboard.**

NEUTRAL CURVE DRAFTS

Neutral curve drafts allow for the creation of a draft surface by pivoting around a datum curve or edge. They allow nonplanar surfaces and datum curves to serve as the draft hinge. There are three split options available (Figure 5–14):

NO SPLIT

The No Split option does not allow for a split of the drafted surface at the neutral curve (Figure 5–14).

SPLIT AT DRAFT HINGE

The Split at Draft Hinge option splits a drafted surface at a neutral curve (Figure 5–14). Drafted surfaces can be created on both sides or just one side of the neutral curve.

SPLIT BY SPLIT OBJECT

The Split by Split Object option creates a drafted surface out of a user-sketched section or a selected surface feature (Figure 5–14). The sketched section is pivoted around the neutral curve. The user selects the neutral plane and a surface to sketch upon. A sketcher environment provides the user with the requirement of sketching the surface to be drafted.

Neutral curve draft surfaces created with the Split at Draft Hinge and Split by Split Object options allow for draft surfaces to be created on one or both sides of the splitting surface. With the Draft Sides Dependently option, the surfaces on both sides of the splitting curve or surface can have the same draft angle. The Draft Sides Independently option allows for the specification of different draft angles for both sides of the draft hinge.

CREATING A NEUTRAL CURVE DRAFT

This step-by-step guide will detail the construction of a neutral curve draft (Figure 5–15). The part for this guide (*draft3.prt*) is available on the book's Web page.

STEP 1: Select the Draft option on the engineering features toolbar.

STEP 2: Using the Control key, pick surfaces to draft.

STEP 3: Select the Draft Hinge reference box.

Draft options available include Neutral Plane for pivoting around a plane or surface and Neutral Curve for pivoting around a datum curve or edge. In this guide, you will select a datum curve.

STEP 4: On the workscreen, select the datum curve to serve as the draft hinge (or neutral curve).

The draft hinge defines the pivot point for rotating the drafting surfaces.

STEP 5: Select the Pull Direction reference box.

The pull direction plane has to lie perpendicular to the draft surfaces.

STEP 6: Select the Pull Direction plane.

STEP 7: On the model, select a plane to serve as the pull direction.

The selected pull direction plane will serve to define the direction of rotation for the drafted surfaces.

STEP 8: Open the Split menu option on the dashboard (Figure 5–16).

STEP 9: Select the SPLIT BY DRAFT HINGE option on the References menu.

STEP 10: Select the Draft Sides Independently option on the References menu.

Available side options:

- **Draft Sides Independently** Surfaces on each side of the draft hinge have independent draft angles.

- **Draft Sides Dependently** Surfaces on each side of the draft hinge have the same draft angle.

- **Draft First Side Only** The first side surfaces are drafted.

- **Draft Second Side Only** The second side surfaces are drafted.

STEP 11: Enter draft angles for both sides of the neutral curve.

STEP 12: Manipulate the reference angles and pull direction angle to receive the desired draft.

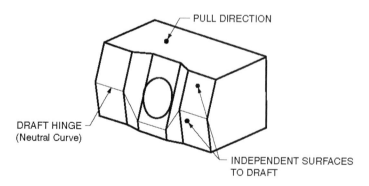

Figure 5-15   Neutral Curve Split at Curve Draft

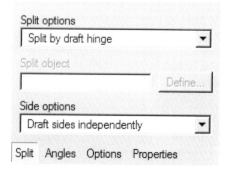

Figure 5-16   References Menu Options

STEP 13:   Preview the draft.

STEP 14:   Reselect the Preview option.

STEP 15:   ✓ Select the Build Feature option on the dashboard.

SHELLED PARTS

The Shell command removes a selected surface from a part and hollows the part with a user-defined wall thickness. The wall is created around the outside surfaces of the part, with any features created before the shell feature being included.

> **MODELING POINT**   Commonly, shell commands in CAD applications provide a uniform wall thickness for the entire shell feature. Pro/ENGINEER's Shell command has an option for providing multiple wall thickness for one shell feature. As shown in Figure 5–17, to perform a shell with multiple wall thicknesses, use the Non-Default Thickness option under the dashboard's References menu.

Perform the following steps to create a shelled part:

STEP 1:   ▢ Select the Shell option on the toolbar.

STEP 2:   Pick surfaces to remove.

The Shell command functions by removing selected surfaces from the part. If necessary, use the Control key to pick multiple surfaces. Figure 5–18 shows the shell with the top surface removed.

STEP 3:   Enter a Shell wall thickness.

STEP 4:   Use the References >> Non-Default Thickness option to create multiple wall thicknesses for your shell (Optional Step).

To perform this operation, you must pick surfaces that lie adjacent to the removed surface or surfaces.

STEP 5:   ✓ Build the feature.

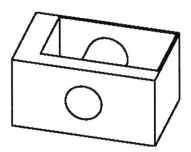

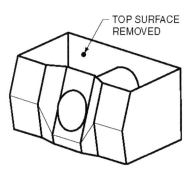

Figure 5–17   Shell with Multiple Wall Thicknesses

Figure 5–18   A Shelled Feature

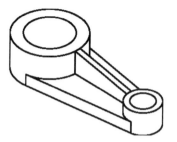

Figure 5-19   A Rib on a Part

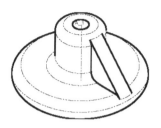

Figure 5-20   A Rib on a Revolved Feature

RIBS

A Rib is a thin web feature created between features on a part (Figure 5–19). It can be revolved or straight. Ribs are similar to protrusions that are extruded in both directions from a sketching plane. Unlike most extruded protrusions, the section of a rib must be open. Additionally, the ends of a section of a rib must be aligned with existing part surfaces. Ribs sketched on a through/axis datum plane and referenced to a surface of revolution will form a conical surface on the top of the rib (Figure 5–20).

CREATING A RIB

Figure 5–21 shows a part before and after a rib has been created. The part (*rib.prt*) shown in the figure is available on the book's Web page. Perform the following steps to create this rib:

STEP   1:   Select the Sketch tool.

STEP   2:   Select a datum plane as the sketching plane for the rib (Figure 5–22).

A rib must be sketched on a datum plane. In Figure 5–22, datum plane FRONT intersects the middle of the two cylindrical features and is used in this example.

STEP   3:   Orient the sketching environment.

STEP   4:   Select the SKETCH option on the Section dialog box.

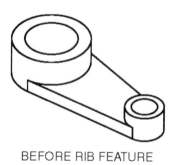

BEFORE RIB FEATURE

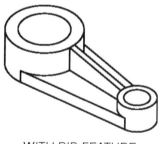

WITH RIB FEATURE

Figure 5-21   Creating a Rib

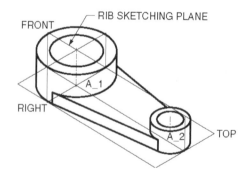

Figure 5-22   Rib Sketching Plane

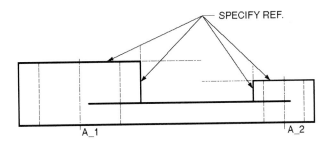

Figure 5-23 Specifying References for a Rib

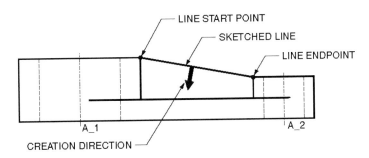

Figure 5-24 Sketching the Rib

STEP 5: **Within the sketcher environment, specify references to meet design intent (Figure 5–23).**

The sketched entity defining the rib will be aligned with each selected edge.

STEP 6: **Sketch the outline of the rib feature (Figure 5–24).**

Sketch a line from the endpoints shown in Figure 5–24. This will be the only entity required to define the section.

STEP 7: ✔ **Select the Continue option to exit the sketcher environment.**

STEP 8: ◿ **Starting with the part shown in Figure 5–21, select the Rib option of the toolbar.**

STEP 9: **Enter the thickness for the rib.**

The thickness provided will extrude in both directions from the sketching plane.

STEP 10: ╱ **The direction of extrusion should point toward the part. If necessary, reverse the direction with the REFERENCES >> FLIP option.**

STEP 11: ✔ **Build the feature.**

COSMETIC FEATURES

Cosmetic features enhance the display of parts without complicated geometric features that require regeneration. Since cosmetic features are considered part features, they can be re-defined and modified. Four cosmetic features are available: sketched, threads, grooves, user-defined, and ECAD areas.

SKETCHED COSMETIC FEATURES

Sketched cosmetic features are useful for including names and logos on a part. These cosmetic features are sketched on a part surface using normal sketching environment techniques (Figure 5–25).

COSMETIC GROOVES

Cosmetic grooves are sketched sections projected onto a part surface. The section is sketched with normal sketcher tools. The projected cosmetic feature has no defined depth. An example of a cosmetic groove is shown in Figure 5–25. In this example, the cosmetic groove was sketched on the same plane as the cosmetic sketch and projected on to the receiving surfaces.

COSMETIC THREADS

Realistic threads can be created within Pro/ENGINEER as a helical cut. Despite this, often it is satisfactory to only symbolically display a thread. The Cosmetic Thread option allows for the creation of thread symbols that correspond to a simplified thread representation (Figure 5–26).

The parameters that define a cosmetic thread resemble parameters associated with true threads. As an example, internal and external threads can be defined as well as the major and minor diameters. The following are definable parameters for a cosmetic thread:

• **Major Diameter** This parameter allows for the defining of the thread's major diameter. For external threads, the default value is 10 percent smaller than the thread surface diameter; for internal threads, the default value is 10 percent larger.

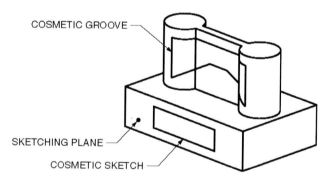

Figure 5-25   Sketched Cosmetic Feature

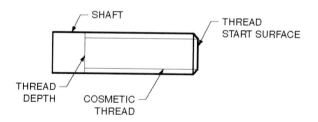

Figure 5-26   Cosmetic Threads

- **Threads Per Inch** This parameter defines the number of threads-per-inch.
- **Thread Form** This parameter allows for the selection of a thread form. An example would be the attribute UN (Unified National).
- **Thread Class** The class of a thread is defined with this parameter. Examples include fine, extra fine, and coarse.
- **Placement** This parameter defines an external or internal thread.
- **Metric** This parameter sets the thread to metric (true) or not to metric (false).

CREATING A COSMETIC THREAD

Perform the following steps to create a cosmetic thread:

STEP 1: **Select INSERT >> COSMETIC on the menu bar.**

STEP 2: **Select the THREAD option.**

Selecting the Thread option will display the Cosmetic Thread dialog box. This dialog box displays the parameters required for the creation of a cosmetic thread.

STEP 3: **Select the Thread Surface.**

On the workscreen, select the cylindrical surface upon which to place the threads. Only one surface can be selected.

STEP 4: **Select the starting surface for the thread (Figure 5–26).**

The starting surface may be a part surface, quilt, or datum plane.

STEP 5: **Choose the direction of thread creation.**

A red arrow displays the default thread direction. Select Okay to accept the direction or select Flip to change the direction.

STEP 6: **Select a thread depth specification then provide appropriate depth information.**

Depth specifications include Blind, UpTo PntVtx, UpTo Curve, and UpToSurface. After selecting a specification, provide the necessary information that corresponds to the selected specification.

STEP 7: **Enter a thread diameter.**

The default diameter is displayed. For external threads, the diameter is 10 percent smaller than the thread surface diameter. For internal threads, the diameter is 10 percent larger than the thread surface diameter.

STEP 8: **Select one or more options from the Feature Parameters menu.**

The following options are available on the Feature Parameters menu:

- **Retrieve** This option allows for the retrieval of an existing thread parameter.
- **Save** This option allows a defined thread parameter to be saved for later use.
- **Mod Params** This option allows for the modification of a thread's parameters.
- **Show** This option shows parameters set for a thread.

STEP 9: **Select DONE/RETURN on the Feature Parameters menu.**

STEP 10: **Select PREVIEW on the Cosmetic Thread dialog box then select OK.**

PATTERNED FEATURES

The Pattern tool is used to create multiple instances of a feature. Instances of a pattern are copies of the feature. In most cases, an instance is an exact duplicate of the parent feature. The two most common pattern types are linear and circular (Figure 5–27), but ENGINEER's pattern tool has been significantly enhanced with the addition of other configuration options.

PATTERN TYPES

Seven types of patterns are available: Dimension, Direction, Axis, Reference, Curve, Table and Fill.

- **Dimension Patterns** Dimension patterns are created by varying one or more specific dimensions that define the pattern's parent feature. Relations can be used within leader dimensions to create innovated instance increments.

- **Direction Pattern** Direction patterns are defined by selecting a datum plane, planar surface, axis, or coordinate system axis. The selected reference defines the direction of the pattern. This option only allows for linear patterns.

- **Axis Pattern** Axis patterns are used to create a circular pattern around an axis.

- **Reference Pattern** Reference patterns are created by referencing a previously created pattern. The leader dimensions from the reference pattern govern the new pattern.

- **Curve Pattern** The path of a curve pattern follows the direction of a sketched curve.

- **Table Pattern** In a Table pattern, instances of a pattern receive their parametric definitions from a table (Figure 5–28). They are useful for arrays of features that don't have linear or circular patterns.

- **Fill Pattern** In a Fill pattern, instances of a pattern fill the interior of a sketched region. The sketched region is created within the pattern command (Figure 5–29).

PATTERN OPTIONS

Three pattern options are available within Pro/ENGINEER. The option to use is based on the complexity of the part. Less complex patterns allow for more assumptions during the pattern construction. This allows Pro/ENGINEER to regenerate the pattern quicker. The following pattern options are available:

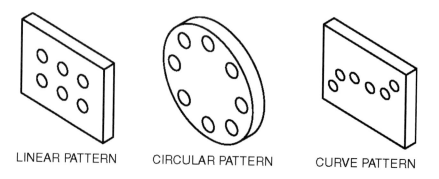

LINEAR PATTERN   CIRCULAR PATTERN   CURVE PATTERN

Figure 5-27   Pattern Configurations

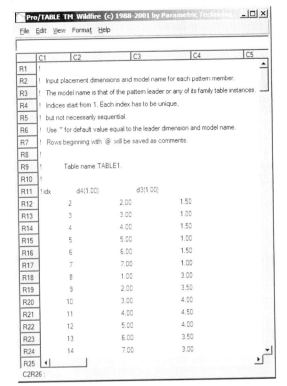

PATTERN TABLE PATTERN DEFINITION

Figure 5-28 Table Pattern Example

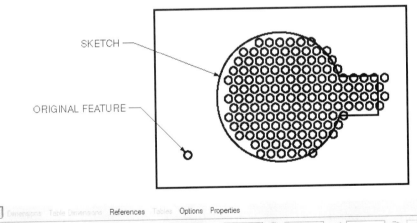

Figure 5-29 Fill Pattern Example

IDENTICAL PATTERNS

Identical patterns are the least complex and allow for the most assumptions. Identical pattern instances must be of the same size and must lie on the same placement surface. Identical pattern instances cannot intersect other features, instances, or the edge of the placement plane.

VARYING PATTERNS

Varying patterns are more complex than identical patterns. Varying pattern instances can vary in size and can lie on different placement surfaces. As with identical patterns, varying pattern instances cannot intersect other instances.

GENERAL PATTERNS

General patterns are the most complex and require longer to regenerate. With general patterns, no assumptions are made during the construction process. Instances can intersect other instances and placement plane edges. Additionally, instances can vary in size and lie on different surfaces.

> **MODELING POINT**   When creating a feature that will be patterned, placement of dimensions defining the feature is critical. As an example, when creating a linear pattern, the dimensioning scheme defining the location of the feature will be used to create the pattern. For angular patterns, an angular dimension must exist.

CREATING A LINEAR DIMENSION PATTERN

Dimension patterns are created by varying linear and/or angular dimensions. A feature can be patterned unidirectionally or bidirectionally. Additionally, more than one dimension of a feature can be varied during the process. The following is a step-by-step approach for patterning the cylindrical feature shown in Figure 5–30. The part used in this guide (*pattern1.prt*) is available on the book's Web page. As shown, the 1.75-in and 1.5-in dimensions locating the feature will be patterned, as will the 2.0-in and 1.25-in dimensions defining the height and the diameter.

STEP   1:   **On the model tree, pick the feature to pattern.**

On the workscreen or on the model tree, select the cylindrical feature (Figure 5–30). With the Pattern command, only one feature can be patterned at a time.

STEP   2:   ▦   **Select the Pattern tool on the toolbar.**

STEP   3:   **Select DIMENSION on the dashboard as the pattern type (Figure 5–31).**

Dimension is selected by default. Other options include Table, Direction, Axis, Reference, Fill, and Curve.

STEP   4:   **Using the Control key, pick the 1.50-in leader dimension for patterning in the first direction (Figure 5–31).**

On the workscreen, select the first dimension to vary. This pattern will copy the feature in the first direction.

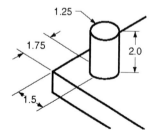

Figure 5–30
The Feature to be Patterned

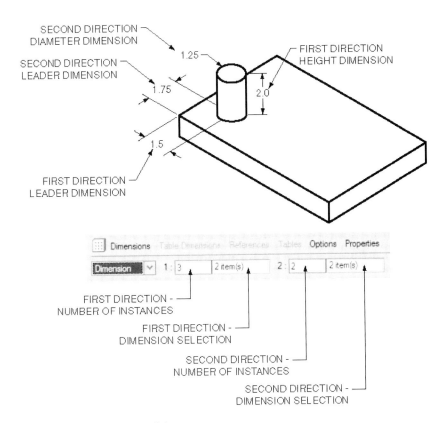

Figure 5–31 Dimension Selection

STEP 5: Enter 2.75 as the dimension increment value.

STEP 6: Using the Control key, pick the height dimension for varying (Figure 5–31).

The height value will be varied in the first direction.

STEP 7: Enter 0.25 as the dimension increment value.

STEP 8: On the dashboard, enter 3 as the number of instances in the first direction (see Figure 5–31).

Three instances of the feature will be created in the first direction.

STEP 9: Pick inside the Second Direction Dimension Selection edit box as shown in Figure 5–31.

Selecting the Second Direction Dimension Selection option will allow you to select dimensions to define the second direction of the pattern.

STEP 10: Using the Control key, pick the 1.75-in leader dimension for varying in the second direction.

STEP 11: Enter 2.50 as the dimension increment value.

STEP 12: Using the Control key, pick the cylinder diameter dimension for varying.

You will also vary the diameter dimension in the second direction of the pattern.

STEP 13: Enter 0.500 as the dimension increment value.

STEP 14: Enter 2 as the number of instances in the second direction.

Two instances in the second direction, in combination with three instances in the first direction, will create a total of six instances. Because of the

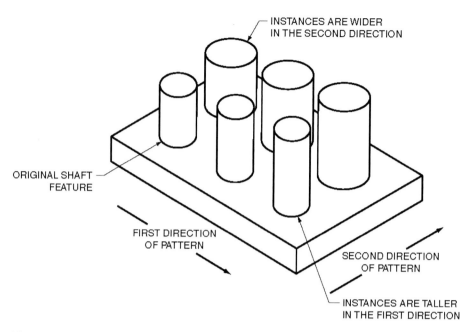

INSTANCES ARE WIDER
IN THE SECOND DIRECTION

ORIGINAL SHAFT
FEATURE

FIRST DIRECTION
OF PATTERN

SECOND DIRECTION
OF PATTERN

INSTANCES ARE TALLER
IN THE FIRST DIRECTION

Figure 5–32   Pattern Feature

variations selected for the height and the diameter dimensions of the cylinder, each instance of the pattern will be a different size.

STEP 15:   **Select OPTIONS >> VARYING as the Pattern option.**

Other pattern options include Identical and General. Unlike the Identical option, the Varying option allows dimensional values on each pattern instance to vary. Also unlike Identical, instances can lie on different placement surfaces.

STEP 16:   **Select the Build Feature option.**

After selecting the Build Feature option, the pattern will appear as shown in Figure 5–32.

SUMMARY

While the basic tools such as extrude and revolve are primary for creating features, other options are available that enhance the power of Pro/ENGINEER's modeling capabilities. As an example, if a part has multiple instances of a feature, it would be time-consuming to individually model each feature. In this case, Pro/ENGINEER provides the Pattern command to create a rectangular or polar array of the feature. Feature creation tools such as Round, Chamfer, Shell, and Draft are available for unique modeling situations. It is important to understand the capabilities of Pro/ENGINEER's various construction tools and to know when one is appropriate over another.

FEATURE CONSTRUCTION TUTORIAL 1

This tutorial exercise will provide instruction on how to model the part shown in Figure 5–33. Within this tutorial, the following topics will be covered:

- Creating a new object.
- Creating an extruded protrusion.
- Creating a simple round.
- Creating a straight chamfer.
- Creating a standard coaxial hole.
- Creating an advanced round.
- Creating a shell feature.

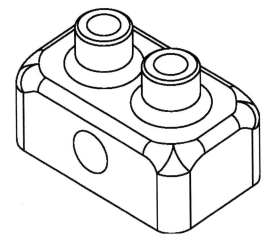

Figure 5–33   Finished Model

STARTING A NEW OBJECT

This segment of the tutorial will establish Pro/ENGINEER's working environment.

STEP 1:   **Start Pro/ENGINEER.**

STEP 2:   **Set Pro/ENGINEER's Working Directory.**

From the File menu, select the Set Working Directory option then establish an appropriate working directory for the part.

STEP 3:   **Start a NEW Pro/ENGINEER part and name it *construct1*.**

From the File menu, select New, then create a new part file named *construct1*. Use the default template file as supplied by Pro/ENGINEER. Using Pro/ENGINEER's default template file, your part model should start with a set of default datum planes, the default coordinate system, and *inch_lbm_second* as the units.

CREATING THE BASE GEOMETRIC FEATURE

This segment of the tutorial will create the extruded feature shown in Figure 5–34.

STEP   1:   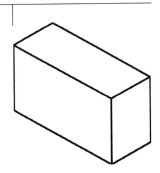   **Select the Sketch option on the toolbar.**

STEP   2:   **Select datum plane FRONT as the sketching plane, then select SKETCH on the Sketch dialog box.**

STEP   3:   **If necessary, close the References dialog box.**

When no existing geometric features exist for a part, Pro/ENGINEER will automatically specify datum planes RIGHT and FRONT as references. These are the only required references for this model.

Figure 5–34
The Extruded Base Feature

STEP 4: Using the LINE icon, sketch the section shown in Figure 5–35.

As shown in Figure 5–35, sketch the Line entities aligned with datum planes RIGHT and TOP. When sketching, do not worry about the size of the entities being sketched. Entity size parameters will be modified in another step. Allow Intent Manager to apply the vertical and horizontal constraints as shown.

MODELING POINT   When sketching, your left mouse button is used to pick entity points on the workscreen and your middle mouse button is used to cancel commands.

STEP 5: Use the MODIFY icon to match the dimension values shown in Figure 5–35.

Select Modify, then select the dimension value to change. Follow this by entering a new value for the dimension.

STEP 6: Select the Continue icon when the section is complete.

STEP 7: Select the Extrude option.

STEP 8: Enter blind extrude depth of 2.00.

STEP 9: Preview the feature.

After selecting the Preview option, shade and dynamically rotate the object. It is not too late to change a parameter defining the feature. Selecting the parameter on the dashboard will allow the selected parameter to be modified. As an example, the depth of the extrusion can be changed by redefining the depth parameter.

STEP 10: If the protrusion is correct, build the feature.

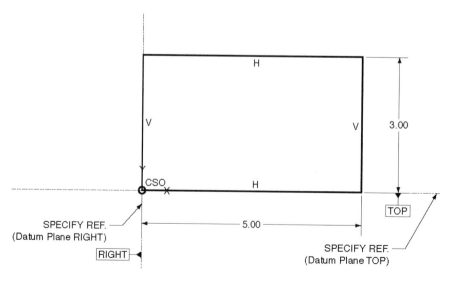

Figure 5–35   Section for Extrusion

ADDING EXTRUDED FEATURES

An extruded protrusion can be sketched on an existing part surface. This section of the tutorial will create the protrusion shown in Figure 5–36.

STEP 1: Select the Sketch option on the dashboard.

STEP 2: Select the surface shown in Figure 5–36 as the sketching plane.

As shown in Figure 5–36, select the top surface of the existing part to use as a sketching plane. Observe the dimensions of the part; care should be taken not to select one of the side surfaces.

STEP 3: Select the SKETCH option on the Section dialog box.

STEP 4: In the sketching environment, specify the references shown in Figure 5–37.

STEP 5: On the Toolbar, turn off the display of datum planes.

STEP 6: On the Toolbar, select No Hidden as the model display.

Sketching a new feature on an existing part surface is usually easier with No Hidden or Hidden set as the model display style.

STEP 7: Using the CIRCLE icon, sketch the two circles shown in Figure 5–37.

Circle option lets you define a circle through the selection of the circle's center point, then drag the size of the circle. Allow Intent Manager to create a Horizontal Alignment constraint between the centers of the two circles.

STEP 8: Use the DIMENSION icon to match the dimensioning scheme shown in Figure 5–37.

To dimension the location of a circle, with the left mouse button, select the center of the circle, then the locating edge. Place the dimension with the right mouse button. Diameter dimensions are created by double picking the circle with the left mouse button, followed by placing the dimension with the middle mouse button.

MODELING POINT An alternative to the two separate diameter dimensions as shown in Figure 5–37 would be to create an Equal-Radius constraint between the two circles. With an Equal-Radius constraint, one dimension will control the size of both circles. This is a common way to capture design intent. An Equal-Radius constraint can be created through the Constraint icon and the Equal option.

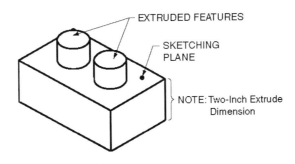

Figure 5–36 Extrude Features

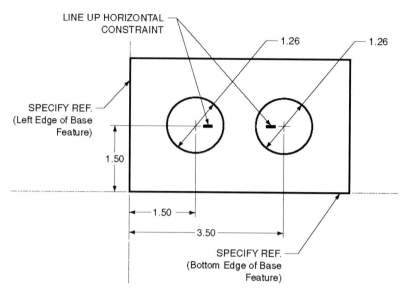

Figure 5–37   Sketched Regenerated Section

STEP 9:   Using the Pick icon, double-pick dimensions and modify their values to match Figure 5–37.

STEP 10:   Using the Control key, select the D character on the keyboard.

The combination of the Control key and the D character will place the model in its default orientation.

STEP 11:   Select the Continue icon on the Sketch toolbar when the section is complete.

STEP 12:   Select the Extrude option.

STEP 13:   Enter a Blind depth of 1.00.

Select Blind as the depth option then enter a value of 1.00.

STEP 14:   Build the feature.

CREATING A ROUND FEATURE

This segment of the tutorial will create the simple round feature shown in Figure 5–38.

STEP 1:   Select the Round option on the toolbar.

The Round option is available to create rounds and fillets. A round is considered a part feature.

STEP 2:   Using the Control key, pick the two edges at the base of the circular protrusion (Figure 5–39).

STEP 3:   Enter 0.25 as the radius of the rounds.

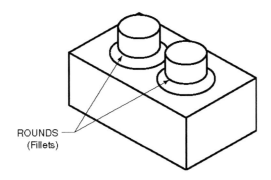

Figure 5-38 Part with Simple Rounds

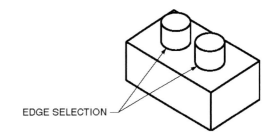

Figure 5-39 Round Edge Selection

STEP 4: Preview the rounds.

STEP 5: Build the round feature.

Your rounds should look like Figure 5–38.

CREATING A CHAMFER

Pro/ENGINEER provides the option to create either a corner chamfer or an edge chamfer. This segment of the tutorial will create the edge chamfer shown in Figure 5–40.

STEP 1: Select the Chamfer option on the toolbar.

The Chamfer option on the toolbar creates an edge chamfer. A Corner Chamfer option, which creates an angled surface between three planes, is available under the Insert menu.

STEP 2: As shown in Figure 5–41, use the Control key to pick the edges on the top of the two circular protrusions.

STEP 3: On the dashboard, select 45 x D as the dimensioning scheme (Figure 5–42).

The 45 x D option will create a 45 degree angle with a user-defined distance.

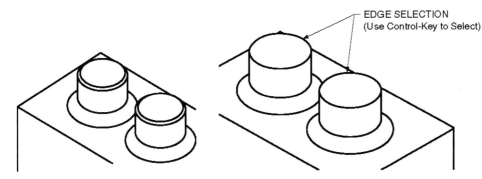

Figure 5-40 Chamfer Features   **Figure 5-41** Edge Selection

Figure 5-42 Chamfer Definition

STEP 4:   Enter 0.06 as the chamfer dimension.

STEP 5:   ✓ Select the Build Feature option on the dashboard.

CREATING A STANDARD COAXIAL HOLE

This segment of the tutorial will create the two coaxial threaded holes shown in Figure 5–43. The specification will be a standard 0.75-in unified national course internal thread.

STEP 1:   ⊥ Select the Hole tool on the toolbar.

STEP 2:   Change the Selection Filter to Axis.

Note: The Selection Filter option is available in the lower right-hand corner of your workscreen. Smart is the default selection option. The primary reference for a coaxial hole is an axis. Filtering to select an axis will ensure that you pick the correct reference.

STEP 3:   As shown in Figure 5–44, select the first axis as the Primary Reference.

STEP 4:   Change the Selection Filter back to ALL.

STEP 5:   On the dashboard, select the PLACEMENT menu option (Figure 5–44).

STEP 6:   Using the control-key, pick the Placement Plane show in Figure 5–44.

Using the control key will allow you to pick the secondary reference without undefining the prevousily picked primary reference.

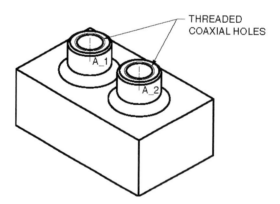

Figure 5–43   Coaxial Holes

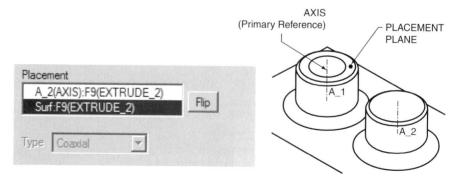

Figure 5–44   Hole Placement Options

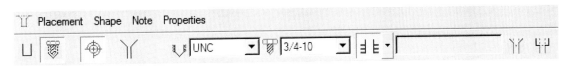

Figure 5-45 Hole Placement References

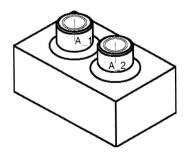

Figure 5-46 Two Coaxial Holes

STEP 7: ⊔ Select the Standard hole option on the hole dashboard.

Standardized thread specification tables are used to define standard holes. As an example, a 3/4-in nominal size hole with Unified National Coarse threads has 10 threads per inch. Pro/ENGINEER has three preexisting hole tables: UNC, UNF, and ISO. Each table provides parameters needed to detail a threaded hole.

STEP 8: Select UNC and 3/4–10 as the standard hole screw size (Figure 5–45).

STEP 9: ⊒ ⏚ Select Through All on the dashboard as the hole depth (Figure 5–45).

STEP 10: ⊕ Select the Tap Drill option on the Dashboard (Figure 5–45).

STEP 11: Select the SHAPE menu option then select the THRU THREAD option.

The Through All option will construct the hole through the entire part. Similarly, the Thru Thread option will construct the thread through the entire length of the hole.

STEP 12: Ensure that the Countersink and Counterbored options are not active on the dashboard (Figure 5–45).

STEP 13: ✓ Select the Build Feature option.

STEP 14: ▱ On the Datum Toolbar, turn off the display of 3D NOTES.

Deselecting the 3D Annotations (Notes) environmental option will turn off the display of the standard hole's thread representation note. This option can also be set with the configuration file option *model_note_display*.

STEP 15: Repeat Steps 1 through 13 to create the second standard coaxial hole (Figure 5–46).

STEP 16: Save your part.

CREATING A LINEAR HOLE

The Linear Hole Placement option will locate a hole from two reference edges. This tutorial will create the linear hole shown in Figure 5–47.

STEP   1:   Select the Hole option on the toolbar.

STEP   2:   Select Straight and Simple on the dashboard as the Hole Type.

STEP   3:   Enter 1.00 as the hole's diameter (see Figure 5–48).

STEP   4:   Select Through All as the hole depth.

STEP   5:   Select the front surface of the part as the primary reference (see Figure 5–49).

STEP   6:   Drag the hole's first placement handle to the first linear reference edge shown in Figure 5–49.

When you are placing a linear hole, Pro/ENGINEER will provide placement handles for locating the hole and for repositioning the hole. These placement handles appear as small white dots connected to the hole and can be repositioned with the mouse. A linear placement handle will convert to a dimension when it comes in contact with a valid placement reference (e.g., plane, edge, or datum).

STEP   7:   Drag the hole's second placement handle to the second linear reference edge shown in Figure 5–49.

DESIGN INTENT   When you are utilizing the Linear Placement option and the placement plane shown in Figure 5–49, any edge of the Placement plane could be used to locate the hole. Since the design intent requires the hole to be located from the edges shown in the figure, these references should be the edges selected during part modeling. Pro/ENGINEER is a feature-based/parametric modeling system. These types of systems allow for the incorporation of design intent into a model. Placing a hole is one example of where design intent can be incorporated. In this example, placing the hole from these edges incorporates design intent.

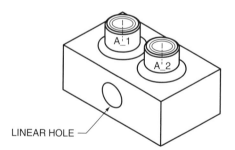

Figure 5–47   Linear Hole Placement

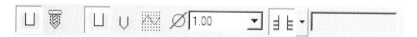

Figure 5–48   Straight Hole Parameters

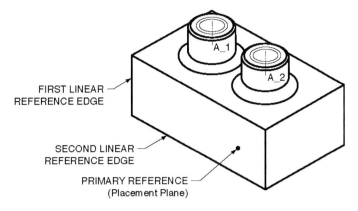

Figure 5–49   Hole References

STEP 8:   For the first location dimension, change the value to 2.50.

STEP 9:   For the second location dimension, change the value to 1.00.

STEP 10:   Select the Build Feature option to create the hole.

STEP 11:   Save your part.

CREATING AN ADVANCED ROUND

The advanced round created in this segment of the tutorial will consist of two round sets. The first round set will have a radius value of 0.500, while the second round set will have a radius value of 0.235. The result of this round feature is shown in Figure 5–50.

STEP 1:   Select the Round option on the toolbar.

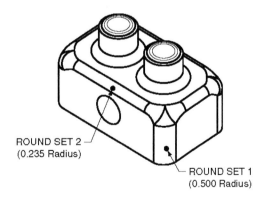

Figure 5–50   Advanced Round Feature

STEP 2:   Using the Control key, pick the four vertical corners of the part (Figure 5–51).

INSTRUCTIONAL NOTE   The fourth corner is hidden in Figure 5–51. Use the Activate Query List option (right mouse button pop-up menu) to pick this edge, or dynamically rotate the part. Don't release the Control key during selection.

STEP 3:   Enter 0.500 as this round set's radius value.

STEP 4:   Not using the Control key, pick one of the four edges shown in Figure 5–52.

Make sure you don't use the Control key to make this selection. Releasing the Control key and picking a reference will create a new round set.

STEP 5:   Using the Control key, pick the remaining three edges that are shown in Figure 5–52.

STEP 6:   Enter 0.235 as this round set's radius value.

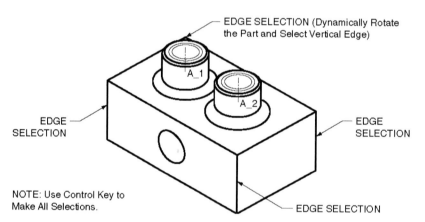

Figure 5–51   Round Set One Edge Selection

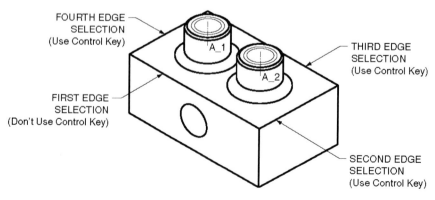

Figure 5–52   Round Set Two Edge Selection

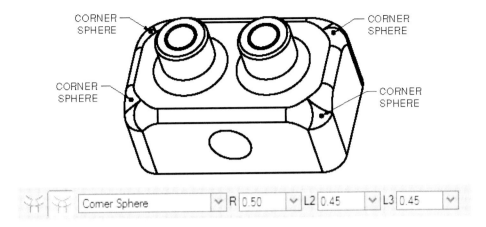

Figure 5-53   Corner Sphere Transitions

Step 7:   Select the **Round Transition** option on the dashboard.

The default transition created between two round sets is a Round Only type. You will change each of the transitions to a Corner Sphere.

Step 8:   Pick one of the transitions in Figure 5–53 and on the dashboard change its transition type to **CORNER SPHERE.**

Step 9:   Change the remaining three transitions to Corner Sphere.

Step 10:   Build the round.

INSERTING A SHELL

This segment of the tutorial will create the shelled feature shown in Figure 5–54. The Shell command removes selected surfaces from a part and provides the remaining surfaces with a user-defined wall thickness.

Step 1:   Select the **Shell** option on the toolbar.

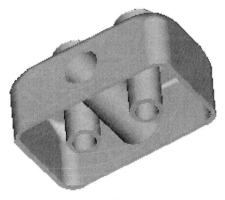

Figure 5-54   A Shelled Feature

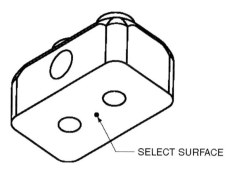

Figure 5–55   Select Shell Surface

STEP 2:   **Select the Surface Shown in Figure 5–55.**

The selected surface will be removed from the part.

STEP 3:   **Enter 0.125 as the wall thickness.**

STEP 4:   **Preview the Shelled feature then select the Play option on the dashboard.**

STEP 5:   **Select the Build Feature option on the dashboard.**

STEP 6:   **Save your part.**

FEATURE CONSTRUCTION TUTORIAL 2

This tutorial exercise will provide instruction on how to model the part shown in Figure 5–56. Within this tutorial, the following topics will be covered:

- Creating a new object.
- Creating an offset datum plane.
- Creating an extruded protrusion.
- Creating a straight coaxial hole.
- Creating a linear straight hole.
- Creating a linear pattern.
- Creating a straight chamfer.
- Creating a cut feature.
- Creating a rib.
- Creating a simple round.

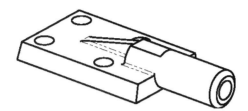

Figure 5–56   Finished Part

STARTING A NEW PART

This segment of the tutorial will establish Pro/ENGINEER's working environment.

STEP 1:   **Start Pro/ENGINEER.**

STEP 2:   **Establish an appropriate working directory.**

STEP 3:   **Create a new part file with the name *CONSTRUCTION2* (use Pro/ENGINEER's Default template).**

CREATING THE BASE GEOMETRIC FEATURE

Model the base geometric feature. The base feature is constructed from an 8- by 5-in rectangular section extruded in one direction a blind distance of 1 in. Sketch the section on datum plane FRONT. Align the middle of the part with datum plane TOP and the left edge of the part with datum plane RIGHT. Figure 5–57 shows the finished feature and the sketched section.

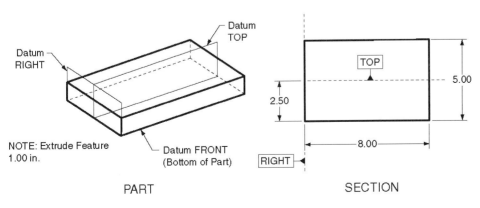

Figure 5–57   The Base Feature

CREATING AN OFFSET DATUM PLANE

This section of the tutorial will create an offset datum plane. As shown in Figure 5–58, the datum plane will be offset from datum plane RIGHT a distance of 6.00 in. A datum plane can be offset from an existing part planar surface or an existing datum plane.

STEP 1:   Select the CREATE DATUM PLANE icon on the Datum Toolbar.

STEP 2:   As shown in Figure 5–58, pick datum plane RIGHT as the surface to offset from.

On the workscreen or on the model tree, select datum plane RIGHT. To make selection easier, change your selection filter to Plane.

STEP 3:   On the Datum Plane dialog box, ensure that OFFSET is selected as the constraint type (Figure 5–58).

STEP 4:   In the Datum Plane dialog box, enter a value that will offset the new datum plane 6.00 in in the direction shown in Figure 5–59.

A positive or negative value can be entered. If your offset direction is in the opposite direction, enter a negative value.

STEP 5:   Select OK on the Datum Plane dialog box.

Notice on the workscreen and on the model tree (Figure 5–60) the addition of datum plane DTM1. Pro/ENGINEER names datum planes in sequential order starting with DTM1. The next two steps in the tutorial will rename datum plane DTM1.

STEP 6:   On the model tree, right-mouse select the DTM1 feature to access the model tree's pop-up menu.

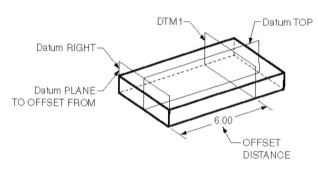

Figure 5–58   Datum Offset

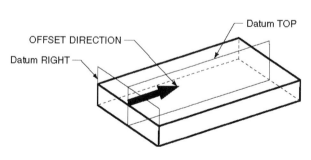
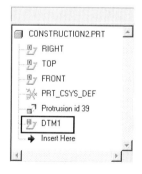

Figure 5–59   Datum Offset Direction

Figure 5–60
Datum Plane Selection

STEP 7:   On the pop-up menu, select the RENAME option.

STEP 8:   In the textbox enter *OFFSET_DATUM* as the new name for datum plane DTM1.

Notice the underscore between *Offset* and *Datum*. On the Model Tree, new feature names cannot have spaces.

MODELING POINT   The default names that Pro/ENGINEER gives features are not descriptive. It is helpful to rename them to allow for ease of identification and selection.

STEP 9:   Save your part.

CREATING AN EXTRUDED PROTRUSION

This section of the tutorial will create the cylindrical protrusion shown in Figure 5–61. This protrusion will consist of a circle entity sketched on the offset datum plane (*Offset_Datum*). The section will be extruded a depth of 6.00 in.

STEP 1:   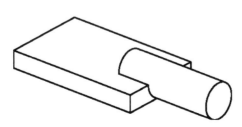 Set up a sketch for an extrusion with datum plane *OFFSET_DATUM* as the sketching plane and the direction of extrusion set as shown in Figure 5–62.

Pick datum plane *Offset_Datum* as the sketching plane (Figure 5–62). Within Pro/ENGINEER, there are two sides to a datum plane: the positive side and the negative side. Using Pro/ENGINEER's default color scheme (see the Color option under the Utilities menu), the positive side of a datum plane is yellow and the negative side is red. When you select a datum as the sketching plane for an extruded protrusion, by default, the feature will be extruded in the positive direction. To extrude in the negative direction, use the Direction menu's Flip option.

STEP 2:   Orient the sketching environment as shown in Figure 5–63.

Use the Top option and select the top of the part.

STEP 3:   On the model display toolbar, select NO HIDDEN as the model's display.

STEP 4:   On the datum display toolbar, turn off the display of datum planes.

Figure 5–61   An Extruded Protrusion

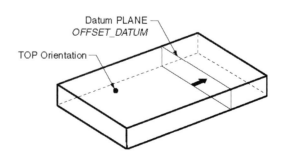

Figure 5–62   Extrude Direction

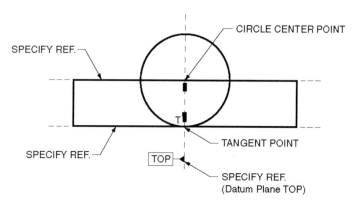

Figure 5–63 Feature's Section

STEP 5: Using the References dialog box (Sketch >> References), specify the three references shown in Figure 5–63.

STEP 6: ○ Sketch a circle entity as shown in Figure 5–63.

The center of the circle is aligned with the top of the part and with datum plane TOP. The bottom edge of the circle is aligned with the bottom of the part (creating a Tangent constraint).

> **DESIGN INTENT** The design intent for this extruded feature requires the diameter of this cylindrical feature to be twice the thickness of the base feature. Also, the design intent requires the center of the feature to be aligned with the top of the part. By sketching the feature in the manner shown in Figure 5–63, the intent of the design will be captured. For this part, if the thickness of the base feature is changed (it is currently set at 1.00 in), the size of the cylindrical feature will change accordingly. This meets the intent for this design.

STEP 7: ✔ Select the Continue icon to exit the sketcher environment.

STEP 8: Select the Extrude option.

STEP 9: Select the Control key and the D character key simultaneously to view the model's default orientation.

STEP 10: Ensure that the extrude direction matches the requirements from Figures 5–61 and 5–62.

The feature should extrude in a manner that will create the part shown in Figure 5–61. If necessary, use the Change Direction option to flip the direction of extrusion.

STEP 11: Construct a blind depth of 6.00 in.

STEP 12: ✓ Select the Build Feature option.

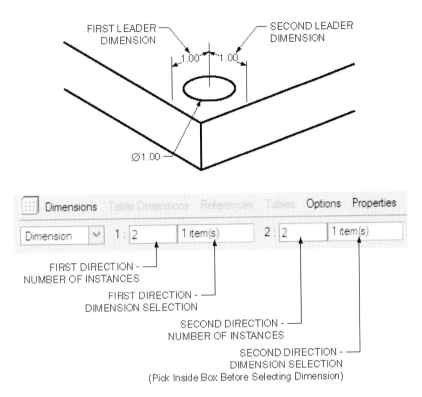

Figure 5–70   Dimension Selection

STEP 1:   On the model tree, pick the previously created hole feature.

STEP 2:   🔲   Select the Pattern option on the toolbar.

STEP 3:   As shown in Figure 5–70, pick the first leader dimension for patterning in the first direction.

Features can be patterned in two directions. In this step, you are selecting a leader dimension that will be used to determine the first direction of pattern. Additionally, this leader dimension will be used to determine the distance between each instance of the pattern.

STEP 4:   Enter 4.00 as the dimension increment value.

Each instance of the pattern in the first direction will be 4.00 in apart.

STEP 5:   On the dashboard, pick inside the Second Direction Dimension Selection edit box as shown in Figure 5–70.

Selecting the Second Direction Dimension Selection option will allow you to select dimensions to define the second direction of the pattern.

STEP 6:   As shown in Figure 5–70, select the second leader dimension for patterning in the second direction.

This tutorial will create a pattern in two directions. Select the second hole placement dimension as shown in Figure 5–70.

STEP 7:   Enter 3.00 as the dimension increment value.

Each instance of the pattern in the second direction will be 3.00 in apart.

STEP 8:   On the Pattern dashboard, ensure that 2 is entered as the number of instances in the second direction (Figure 5–70).

STEP 9:   On the dashboard, select OPTIONS >> IDENTICAL as the Pattern option.

The Identical option requires the most assumptions. The following assumptions must be met:

* The pattern must be on one placement surface.
* The pattern cannot intersect an edge.
* The pattern's instances cannot intersect.

STEP 10:   **Select the Build Feature option.**

STEP 11: **Save the part.**

CREATING A CHAMFER

This segment of the tutorial will create the Straight Chamfer shown in Figure 5–71.

STEP 1:   **Select the Chamfer option on the toolbar.**

Two chamfer options are available: Edge and Corner. The Edge option creates a chamfer between two surfaces, while the Corner option creates a chamfer between three surfaces. The Chamfer option on the toolbar creates an edge chamfer. The Corner chamfer option is available under the INSERT >> CHAMFER option on the menu.

STEP 2: **Pick the feature edge as shown in Figure 5–71.**

STEP 3: **Select 45 x D as the chamfer's dimensioning scheme.**

The 45 x d option will create a chamfer that consists of a 45 degree angle with a user-specified distance.

STEP 4: **Enter 0.25 as the chamfer's dimension value.**

STEP 5:   **Build the feature.**

CREATING A CUT

This section of the tutorial will create the cut feature shown in Figure 5–72.

STEP 1:   **Setup a sketching environment with the sketching plane and top orientation shown in Figure 5–73.**

STEP 2: **While in the sketching environment, on the model display toolbar, select NO HIDDEN as the model's display style.**

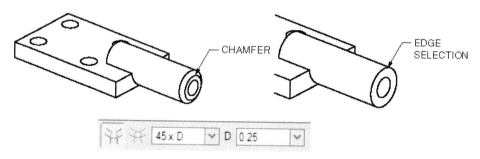

Figure 5–71   Chamfer Feature

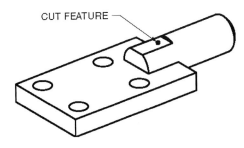

Figure 5-72   Cut Feature

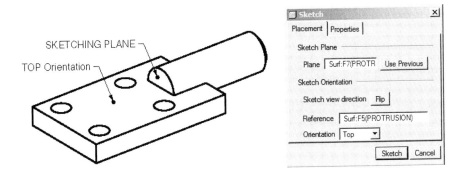

Figure 5-73   Cut Orientation Selection

STEP 3: On the datum display toolbar, turn off the display of datum planes.

STEP 4: Using the References dialog box (Sketch >> References), pick the edge of the circular protrusion as a reference for this section (see Figure 5–74).

STEP 5: Sketch the Line entity shown in Figure 5–74, then use the DIMENSION and MODIFY options to create the dimensioning scheme shown.

STEP 6: Select the Continue option to exit the Sketcher menu.

STEP 7: Select the Extrude option on the toolbar.

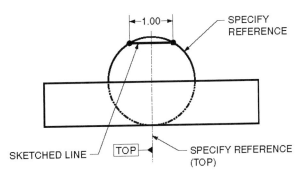

Figure 5-74   Cut Sketch

STEP 8:  Select the Cut (remove material) option on the extrude dashboard.

STEP 9: Create a blind depth of 2.00.

STEP 10: Ensure that the cut extrudes in a direction that will remove material.

STEP 11: Build the feature.

CREATING A RIB

This segment of the tutorial will create the rib feature shown in Figure 5–75. A rib feature is similar to an extruded protrusion. With ribs, the section has to be opened and the extrude direction is both sides by default. In this tutorial, the rib will be sketched on datum plane TOP.

STEP 1: Select the Sketch option on the dashboard.

STEP 2: Select datum plane TOP as the sketching plane (Figure 5–76).

Datum plane TOP can be picked on the workscreen or on the model tree. For parts with multiple features, it is often easier to use the model tree to select a specific feature.

STEP 3: Orient the top of the part toward the top of the workscreen then select SKETCH on the Sketch dialog box (Figure 5–76).

STEP 4: Using the References dialog box, specify the four references shown in Figure 5–77.

STEP 5: Sketch the single LINE shown in Figure 5–77.

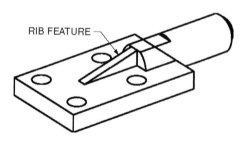

Figure 5–75   Rib Feature

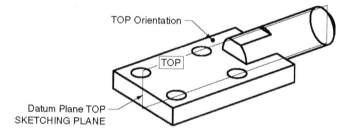

Figure 5–76   Sketch Plane Orientation Selection

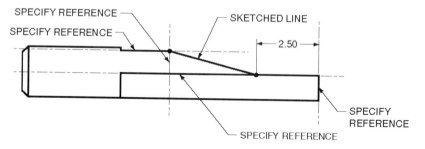

Figure 5–77   The Sketched Section

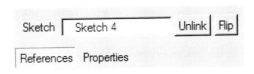

Figure 5–78 Flip Add Material Direction

STEP 6:   Use the DIMENSION and MODIFY options to create the dimensioning scheme shown in Figure 5–77.

With the correct references specified, only one dimension is needed to fully define the section. In this example the design intent requires the end of the rib to be measured from the end of the part.

STEP 7:   Select the Continue icon to exit the sketcher environment.

STEP 8:   Select the Rib option on the toolbar.

STEP 9:   Enter 0.500 as the rib's thickness.

STEP 10:   Use the REFERENCES >> FLIP option to change the direction of material creation (Figure 5–78).

Because of the open section, you have to specify the side of the sketch to create material. Your arrow should point toward the direction where material will be added.

STEP 11:   Build the feature.

CREATING A DRAFT

This section of the tutorial will draft three surfaces as shown in Figure 5–79.

STEP 1:   Select the Draft option on the engineering features toolbar.

STEP 2:   Using the Control key, pick surfaces to draft (Figure 5–80).

Pick the three surfaces shown in Figure 5–80. You will need to rotate the model dynamically or use the Pick From List pop-up menu option to select the third surface. If necessary, access the References >> Draft Surfaces option on the dashboard.

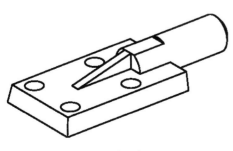

Figure 5–79 Drafted Surfaces

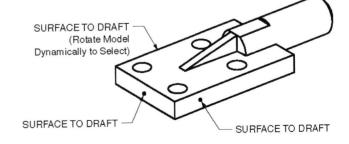

Figure 5–80 Surface Selection

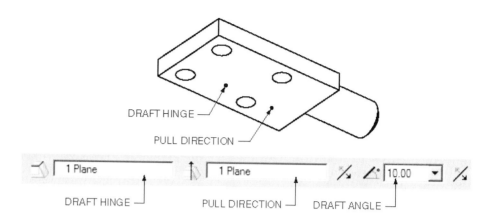

Figure 5-81   Draft Hinge and Pull Direction Selection

STEP 3:   Select the Draft Hinge reference box (see Figure 5–81).

Draft options available include Neutral Plane for pivoting around a plane or surface and Neutral Curve for pivoting around a datum curve or edge.

STEP 4:   As shown in Figure 5–81, pick the bottom of the part to serve as the draft hinge (or neutral plane).

There are two important points to remember in selecting a draft hinge. First, the selected surface must be normal to the surfaces being drafted. Second, the draft hinge defines the pivot point for rotating the drafting surfaces. The draft hinge surface will keep its form and size.

STEP 5:   Select the Pull Direction reference box (Figure 5–81).

The pull direction plane has to lie perpendicular to the draft surfaces. Since this surface is often the same as the draft hinge, Pro/ENGINEER automatically will make the pull direction plane the same as the draft hinge.

STEP 6:   Pick the bottom of the part as the pull direction plane (Figure 5–81).

STEP 7:   Enter 10.0 as the draft angle.

STEP 8:   Preview the draft.

STEP 9:   Reselect the Preview option.

STEP 10:   Select the Build Feature option on the dashboard.

CREATING A ROUND

This segment of the tutorial will create the rounds shown in Figure 5–83. Despite their apparent simplicity, rounds can be one of the most troublesome commands in Pro/ENGINEER. Since it is advisable not to use a round as a reference for the creation of a feature, rounds should be created as the last features of a part.

STEP 1:   Select the Round option on the model tree.

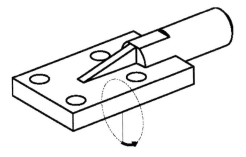

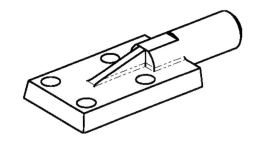

Figure 5-82 Draft Rotation Reference   **Figure 5-83** Rounded Features

STEP 2: **Using the Control key, pick the edges shown in Figure 5–84 and the corresponding edges on the opposite side of the Rib and Cylinder features.**

On the workscreen, pick the edges shown in Figure 5–84. Use one of two methods to pick the corresponding edges on the opposite side of the rib and cylinder features:

- Use the Activate Query List option (right mouse button pop-up menu) to access edges. Ensure that you use the Control key when picking each edge.

- Dynamically rotate the model to pick edges. Ensure that you use the Control key when picking each edge.

STEP 3: **Enter a round radius value of 0.13.**

STEP 4: Build the feature.

STEP 5: **Save your part.**

STEP 6: **Purge old part file by using the FILE >> DELETE >> DELETE OLD option.**

Every time a file is saved in Pro/ENGINEER, a new version of the file is created. The Delete Old option will delete old versions.

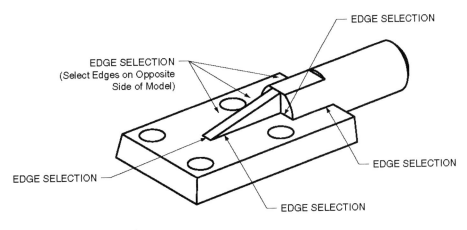

Figure 5-84 Round Selection

PROBLEMS

1. Using Pro/ENGINEER's Part mode, model the part shown in Figure 5–85. Construct the part using the following order of operation:

   a. Use Pro/ENGINEER's default template file.

   b. Sketch the base feature on Datum Plane FRONT.

   c. Model the base protrusion as a both-sides extrusion.

   d. Construct the rounded features within the sketch of the base extrusion.

   e. Model the first hole as a linear straight hole feature.

   f. Pattern the hole feature with the general Pattern option.

   g. Construct the rib feature by sketching on datum plane FRONT.

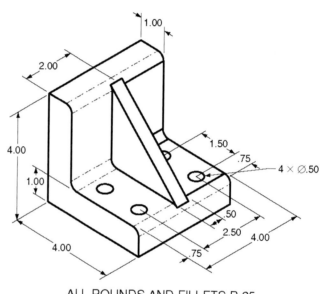

ALL ROUNDS AND FILLETS R.25

Figure 5–85   Problem 1

2. Using Part mode, model the part shown in Figure 5–86.

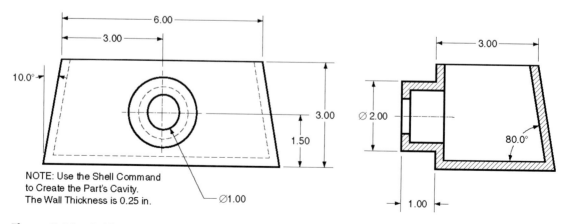

Figure 5–86   Problem 2

3. Use Pro/ENGINEER to model the part shown in Figure 5–87. This part will be used within an assembly model problem later in this textbook.

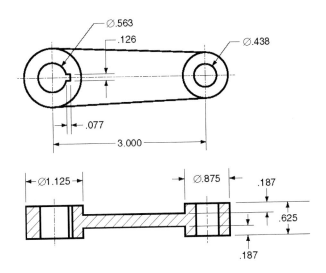

Figure 5–87 Problem 3 (Arm Part)

4. Model the part shown in Figure 5–88.

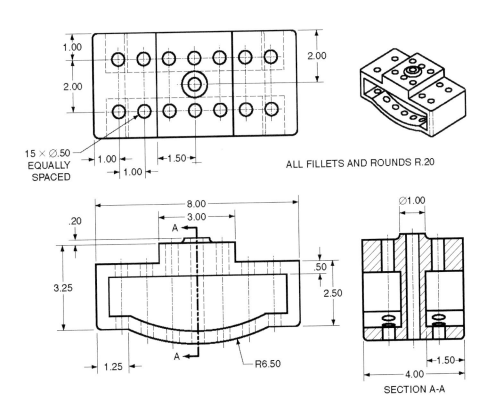

Figure 5–88 Problem 4

QUESTIONS AND DISCUSSION

1. Describe the five different methods for placing a hole.

2. What is the difference between a straight hole, a sketched hole, and a standard hole? What are some uses of a sketched hole?

3. Describe methods for avoiding round regeneration conflicts.

4. How does a neutral plane draft differ from a neutral curve draft?

5. What is the purpose of a neutral plane?

6. Compare and contrast an extruded protrusion feature with a rib feature.

7. Describe the assumptions associated with the following pattern categories:

    • Identical

    • Varying

    • General

8. In regard to the Pattern command, what is a leader dimension?

6

REVOLVED FEATURES

Introduction

Revolved feature construction techniques are common within Pro/ENGINEER. The most obvious revolved technique is the Revolve command. Other revolved feature construction techniques include the sketched hole option and Shaft, Flange, and Neck features. Upon finishing this chapter, you will be able to:

- Construct a sketched hole.
- Model a shaft by utilizing the Shaft command.
- Model a flange by utilizing the Flange command.
- Cut a neck with the Neck command.
- Pattern a radial hole.
- Construct a revolved protrusion.
- Construct a revolved cut.

DEFINITIONS

Axis of revolution	The axis around which a section is revolved. Within Pro/ENGINEER, revolved features require a user-sketched centerline. This centerline serves as the axis of revolution.
Through/axis datum plane	A datum plane constructed through an axis.

REVOLVED FEATURE FUNDAMENTALS

A revolved feature is a section that is rotated around a centerline. For any type of revolved feature, within the sketcher environment, the user sketches the profile of the section to be revolved and the centerline to revolve about (Figure 6–1). Revolved features can be positive space or negative space. A revolved protrusion is an example of a positive space feature, while its negative space counterpart is the revolved cut. The Flange command is another example of a revolved positive space feature. Its counterpart negative space feature is the Neck command.

SKETCHING AND DIMENSIONING

As shown in Figure 6–1, the geometry of a revolved feature must be sketched on one side of the centerline and the section must be closed. A sketch does not require a center, but if multiple centerlines exist in a sketch, the first one sketched will serve as the axis of

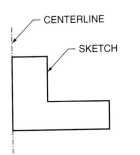

Figure 6–1 Revolve Sketch

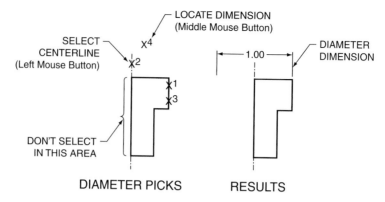

Figure 6–2　Diameter Dimensions

revolution. Entities that lie on the axis of revolution will not serve as a replacement for the centerline, nor will centerlines close a section.

Revolved features are often used to create cylindrical objects such as shafts and holes. Drafting standards require cylindrical objects to be dimensioned with a diameter value. This creates a unique situation within the sketcher environment. As shown in Figure 6–2, to dimension a revolved feature with a diameter value, perform the following steps:

1.　Pick the geometry defining the outside edge of the feature.
2.　Pick the centerline to serve as the axis of revolution.
3.　Pick the geometry defining the outside edge of the feature.
4.　Pick a location for the placement of the dimension text.

The resulting dimension should appear as shown in Figure 6–2.

MODELING POINT　If a diameter dimension is not created, geometry was probably inadvertently selected instead of the required centerline. Pick the centerline on the workscreen where it is clear that a centerline is the only entity residing (Figure 6–2).

REVOLVED PROTRUSIONS AND CUTS

Like the extrude tool, the Revolve command has solid, cut, and surface options. Revolved protrusions are used to create positive space solid features while revolved cuts are used to create negative space features.

REVOLVED FEATURE PARAMETERS

As with extruded protrusions and cuts, a variety of options are available for defining revolved feature parameters.

REVOLVE DIRECTION

The Revolve direction attribute is similar to the Extrude direction attribute. Options are available for selecting a One Side or a Both Sides revolution. The One Side option will revolve the section from the sketching plane in one direction while the Both Sides attribute will revolve the section in both directions from the sketching plane.

ANGLE OF REVOLUTION

The Angle of Revolution parameter is similar to the Extrude option's Depth parameter. This option is used to specify the number of degrees that the section will be revolved about the axis of revolution. The following options are available:

- **Variable** The Variable option is used to specify an angle of revolution. The angular parameter specified is modifiable (Figure 6–3). Use this option to create a 360 degree angle of revolution.

- **To Reference** The Thru Until option revolves a feature up to a selected point, vertex, plane, or surface. The user must specify the reference.

- **Symmetric** The Symmetric option revolves a feature with equal degrees of revolution on both sides of the sketch plane.

CREATING A REVOLVED PROTRUSION

The Revolve option is used extensively for creating base geometric features. The following is a step-by-step guide for creating a revolved protrusion.

STEP 1:   Select the Sketch Tool icon (Figure 6–5).

STEP 2:   On the workscreen, pick a suitable sketching plane.

Notice on your screen and in Figure 6–4 how the Sketch dialog box is used to manage the sketching plane and the sketching plane orientation.

STEP 3:   Orient the Sketching Plane.

Notice in Figure 6–4 how Pro/ENGINEER will provide a default sketch plane orientation. In this example, you are sketching on datum plane FRONT and orienting datum plane RIGHT toward the right of the work-screen. You can see this setup in the Section dialog box's Reference and

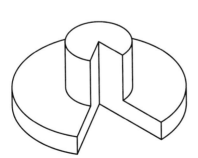

Figure 6–3   Variable Angle of Revolution

Figure 6–4   Section Dialog Box

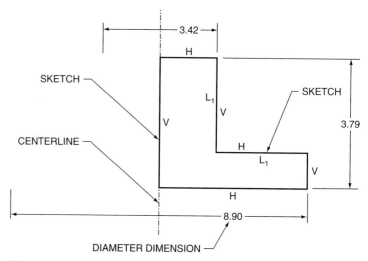

Figure 6–5   Resolve Feature Sketch

Orientation options. You can pick any planar reference to include part surface and you can also orient a plane toward the TOP, BOTTOM, LEFT, or RIGHT.

STEP 4: **Select the SKETCH option on the Sketch dialog box.**

STEP 5: **Use the CENTERLINE option to sketch the centerline (see Figure 6–6).**

Note: With Wildfire 3.0 an axis of revolution can be picked on the workscreen, so a sketched centerline is not a requirement.

For the first geometric feature of a part and when sketching on Pro/ENGINEER's default datum planes, it is recommended that you sketch the centerline of a revolved feature aligned with the edge of a datum plane. Consider design intent when sketching the centerline.

MODELING POINT   A centerline does not have to be the first entity sketched for a revolved feature. If a centerline is not present when you select the Continue icon to exit the sketcher environment, Pro/ENGINEER will provide a warning message.

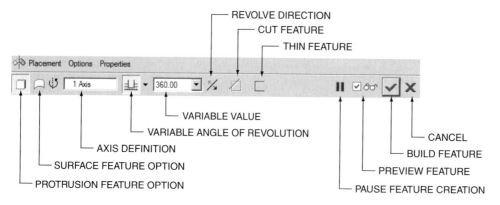

Figure 6–6   Revolve Dashboard

Figure 6–7 Additional Depth Options

STEP 6:   Sketch the Section (Figure 6–6).

Use appropriate sketching tools to create the section. A revolved section has to be closed and must lie completely on one side of the centerline. Sketch the section according to design intent.

STEP 7:   Select the Continue icon when the section is complete.

STEP 8:   Select the Revolve tool on Pro/ENGINEER's Base Feature toolbar.

Remember that you are working from the left to the right on the dashboard.

STEP 9:   Select between a solid protrusion or a surface feature (see Figure 6–4).

Remember that you are working from the left to the right on the dashboard.

STEP 10:   Select an Angle option and, if required, define angle references.

Angle options available include Variable, Through Until, and Symmetric. The references you define depend on the option selected. As an example, if you select Variable, you will have to enter a variable angle. If you select Through Until, you will have to select a feature to serve as the bounding reference.

The Options selection on the dashboard provides additional Angle options (Figure 6–7). For symmetrical revolutions, a user can enter a two-sided variable revolution.

STEP 11:   Select a material creation direction.

This option determines the direction that the feature will extrude from the sketching plane. This option is usually relevant only if a variable angle less than 360 degrees is defined.

STEP 12:   For thin solids, select the thin icon (optional step).

By default, the thin option is not selected. Unlike the solid option, thin features protrude or cut with a user-defined wall thickness. This wall thickness is defined after exiting the sketching environment.

STEP 13:   Preview the feature.

At this point, you can still modify any definitions established for the feature. Definition parameters are shown along the dashboard.

STEP 14:   Build the feature.

Select the Build Feature option on the dashboard to finish the feature.

HOLE OPTIONS

Pro/ENGINEER provides three hole types: simple (straight), sketched, and standard. Simple holes have a straight profile throughout the length of the feature. Sketched holes are used to create unique profiles, such as exist with counterbored and countersunk holes. In addition to the three types of holes, Pro/ENGINEER provides five placement options: Linear, Coaxial, Radial, Diameter, and On Point. The Linear option is used to locate a hole from two reference edges, while the Coaxial option is used to locate a hole's centerline coincident with an existing axis. The Radial option is used to locate a hole at a distance from an axis and at an angle to a reference plane (Figure 6–8). The hole's distance from the reference axis is defined by a radius value. Like the Radial option, the Diameter option is used to locate a hole at a distance from an axis and at an angle to a reference plane (Figure 6–9). With the Diameter option, the hole's distance from the reference axis is defined by a diameter value.

SKETCHED HOLES

The Sketched Hole option requires the user to sketch the profile of a hole (Figure 6–10). Most normal sketching tools can be used. Unlike normal feature sketching environments, the hole sketcher environment does not provide an option for specifying references. A sketched hole is created originally independent of any specific part features and later placed according to the hole placement option being used (e.g., Linear, Coaxial, Diameter, Radial, or On Point).

   When you are sketching a hole, a vertical, user-created centerline is required. Within the sketcher environment, use the Centerline icon to create this entity. All sketched entities must be created on one side of the centerline and must be closed. A geometric entity may be placed on top of the centerline, but the centerline cannot serve as an element of the hole profile. Additionally, one sketched entity must lie perpendicular to the centerline. This en-

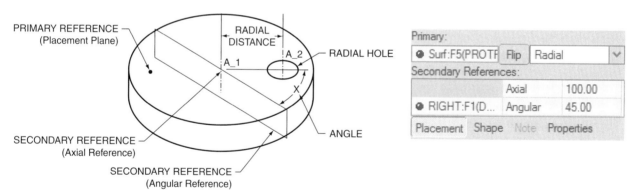

Figure 6–8   Radial Hole Placement

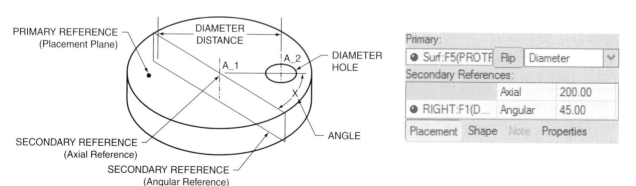

Figure 6–9   Diameter Hole Placement

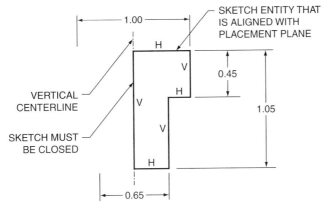

Figure 6-10   Sketching a Counterbored Hole

tity will be aligned with the placement plane when placing the hole. For sketched holes with multiple perpendicular lines, the uppermost line within the sketcher environment will serve as the placement reference.

CREATING A SKETCHED HOLE

Perform the following steps to create a sketched hole:

STEP 1:   Select the Hole option on the toolbar.

STEP 2:   Select the hole's primary reference.

Place the hole according to requirements for a Linear, Coaxial, Radial, Diameter, or On Point hole. The following are the primary references for each placement type:

- **Linear** Placement plane.
- **Coaxial** Axial reference.
- **Radial** Placement plane.
- **Diameter** Placement plane.
- **On Point** Datum point reference.

STEP 3:   Ensure that Simple is selected as the hole type on the dashboard.

A sketched hole is a subtype of a simple hole.

STEP 4:   On the dashboard, select SKETCHED as the Hole Type (Figure 6–11).

STEP 5:   Under the Placement menu, select the hole's placement type (e.g., Linear, Coaxial, Diameter).

STEP 6:   Under the Placement menu, use the Control key to pick offset references as required by the hole's placement type.

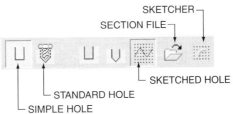

Figure 6-11   Sketched Hole Option

STEP 7:   On the dashboard, select the Sketcher tool (Figure 6–11).

Once the Sketcher icon is selected, Pro/ENGINEER will launch a sketcher environment for the creation of the hole's profile. The Section File option allows predefined section files (*.sec) to be imported as the hole's sketch.

STEP 8:   Within the sketcher environment, use the Centerline option to create a vertical centerline (Figure 6–10).

All revolved features require a centerline within the sketch. For a sketched hole, this centerline must be vertical.

STEP 9:   Sketch the profile of the hole (Figure 6–10).

Sketch entities of the hole must be created on one side of the centerline and must be completely closed. The centerline will not serve as an entity of the sketch.

One sketched line entity must be created perpendicular to the centerline. This entity will be used to align the hole with the placement plane. For multiple perpendicular entities, the uppermost line in the sketcher environment will serve this purpose.

STEP 10:   Dimension the sketch to meet design intent.

Use the Dimension option to dimension the sketch to meet design intent. Since holes are defined with diameter values, use the diameter dimensioning technique described previously in this chapter.

STEP 11:   Modify dimension values.

STEP 12:   Select the Continue icon to exit the sketcher environment.

STEP 13:   Select the Built Feature icon on the dashboard.

RADIAL HOLE PLACEMENTS

The Radial and Diameter hole placement options are used frequently with the Pattern command to create a radial pattern of holes (Figure 6–12). The Radial and Diameter options place a hole at a user-specified distance from an existing axis and at an angle to a reference plane. With patterned holes, this angular dimension is used as the leader dimension for creating the pattern.

CREATING A STRAIGHT DIAMETER (RADIAL) HOLE

Perform the following steps to place a straight diameter hole (Refer to Figure 6–13):

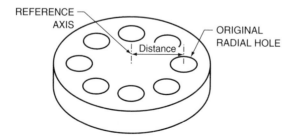

Figure 6–12   A Patterned Hole

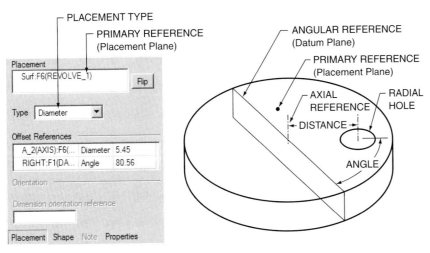

Figure 6–13   Radial Hole Creation

STEP 1: ⊔ Select the Hole option on the toolbar.

STEP 2: ⊔ Ensure that Simple is selected as the hole type on the dashboard.

A sketched hole is a subtype of a straight hole.

STEP 3: Select the hole's placement plane.

The primary reference for a diameter or radial placed hole is the hole's placement plane.

STEP 4: Under the Placement menu (Figure 6–13), select Diameter (or Radial) as the hole's placement type.

Other options available include Linear and Coaxial. The Linear option will locate a hole from two reference edges while the Coaxial option will place the hole's centerline coincident with an existing axis.

STEP 5: Within the Placement menu, select within the Offset References field.

STEP 6: On the workscreen, pick the axial reference for the hole.

STEP 7: Enter the Diameter distance value for the radial hole.

In the Diameter textbox, enter the distance value for the diameter hole. The diameter value will be twice a corresponding radial value. As an example, if the hole is to be located 3 in from the axial reference, then you should enter 6 as the diameter value. This option is often used to create a bolt-circle hole pattern.

STEP 8: Using the Control key, pick a reference plane, then enter an angular value from the plane.

To select multiple secondary references, you must use the Control key when picking references. Since Diameter and Radial placement options are defined with two secondary references, this is a requirement. Pick an existing planar surface, then enter an angular value. The hole will be located from the reference plane at the specified angle. A datum plane or planar surface normal to the hole's placement plane can be selected.

STEP 9: On the dashboard, enter the hole's Diameter value.

STEP 10: On the dashboard, select the hole's depth type (and value if required).

STEP 11: ✓ Select the Build Feature checkmark.

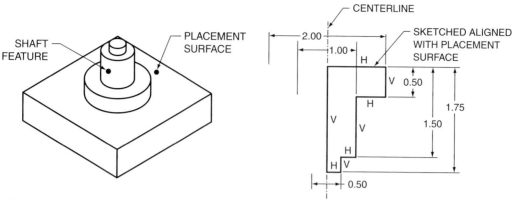

Figure 6-14   The Shaft Option

THE SHAFT COMMAND

The counterpart to the Sketched Hole option is the Shaft command. While the Sketched Hole option creates a negative space feature, the Shaft command creates a positive space feature (Figure 6–14). The techniques and options for creating a shaft are the same as for creating a hole. As an example, shafts require a vertical centerline with entities sketched on one side only. Additionally, placement options for a shaft are the same as for a hole. Compared to the sketched hole option, one concern does exist. As with sketched holes, the uppermost line in the sketcher environment constructed perpendicular to the centerline will be aligned with the placement plane. Because of this, it is often necessary to sketch a shaft's section upside down. To construct a shaft, perform the same steps for creating a sketched hole. The configuration file option *allow_anatomic_features* has to be set to YES for the Shaft command to show on the Insert >> Advanced menu.

THE FLANGE AND NECK OPTIONS

A flange is a revolved positive space feature created around an existing revolved part or feature (Figure 6–15). A Neck is a revolved negative space feature created around an existing revolved part or feature. The sketching plane for both a flange feature and a neck feature must be a Through >> Axis datum plane. As shown in Figure 6–16, when sketching a Flange or a Neck, the ends of the sketched section must be aligned with the surface of an existing revolved feature. The configuration file option *allow_anatomic_features* has to be set to YES for the Flange and Neck commands to show on the Insert >> Advanced menu.

Perform the following steps to create either a flange or a neck:

STEP 1:   **Select INSERT >> ADVANCED >> FLANGE (or NECK) on the menu bar.**

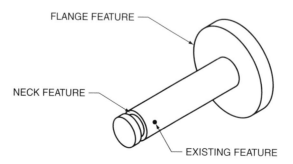

Figure 6-15   Flange and Neck Features

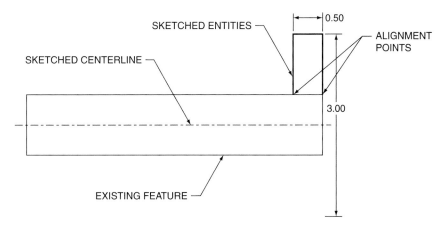

Figure 6–16   The Sketched Flange Section

STEP 2:   On the Options menu, specify the number of degrees of revolution.

STEP 3:   Select ONE SIDE >> DONE on the options menu.

STEP 4:   Select or create a datum plane that lies through the axis of revolution of a part or feature.

The plane for sketching a flange or neck section must be a Through >> Axis datum plane (Figure 6–17). If one does not exist, create one with the Make Datum option.

STEP 5:   Accept the view direction, then orient the sketcher environment.

STEP 6:   Use the CENTERLINE icon to create a centerline to revolve the flange or neck around (Figure 6–16).

Align the centerline with the axis of the existing revolved feature. The existing axis will not serve as the centerline.

STEP 7:   Sketch the section of the flange or neck feature.

As shown in Figure 6–16, the section must be aligned with the surface of the existing revolved part or feature. Dimension the feature according to design intent.

STEP 8:   Select the Continue option on the sketcher toolbar to create the flange or neck.

Compared to most feature creation processes, the Neck and Flange options are unique. They do not utilize a Feature Definition dialog box.

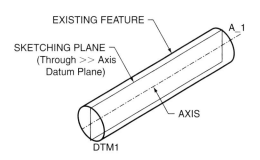

Figure 6–17   Selecting a Datum Plane

ROTATIONAL PATTERNS

The Pattern command can be used to create radial patterns of features. Two options are available to serve this purpose: dimension and axis. The dimension pattern option requires the selection of an angular dimension that will be varied throughout the pattern. The Axis option requires the selection of an existing that will define the center of the pattern.

HOLE RADIAL PATTERN

The Pattern command can be used to create an angular copy of a radial hole (Figure 6–18). The Hole-Radial option places a feature by entering a distance from a selected axis and by entering an angle to a reference plane. This angle can be used as the leader dimension in the pattern creation process.

ROTATIONAL PATTERN

The Pattern option can be used to create a rotational copy of a sketched feature (Figure 6–19). The feature's section has to reference a Through/Axis datum plane created with the Through and Angle constraint options or the feature has to be sketched on a Through/Axis datum plane. The angular dimension defining the sketching plane is used to pattern the feature.

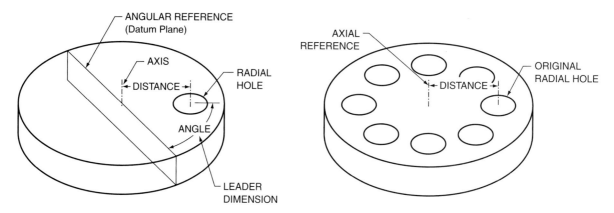

Figure 6–18   Radial Pattern

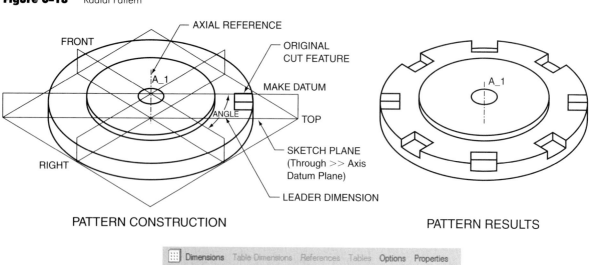

Figure 6–19   Rotational Pattern

AXIS PATTERN

The Axis pattern option is used to create a rotational pattern around an axis.

INSTRUCTIONAL NOTE For more information on patterns, see Chapter 6.

CREATING A RADIAL PATTERN

The following is a step-by-step approach for creating a radial pattern in one direction. Two directional patterns can be used with rotational patterns too. Additionally, dimensions other than the leader dimension can be varied.

STEP 1: **On the model tree, pick the feature to pattern.**

On the workscreen or on the model tree, select the feature to pattern. With the Pattern command, only one feature can be pattern at a time.

STEP 2: **Select the Pattern option on the toolbar.**

The Pattern command is also available under the Insert menu and on the mouse's pop-up menu.

STEP 3: **Select Dimension on the dashboard as the pattern type (Figure 6–19).**

Dimension is selected by default. Other options include Direction, Table, Reference, Curve, and Fill.

STEP 4: **Select an angular leader dimension for use in varying the feature (see Figures 6–18 and 6–19).**

To create a rotational pattern of a feature, the feature has to have an angular dimension. Holes and shafts placed with the Radial or Diameter option incorporate a reference angle dimension to locate the hole. Features sketched on an on-the-fly datum plane can be patterned if the datum plane is defined with the Angle constraint.

STEP 5: **Enter the dimension increment value.**

The pattern will be incremented in the first direction by the amount entered.

STEP 6: **On the dashboard, enter the number of instances in the pattern (see Figure 6–19).**

STEP 7: **Select a Pattern Option on the Options menu.**

On the dashboard's Options menu, select Identical, Varying, or General. Identical patterns require the most assumptions. With this option, instances of a pattern cannot intersect other instances, or the edge of the placement plane. Additionally, an identical pattern has to exist on one placement plane only. The General option does not require any assumptions, but takes longer to regenerate.

STEP 8: **Build the feature.**

DATUM AXES

Datum axes are used as references for the creation of features. As an example, they can be used to place a coaxial hole or a radial hole. Additionally, they are used often to create datum planes. When holes, cylinders, and revolve features are created, datum axes are created automatically. Datum axes created separately from a part feature are considered to be features. They are named in sequential order on the model tree starting with A_1. The following is a list of available constraint options.

THROUGH

The Through constraint option can be used to create a datum axis through an existing edge, datum point, or vertex. It can also be used to create a datum axis through the center of a cylindrical feature (hole, cylinder, etc.).

NORMAL

The Normal option creates a datum axis perpendicular to an existing plane. This plane can be a part or datum plane. Dimensional offset references are required to locate the axis or the Normal constraint option can be used in combination with the Through constraint option. Dimensional offset references are created by selecting two part edges and then providing dimensional values.

TANGENT

The Tangent option places a datum axis tangent to a curve or edge at a selected point. The point has to exist prior to the use of this option.

CREATING DATUM AXES

Datum axes can be placed on parts or assemblies. Perform the following steps to create a datum axis in Part mode:

STEP 1: On the datum toolbar, select the CREATE DATUM AXIS icon.

STEP 2: Using the Control key, pick one or two references to define the datum axis.

Pick references based on the datum constraint option.

STEP 3: If necessary, select offset references to define a datum axis.

Often, one reference will not define the location of a datum axis. As an example, placing an axis normal to a plane requires the selection of two offset references for locating the axis on the plane.

STEP 4: Select the OK option on the Datum Axis dialog box.

The datum axis will be created when you select OK.

SUMMARY

Some of the most common features created within Pro/ENGINEER are revolved features. Options that utilize a form of a revolved feature include Revolved Protrusion, Revolved Cut, Sketched Hole, Shaft, Flange, and Neck. One of the requirements of any revolved feature is the sketching of a centerline. Within the sketcher environment for a feature option, entities of the sketch must lie completely on one side of the centerline. In addition to the typical options that are revolved, the Pattern command allows for the creation of rotational instances of a feature.

REVOLVED FEATURES TUTORIAL

This tutorial exercise provides step-by-step instruction on how to model the part shown in Figure 6–20. This tutorial will cover:

- Creating a revolved protrusion.
- Creating a radial sketched hole.
- Creating a hole radial pattern.
- Creating a revolved cut.
- Modifying the number of holes in a pattern.

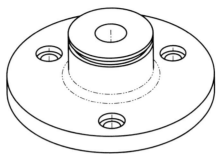

Figure 6–20   Finished Part

CREATING A REVOLVED PROTRUSION

The first section of the Revolved Features Tutorial will create the base revolved protrusion shown in Figure 6–21.

STEP 1: **Start Pro/ENGINEER.**

STEP 2: **Establish an appropriate working directory.**

STEP 3: **Start a new part file named *revolve1* (use the default template file).**

Use either the File >> New option or the New icon to start a new part file named *revolved1*.

STEP 4: **Define the following sketching environment.**

- Datum plane FRONT as the sketching plane.
- Datum plane RIGHT oriented to the right.

STEP 5: **Select SKETCH on the Sketch dialog box to enter the sketching environment.**

STEP 6: **Close the References dialog box.**

STEP 7: **Use the Centerline option to sketch a vertical centerline aligned with datum plane RIGHT (Figure 6–22).**

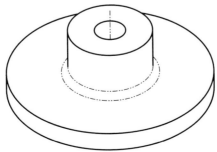

Figure 6–21   Revolved Protrusion

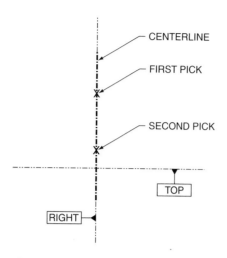

Figure 6-22   Sketching a Centerline

The Centerline icon can be found under the Line icon. Revolved features require a user-created centerline in the sketching environment. The sketched section will be revolved around this centerline. With revolved protrusions and cuts, this centerline can be created at any angle.

STEP 8:   Use the LINE option to sketch the section shown in Figure 6–23.

Sketch the section to the right of the centerline. Sketch all lines either horizontal or vertical as shown in the figure. Align the bottom of the sketch with datum plane TOP.

STEP 9:   Use the CIRCULAR FILLET option to create the fillet shown in Figure 6–23.

The Circular Fillet option requires the selection of two nonparallel entities to fillet between.

STEP 10:   Apply the dimensioning scheme shown in Figure 6–24. (Do not modify the dimension values within this step.)

Use the Dimension option to create the dimensioning scheme shown in Figure 6–24. Pro/ENGINEER does not know the design intent for a feature. As a result, the dimensions created by Intent Manager may not match those necessary for the design.

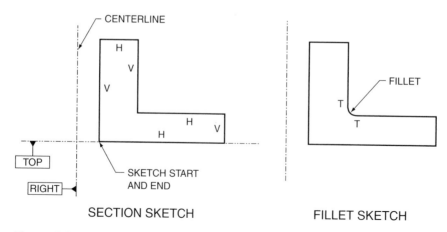

SECTION SKETCH   FILLET SKETCH

Figure 6-23   Sketching the Section

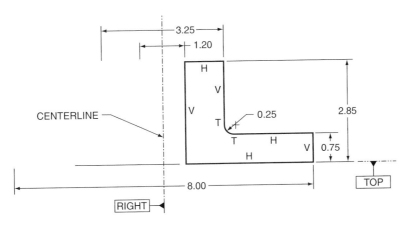

Figure 6-24   The Dimensioning Scheme

To create the diameter dimensions, perform the following four selections (Figure 6–25):

1. Pick the outside edge of the entity (left mouse button).
2. Pick the centerline (left mouse button).
3. Pick the outside edge of the entity (left mouse button).
4. Place the dimension (middle mouse button).

STEP 11:   Select the Modify option.

STEP 12:   **On the workscreen, select the dimension defining the height of the flange (the 2.85 dimension in Figure 6–24).**

STEP 13:   **On the workscreen, select the remaining five dimensions.**

After selecting the six dimensions defining the section, your Modify Dimension dialog box should appear similar to Figure 6–26.

STEP 14:   **Check the LOCK SCALE option on the dialog box (Figure 6–26).**

STEP 15:   **On the Modify Dimension dialog box, modify the flange's height dimension to have a value of 2.85 (select the ENTER key after entering value).**

If you get unexpected results from your dimension modification, use the Undo to correct any errors.

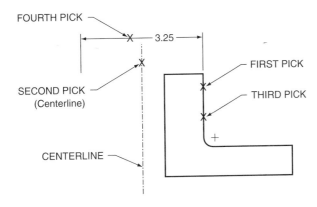

Figure 6-25   Diameter Dimension

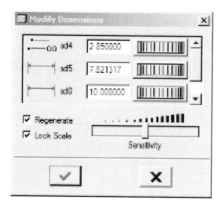

Figure 6–26   Modify Dimensions Dialog Box

STEP 16: Uncheck the LOCK SCALE option.

STEP 17: On the dialog box, modify the remaining dimension values to match Figure 6–24.

STEP 18: When all dimension values match the illustration, select the Regenerate checkmark on the dialog box.

STEP 19: ✓ Select the Continue icon to exit the sketcher environment.

Do not select the Continue option until your section matches Figure 6–24.

STEP 20: ⟨⟩ Select the Revolve option on the toolbar.

STEP 21: On the dashboard, ensure that the Variable Angle of Revolution icon is selected (Figure 6–27).

STEP 22: On the dashboard, enter 360 as the variable angle of revolution.

STEP 23: ✓ Build the feature.

Your feature will look similar to Figure 6–28. The illustration shown displays the part with Tangent Edges set to Phantom and the Default Orientation set to Isometric. To change the tangent edge display and default orientation of a part, make the adjustments to the Environment dialog box (Tools >> Environment), as shown in Figure 6–29.

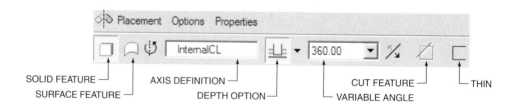

Figure 6–27   Revolve Tool Dashboard

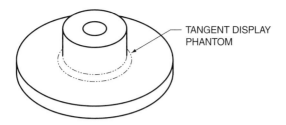

TANGENT DISPLAY
PHANTOM

Figure 6–28   Finished Feature

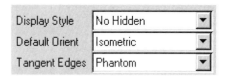

Figure 6–29   Environment Options

CREATING A DIAMETER PLACED SKETCHED HOLE

This segment of the tutorial will create the sketched hole shown in Figure 6–30.

STEP   1:   Select the Hole option on the toolbar.

STEP   2:   **Select the hole's placement plane (Figure 6–31).**

The primary reference for a diameter or radial placed hole is the hole's placement plane.

STEP   3:   **Under the Placement menu (Figure 6–31), select Diameter as the hole's placement type.**

Other options available include Linear and Radial. The Linear option will locate a hole from two reference edges. The radial option will create a radially placed hole with a radius value.

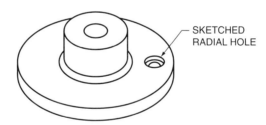

SKETCHED
RADIAL HOLE

Figure 6–30   Finished Hole Feature

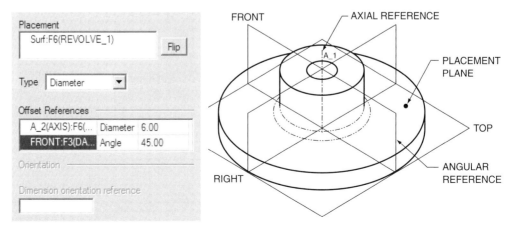

Figure 6–31   Hole Placement References

STEP 4:   Within the Placement menu, select within the Offset References field.

STEP 5:   On the workscreen, pick the center axis of the revolved feature (Figure 6–31).

STEP 6:   On the Placement menu, enter 6.0 as the Diameter distance value for the diameter hole.

In the Diameter textbox, enter 6.0 as the distance value for the diameter hole. The diameter value will be twice the corresponding radial value. As an example, if the hole is to be located 3 in from the axial reference, then you should enter 6 as the diameter value. This option is often used to create a bolt-circle hole pattern.

STEP 7:   Using the Control key, pick datum plane FRONT as the angle reference, then enter 45.0 as the angular value from the plane.

To select multiple secondary references, you must use the Control key when picking references. Since Diameter and Radial placement options are defined with two secondary references, this is a requirement. Pick an existing planar surface then enter an angular value. The hole will be located from the reference plane at the specified angle. A datum plane or planar surface normal to the hole's placement plane can be selected.

MODELING POINT   Creating a radial or diameter hole is the first step in modeling a rotationally patterned hole. The angular dimension created with a radial hole placement option is used as the leader dimension for patterning the hole around the center axis.

STEP 8:   ⊔   Ensure that Simple is selected as the hole type on the dashboard.

A sketched hole is a subtype of a straight hole.

STEP 9:   ▦   On the dashboard, change your hole type from Straight to SKETCHED (Figure 6–32).

In this tutorial, you will sketch the section for the hole feature. The Sketched hole dashboard provides an option for opening and existing section file to serve the same purpose.

STEP 10:   ▦   Select the Sketched Hole Tool on the dashboard.

The Sketched option will require you to sketch the hole's cross section. After you have selected this option, Pro/ENGINEER will open a sketcher environment.

STEP 11:   ┊   Within the sketching environment, use the Centerline option to create the vertical centerline shown in Figure 6–33.

A vertical centerline is a requirement for a sketched hole feature. The section will be revolved around the centerline.

Figure 6–32   Sketched Hole Dashboard

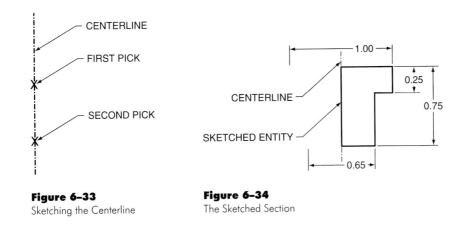

Figure 6–33
Sketching the Centerline

Figure 6–34
The Sketched Section

STEP 12: **Sketch the section shown in Figure 6–34.**

Use the Line option to sketch the section. For this specific section, sketch each entity either horizontally or vertically.

Section sketched holes have to be completely enclosed. As shown in Figure 6–34, the section will be sketched on top of the previously created centerline. *The centerline will not serve as an entity to close the section.* You will receive an error message when exiting the sketcher environment if the section is not closed.

STEP 13: **Apply the dimensioning scheme as shown in Figure 6–34.**

Use the Dimension option to match the dimensioning scheme shown in Figure 6–34. Holes are dimensioned with diameter values.

STEP 14: **Modify dimension values to match Figure 6–34.**

STEP 15: **Select the Continue icon when the section is correct.**

Using the Diameter option, the hole will be located at a distance (3 in) from datum axis A_1 and at an angle to datum plane FRONT (45 degrees).

STEP 16: **Select the Build Feature icon on the dashboard.**

STEP 17: Save your part.

CREATING A RADIAL HOLE PATTERN

This segment of the tutorial will create a radial pattern of the hole created in the previous section of this tutorial. The 45 degree angular reference dimension used to locate the hole will be used as the leader dimension within the pattern. The finished pattern is shown in Figure 6–35.

STEP 1: Select the Hole feature on the model tree.

MODELING POINT The Pattern command will create a rotational or linear pattern of a single feature. To create a pattern of multiple features, each feature must be grouped by using the Group command. Within the Group menu is a separate Pattern option.

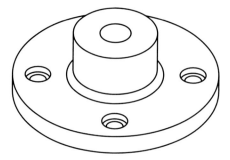

Figure 6–35   Finished Pattern

STEP 2:      **Select the Pattern option on the toolbar.**

The Pattern command is also available under the Insert menu and on the mouse's pop-up menu.

STEP 3:   **Select Dimension on the dashboard as the pattern type (Figure 6–36).**

Dimension is selected by default. Other options include Table, Reference, and Fill.

STEP 4:   **As shown in Figure 6–36, pick the 45 degree angular dimension as the leader dimension for use in varying the feature.**

The 45 degree angular dimension will be used as the leader dimension within the patterning process. This dimension will be varied to create the pattern.

STEP 5:   **Utilizing the on-screen dimension textbox, enter 90 as the dimension increment value.**

The leader dimension will be varied 90 degrees per instance of the hole feature. In other words, each hole within the pattern will be 90 degrees apart.

STEP 6:   **On the dashboard, enter 4 as the number of instances in the pattern (see Figure 6–36).**

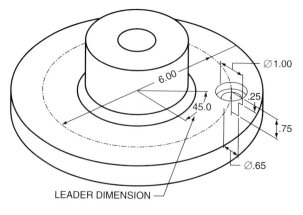

Figure 6–36   Pattern Definition

Step 7: **Under the Options menu, select IDENTICAL as the pattern option.**

The Identical option allows several assumptions in the pattern creation process. Features patterned with this option must lie on the same placement plane, instances cannot intersect the placement plane's edge, and instances cannot intersect other instances. To allow instances to be placed on different planes and to allow instances to intersect an edge, use either Varying or General. To allow instances to intersect, use the General option.

Step 8: **Build the feature.**

Step 9: **Save the part.**

CREATING A REVOLVED CUT

This segment of the tutorial will create the Revolved Cut feature shown in Figure 6–37.

Step 1: 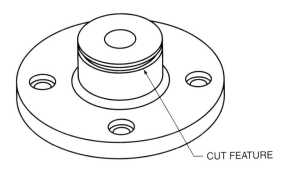 **Select the Sketch option on the dashboard.**

Step 2: **Using the Sketch dialog box, select datum plane FRONT as the Sketching Plane (Figure 6–38), then orient datum plane RIGHT to the right.**

Step 3: **Select SKETCH on the dialog box.**

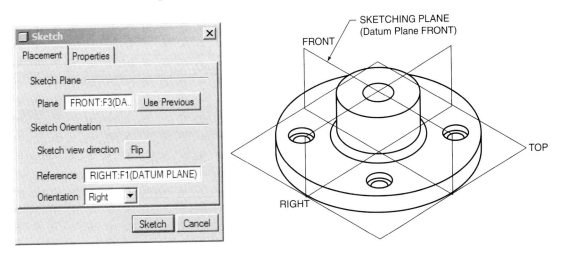

Figure 6–37   Revolved Cut Feature

Figure 6–38   Sketching Plane

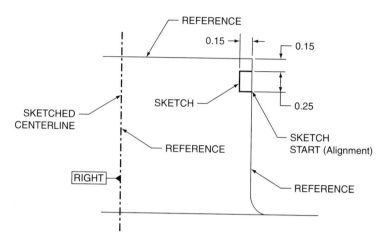

Figure 6–39   Section Creation

STEP 4:   **Within the sketcher environment, specify the three references shown in Figure 6–39.**

Use the References dialog box (Sketch >> References) to specify the three references shown in the illustration.

STEP 5:   **Sketch the section shown in Figure 6–39.**

Sketch the section as shown. To better view the entities being sketched, use a zoom option to match the zoom extents shown in Figure 6–39. By aligning with the outside edge of the existing flange, a closed section is not necessary. After sketching the section, apply the dimensioning scheme shown, then modify the dimension values accordingly.

STEP 6:   Apply the dimensioning scheme and modify values to match Figure 6–39.

STEP 7:   Create a centerline aligned with datum plane RIGHT (see Figure 6–39).

Sketch a vertical centerline aligned with datum plane RIGHT. This centerline will serve as the required axis of revolution.

STEP 8:   Select the Continue icon to exit the sketcher environment.

STEP 9:   Select the Revolve option on the toolbar.

STEP 10:   Select the Remove Material (Cut) option on the dashboard.

Like the extrude command, the revolve command defaults to the creation of a positive space protrusion. The Cut option for negative space features and the surface option are also available.

STEP 11:   On the dashboard, select Variable as the Revolve option.

STEP 12:   Enter 360 as the variable angle value.

STEP 13:   Build the feature.

STEP 14: Select the keyboard shortcut combination *Control D* to view the model's default orientation.

STEP 15: Save the model.

MODIFYING THE NUMBER OF HOLES

In this segment of the tutorial you will use the Edit option on the pop-up menu to modify the number of holes around the bolt circle pattern. The final pattern will appear as shown in Figure 6–40.

STEP 1: Expand the PATTERN feature on the model tree to reveal the instances of the pattern (Figure 6–41).

STEP 2: As shown in Figure 6–41, select one of the three patterned instances of the original sketched hole.

Select one of the patterned instances of the hole, not the original. As shown, an instance of a rotated pattern will show the original leader dimension value (45 degrees for this part) and the increment value (90 degrees). Also shown is the number of instances of the patterned hole (4 holes).

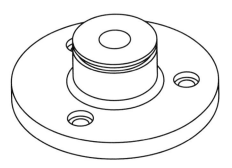

Figure 6–40   Finished Part

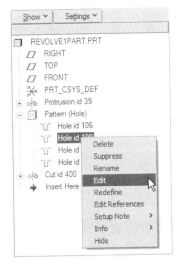

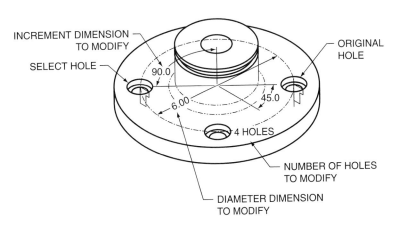

Figure 6–41   Pattern Parameters

Step 3:   Select the EDIT command on the right-mouse button's pop-up menu (Figure 6–41).

Notice the other commands available on the right mouse button's pop-up menu.

Step 4:   Double-pick the 90 degree value and modify it to equal 120.

Step 5:   Double-pick the 4 HOLES parameter and modify it to equal 3.

When modifying the number of instances of a pattern, you have to pick the actual text defining the value. In this case, select the number 4.

Step 6:   Select the 6.00 diameter dimension and modify it to equal 5.50.

Step 7:   Regenerate the part.

Your part should appear similar to Figure 6–40.

Step 8:   Save your part.

Step 9:   Purge old versions of the part by using the FILE >> DELETE >> OLD VERSIONS option.

SHAFT TUTORIAL

This tutorial exercise provides step-by-step instruction on how to model the part shown in Figure 6–42. One of the features of this part will be a shaft. Since the Shaft command by default is not available on any menu or toolbar, the configuration file option *allow_ anatomic_features* has to be set to YES to display this command. Setting this option will be the first section in this tutorial. Chapter 2 provides more information on setting other configuration file options. This tutorial will cover:

- Setting configuration options.

- Creating a base extruded protrusion.

- Creating a shaft feature.

- Creating a cut feature utilizing an on-the-fly datum plane.

- Patterning a cut feature.

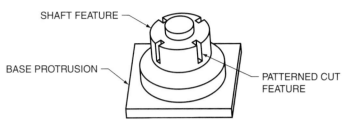

Figure 6–42   Final Part

SETTING CONFIGURATION OPTIONS

The Flange, Neck, Shaft, and Slot commands are not available by default on Pro/ENGINEER's Insert menu. The configuration file option *allow_anatomic_features* set to YES will reveal these commands. Perform the following steps to set this option:

STEP 1: **On Pro/ENGINEER's menu bar, select TOOLS >> OPTIONS.**

Selecting Options will reveal the Options dialog box, which is used to set configurations settings. See Chapter 2 for more information on configuration file options.

STEP 2: **In the Options textbox of the Options dialog box, enter ALLOW_ANATOMIC_FEATURES (Figure 6–43).**

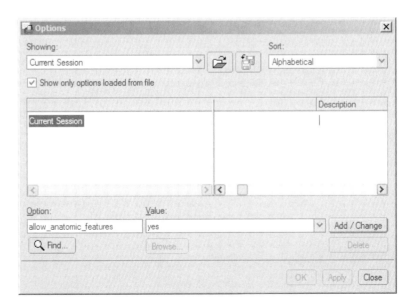

Figure 6–43   Options Dialog Box

Step 3:   Set the value of the *allow_anatomic_features* configuration option to YES.

Step 4:   Select the ADD/CHANGE option.

After selecting this option, notice how the allow_anatomic_features option is added to the configuration settings.

Step 5:   Select APPLY then Close the dialog box.

CREATING THE BASE PROTRUSION

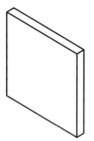

The first segment of this tutorial will create the base feature upon which the shaft feature will be placed. This base feature, shown in Figure 6–44, will be created as an Extruded Protrusion. This tutorial will not utilize Pro/ENGINEER's default datum planes.

Step 1:   Establish an appropriate working directory.

Step 2:   Select FILE >> NEW.

Step 3:   On the New dialog box, *deselect* the Use Default Template (see Figure 6–45).

Step 4:   Enter *SHAFT1* as the name of the part file then select OK.

Figure 6–44
The Base Feature

Step 5:   On the New File Options dialog box, select the *EMPTY* Template file, then select OK.

The Empty template has no default settings, parameters, features, or datums.

INSTRUCTIONAL POINT   This tutorial will NOT utilize Pro/ENGINEER's default datum planes as the first feature of the part. When modeling this feature, notice the steps that are missing in the protrusion process.

Step 6:   Select the Extrude option on the toolbar.

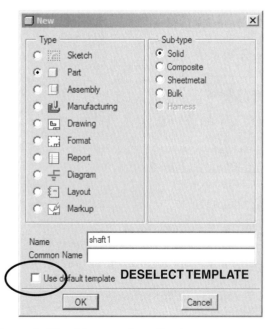

Figure 6–45   Selecting an Empty Template

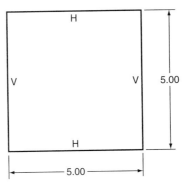

Figure 6-46   The Section for the Base Protrusion

Step 7: **Select PLACEMENT >> DEFINE on the dashboard.**

As the first feature of the part, there will be no option for selecting a sketching plane, nor will there be an option for orientating the sketching environment. Additionally, without a sketching plane, Pro/ENGINEER will allow a one-side protrusion only.

Step 8: **Using appropriate sketching tools, sketch the section shown in Figure 6–46.**

Use either the Line option or the Rectangle option to create the sketch. Modify the dimension values to match the figure.

Step 9: **After the section is complete, select the Continue icon to exit the sketcher environment.**

As with any section, do not select continue until the section is complete.

Step 10: **Enter a variable extrude depth of 0.500.**

Step 11: **Build the feature.**

Step 12: **Save the part.**

Creating a Shaft

A Shaft will be the next feature created in this tutorial (Figure 6–47). The creation of a shaft is similar to the creation of a sketched hole feature. Unlike a sketched hole, a shaft feature has to be sketched upside down. The configuration file option *allow_anatomic_features* has to be set to YES for this command to show on the Solid menu (Tools >> Options).

Step 1: **Select INSERT >> ADVANCED >> SHAFT on the menu bar.**

The Shaft command creates a positive space feature. Its negative space counterpart is the Sketched Hole.

Step 2: **Select LINEAR >> DONE as the Placement option.**

The Linear option will place the shaft from two user-selected edges. This is similar to placing a linear hole.

Step 3: **Within the independent sketcher environment, sketch a vertical centerline as shown in Figure 6–48.**

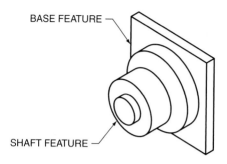

Figure 6–47   The Shaft Feature

All revolved features require a centerline. It is recommended for sketched holes and shafts that the centerline be constructed vertical. Pro/ENGINEER will allow you to exit the sketcher environment with a nonvertical centerline, but you will receive an error message when selecting OK on the Feature Definition dialog box.

STEP 4:   Sketch the section shown in Figure 6–48.

Shafts are normally sketched upside down. With a sketched hole or shaft, the uppermost line in the sketching environment that is perpendicular to the centerline will be the entity that aligns with the placement plane. As a result, the portion of the section that is to be aligned with the placement plane should be sketched at the top of the section.

STEP 5:   Dimension the sketch to match the design intent shown in Figure 6–48.

STEP 6:   Modify the dimensioning scheme and values to match Figure 6–48.

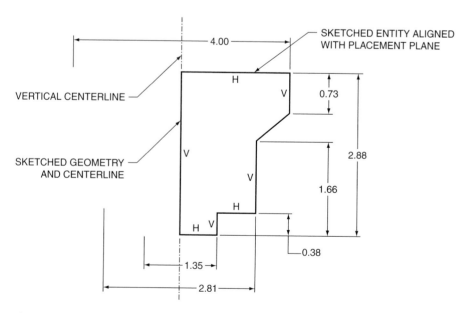

Figure 6–48   The Sketched Section

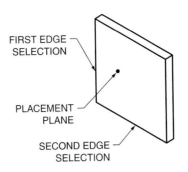

Figure 6–49   Shaft Location

You can use either the Modify option or the Pick option to modify dimension values. To eliminate the possibility of a regeneration error, it is advisable to modify smaller dimension values first.

STEP 7: ✓ **Select the Continue option when the section is complete.**

STEP 8: **Select the top of the base feature as the placement plane (Figure 6–49).**

The placement plane for a shaft feature must be planar. Remember to observe Pro/ENGINEER's message area after each option's selection.

STEP 9: **Select one edge of the base feature as a location reference (Figure 6–49).**

The Linear placement option specified earlier in this tutorial will locate the shaft from two reference edges. This step requires the selection of one edge.

STEP 10: **Enter 2.50 as the reference distance.**

Enter the distance that the centerline of the shaft will lie from the first selected edge.

STEP 11: **Select a second edge of the base feature as a location reference (Figure 6–49).**

STEP 12: **Enter 2.50 as the second reference distance.**

STEP 13: **Select Preview on the Feature Definition dialog box.**

STEP 14: **If the shaft is correct, select OK on the dialog box.**

STEP 15: **Save the part.**

CREATING A CUT

This segment of the tutorial will create the cut feature shown in Figure 6–50. This feature will be sketched on an on-the-fly datum plane. Additionally, the section will be extruded both sides from the sketching plane.

STEP 1: **Select the Sketch option on the dashboard.**

Since no suitable sketching plane currently exists, the next several steps will require you to complete an on-the-fly datum plane to serve this purpose.

STEP 2: **Select the Datum Plane icon on the datum toolbar.**

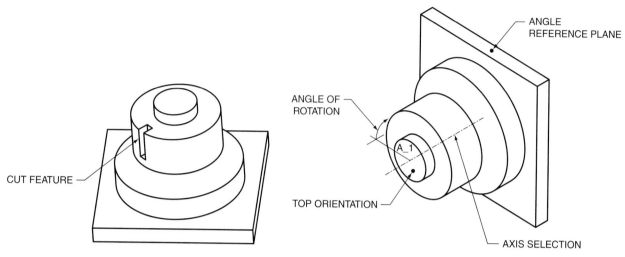

Figure 6–50   Cut Feature

Figure 6–51   Datum Creation

Notice how the Datum Plane dialog box will be displayed concurrently with the Section dialog box. Datum features can be created on the fly in the middle of most modeling processes.

STEP   3:   **Using the Control key, select the axis reference and the plane reference (angle reference plane) shown in Figure 6–51.**

The through/axis datum plane created in this tutorial requires an axis reference and a planar reference. Like most datum references in Pro/ENGINEER, you have to pick each entity with a Control key combination.

STEP   4:   **On the Datum Plane menu, enter 45 degrees as the rotation offset.**

This option will create a datum plane with an angular reference value of 45 degrees. Notice on the workscreen the graphical display of the angle of rotation. A negative value can be entered if necessary. Note, the angle defined in this step will be used as the leader dimension in the rotation pattern created in the next section of this tutorial.

STEP   5:   **Select OK on the Datum Plane menu.**

After selecting OK, notice on the Section dialog box how the previously created datum plane is now the sketching plane for the cut feature under construction.

STEP   6:   **On the Sketch dialog box, pick inside the Reference edit box.**

STEP   7:   **On the workscreen, pick the TOP orientation plane shown in Figure 6–51.**

STEP   8:   **On the Section dialog box, select TOP as the orientation.**

STEP   9:   **Select SKETCH on the Section dialog box.**

STEP 10:   **Specify the two references shown in Figure 6–52, then close the References dialog box.**

STEP 11:   **Use the Rectangle option to create the sketch shown in Figure 6–52.**

After sketching the section, apply the dimensioning scheme shown in the figure, then modify the dimension values to match the figure. Remember, this is an extruded cut. A centerline is not required.

STEP 12:   ✔   **Select the Continue option to exit the sketching environment.**

4. Using Pro/ENGINEER's Part mode, model the part shown in Figure 6–61. Create the base geometric feature as a revolved protrusion.

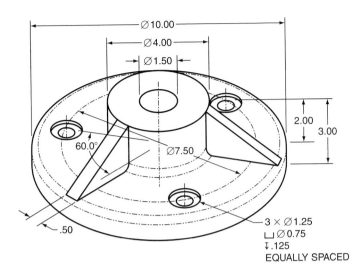

Figure 6-61   Problem 4

QUESTIONS AND DISCUSSION

1. List and describe various revolved features available within Pro/ENGINEER.

2. Describe the necessary requirements for sketching a revolved feature's section.

3. Describe the procedure for creating a diameter dimension within a revolved protrusion's section.

4. What revolved feature requires a vertical centerline?

5. Describe the difference between a hole placed with the Radial option and a hole placed with the Linear option.

6. What dimension type that defines a feature must be available to allow for the creation of a rotational pattern?

7

FEATURE MANIPULATION TOOLS

Introduction

Commands such as Extrude, Revolve, Rib, Datum, and Hole are used to create part features. Pro/ENGINEER provides a variety of tools for manipulating existing features. As an example, multiple features can be combined with the Group option. When a group is created, it can be manipulated with options such as Copy and Pattern. Other manipulation tools covered in this chapter include user-defined features, relations, and family tables. Upon finishing this chapter, you will be able to

- Combine features as a local group.
- Pattern a local group.
- Copy features in a linear direction.
- Mirror features across a plane.
- Copy-rotate features.
- Copy features by specifying new references.
- Add a dimension relationship to a part.
- Create a family table.
- Create a user-defined feature.

DEFINITIONS

Family of parts	A grouping of parts that are similar in shape, size, and geometry.
Family table	A combination of parts that have similar features and geometry but vary slightly in selected items.
Group	A collection of features combined to serve a common purpose.
Relation	An explicit relationship that exists between dimensions and/or parameters.
User-defined feature	A feature that is stored to disk as a group and can be used in other models.
User-defined feature library	A computer directory location where user-defined features can be stored.

GROUPING FEATURES

Most Pro/ENGINEER modification options are utilized to manipulate individual features. Often, it is desirable to manipulate multiple features together. As an example, Figure 7–1 shows a part with a rotationally patterned boss, round, and hole. The normal Pattern

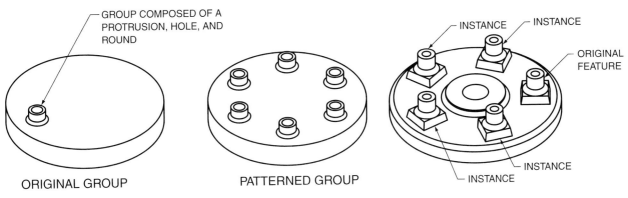

Figure 7–1   A Patterned Group

Figure 7–2   Pattern of Features

command is used on individual features, not multiple features. The grouped boss, round, and hole features shown in the illustration were patterned using the Group >> Pattern command.

GROUPED FEATURES

Several feature operations are available within Pro/ENGINEER. Examples include Copy, Delete, Group, Suppress, and Reorder. Each of these operations is available under the Edit menu. The Group tool will group together preselected features. These features must be adjacent to each other within the part's order of regeneration.

> **MODELING POINT**   The Group, Pattern, Unpattern, and Ungroup options can be used in sequences to copy multiple features in a way that will not require each feature to lose its individual identity. After a group has been patterned, it can be unpatterned with the Unpattern option. After performing an unpattern, each group that made up the pattern can be ungrouped with the Ungroup option.

GROUP TYPES

There are two types of groups: user-defined features and local groups. A user-defined feature (UDF) is a feature that has been grouped and saved to disk, often in a UDF library. A UDF can be retrieved and placed in the current working model or in another model. When a UDF is placed in an object, it becomes a grouped feature on the model tree.

A local group is a combination of features that is available only within the current model. The Group option allows multiple features to be grouped and manipulated. Features that are combined to form a local group must be adjacent to each other in the order of regeneration. Because of this, it is important during the modeling process to place intended group features next to each other on the model tree. To create a local group, perform the following steps:

PATTERNING A GROUP

A common reason to create groups is to pattern multiple features (Figure 7–2). Before Pro/ENGINEER's Wildfire release, this was the exclusive tool used to pattern groups. With the release of Wildfire, the normal Pattern command on the toolbar is available for individual features and groups. With the Pattern command, Local Groups and User-Defined Features can be patterned in the same way that individual features are patterned. Linear and Rotational patterns can be created. When patterning individual features, three pattern options exist: Identical, Varying, and General (see Chapter 5). With patterned groups, the General option is the only option available. The part (*group_pattern.prt*) shown in Figure 7–2 is available on the book's Web page for practice at patterning groups.

COPYING FEATURES

While the Pattern command is used to create multiple instances of a feature in a rotational or linear fashion, the Copy command is used to make a single copy of a feature or features. While the Pattern command creates new instances of a feature by varying dimensions that define the feature, the Copy command creates a copy by changing the placement of references and/or by changing dimension values. Figure 7–3 shows an example of a boss and coaxial hole that have been copied to another location. References required to define the location of the two features include a placement plane and two location edges. These references were changed to create the copy.

COPY OPTIONS

Unlike the Pattern command, multiple features can be simultaneously copied with the Copy command. As shown in Figure 7–4, there are four basic types of copies that can be created.

MIRRORING FEATURES

The Mirror option is used to copy-mirror features about a plane. A mirrored image of the original feature is created.

ROTATIONAL COPIES

Rotate is a suboption under the Move option. Features can be copy-rotated around a datum curve, edge, axis, or coordinate system.

TRANSLATED FEATURES

Translate is a suboption under the Move option. Copied features are translated in a linear direction from the original features. The Same Refs option under the Copy menu creates a form of a translated copy.

NEW REFERENCES

Copies can be created of features by varying dimension values and by selecting new references. Examples of references include placement edges, reference axis, and placement plane. A feature can be copied with completely new references. The Same References (Same Refs) option creates a copy of selected features, but does not allow the selection of alternative references.

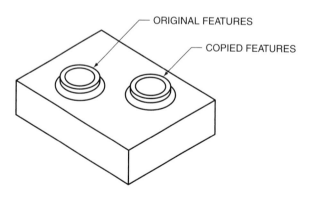

Figure 7–3   Copied Features

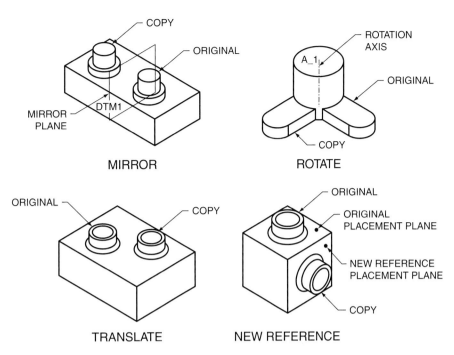

Figure 7-4   Copy Options

INDEPENDENT VERSUS DEPENDENT

Copies of features can be independent or dependent of their parent features. When a feature is copied as dependent, its parent feature's dimension values govern its dimension values. If a dimension is changed in the parent feature, the corresponding dimension is changed in the copy. The Independent option allows copied features to be independent of parent features. When a feature is copied as independent, the copy's dimensions will remain independent of its parent's. The dimensions of the copy can be changed and any changes to the parent's dimensions will not affect the child's dimensions.

SELECTING A MODEL

It is a common procedure to copy a feature from one location on a model to another location. Options are available for copying features from a different version of a model or from a completely different model. The From Different Model (FromDifModel) option is used to copy features from a different model. Because of the change in references required to copy from one model to another, this option is available with the New Refs option only. The From Different Version (FromDifVers) option is used to copy features from a different version of the active model. The New Refs or the Same Refs option may be used.

MIRRORING FEATURES

The Mirror option creates a reflected image of selected features. To construct a mirror, select the features to be copied and then select a mirroring plane. Perform the following steps to create a mirrored feature:

STEP 1:   Select EDIT >> FEATURE OPERATIONS >> COPY.

STEP 2:   Select the MIRROR option.

STEP 3:   **Select either DEPENDENT or INDEPENDENT, then select DONE.**

The Independent option will make the new copy's dimension values independent of any parent features.

STEP 4:   **Select features to be mirrored.**

Features can be selected by picking them on the workscreen or on the model tree. Additionally, the All Feat option can be used to mirror every available feature of a part. The Select option on the Copy menu can be used to select features during later steps.

STEP 5:   **Select DONE to finish selecting features.**

STEP 6:   **Select a Mirroring Plane (Figure 7–5).**

The mirroring plane can be any planar surface or datum plane. If a plane is not available, one can be created with the Make Datum option. The Mirror option does not utilize a Feature Definition dialog box. After the mirroring plane is selected, the copy is created.

ROTATING FEATURES

The Rotate option is located under the Move menu option. It is used to copy features by rotating them around an axis, edge, datum curve, or coordinate system (Figure 7–6). As with other Copy options, multiple features can be copied. Perform the following steps to copy-rotate selected features:

STEP 1:   **Select EDIT >> FEATURE OPERATIONS >> COPY >> MOVE.**

STEP 2:   **Select between INDEPENDENT and DEPENDENT.**

When features are copied as dependent, their dimension values are dependent on the original feature. Any changes made to the parent feature will be reflected in the copy.

STEP 3:   **Select DONE on the Copy Feature menu.**

STEP 4:   **Select features to copy then select DONE.**

STEP 5:   **Select ROTATE on the Move Feature menu.**

STEP 6:   **Select either CRV/EDG/AXIS or CSYS on the GEN SEL DIR menu then select the appropriate entity on the work screen.**

Features can be copy-rotated around a datum curve, edge, axis, or coordinate system. After choosing a rotation type, select the appropriate entity on the workscreen.

STEP 7:   **Select OKAY for direction of Rotation or FLIP to change direction.**

Pro/ENGINEER utilizes the right-hand rule to determine the direction of rotation. With the right-hand rule, point the thumb of your right hand in the

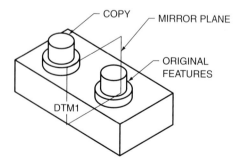

Figure 7–5   Mirroring Features

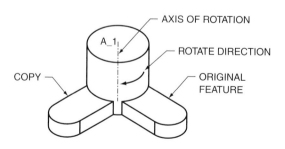

Figure 7–6   Rotating Features

direction of the arrow shown on the workscreen. The remaining fingers of your right hand point in the direction of rotation.

STEP 8: **Enter the degrees of rotation.**

Enter the number of degrees that the selected features will be rotated.

STEP 9: **Select DONE MOVE on the Move Feature menu.**

STEP 10: **Select DONE on the GP VAR DIMS menu.**

Notice in Figure 7–7 the options for checking a dimension. The GP VAR DIMS menu allows for the selection and varying of dimension values during the copy process.

STEP 11: **Select OK on the Feature Definition dialog box.**

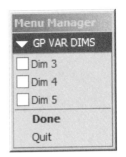

Figure 7–7
Dimension Variation

TRANSLATED FEATURES

Features can be copied in a linear direction using the Translate option (Figure 7–11). Features are copied perpendicular to a selected plane.

STEP 1: **Select EDIT >> FEATURE OPERATIONS >> COPY >> MOVE.**

STEP 2: **Select between INDEPENDENT and DEPENDENT.**

When features are copied as dependent, their dimension values are dependent on the original feature. Any changes made to the parent feature will be reflected in the copy.

STEP 3: **Select DONE on the Copy Feature menu.**

STEP 4: **Select features to copy then select DONE.**

STEP 5: **Select TRANSLATE on the Move Feature menu.**

STEP 6: **Select PLANE on the GEN SEL DIR menu then select a plane on the workscreen (Figure 7–8).**

Select a planar surface or a datum plane. Features will be copied perpendicular to the selected plane. The Make Datum option is available to create an on-the-fly datum plane.

STEP 7: **Accept or FLIP the Translate Direction.**

On the workscreen, Pro/ENGINEER will graphically display the direction of translation.

STEP 8: **Enter the translation value.**

Enter the value that the copied features will be offset from the original features.

STEP 9: **Select DONE MOVE on the Move Feature menu.**

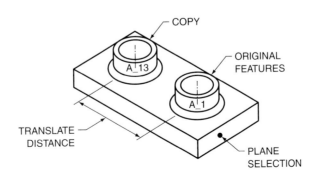

Figure 7–8   Translation

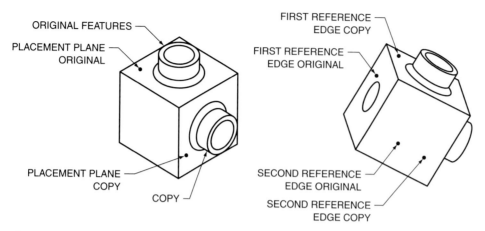

Figure 7–9   New Reference Option

STEP 10:   **Select DONE on the GP VAR DIMS menu.**

As with the Rotate option, the GP VAR DIMS menu allows for the selection and varying of dimension values during the copy process.

STEP 11:   **Select OK on the Feature Definition dialog box.**

COPYING WITH NEW REFERENCES

The New References (New Refs) option copies selected features by specifying new references and by varying feature dimensions (Figure 7–9). Examples of references that can be changed include sketching planes and reference edges. Perform the following steps to copy a feature with new references.

> **INSTRUCTIONAL NOTE**   When Pro/ENGINEER requests the selection of a new reference for a copy (or other options such as Reroute), it highlights the old reference with the section color setting. The section color is set under the View >> Display Settings >> System Colors option. If the default reference color is hard to see, its setting should be changed.

STEP 1:   **Select EDIT >> FEATURE OPERATIONS >> COPY >> NEW REFS.**

STEP 2:   **Select between INDEPENDENT and DEPENDENT, then select DONE.**

STEP 3:   **Select features that will be copied, then select DONE.**

STEP 4:   **Select dimensions to vary (optional step).**

Dimensions defining a feature can be varied during the copy process (Figure 7–10). Pro/ENGINEER will display the dimensions defining the selected features. On the workscreen or on the GP VAR DIMS menu, select the dimensions to vary.

STEP 5:   **Select DONE on the GP VAR DIMS menu (Figure 7–10).**

STEP 6:   **Choose an option for each highlighted reference then perform the appropriate reference selection.**

In sequential order, Pro/ENGINEER will highlight each reference in the established reference (section) color. For each reference, you must perform one of the following options:

• **Alternate**   This option requires the selection of a new reference. Select the appropriate new reference for the copy.

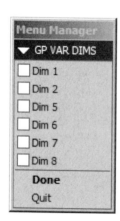

Figure 7–10
Variable Dimension
Menu

- **Same** This option will keep the highlighted reference for the copy.
- **Skip** This option will allow you to skip the definition of a new reference for a copy. This reference will have to be defined later with the Edit Definition option.
- **Ref Info** This option provides information about the reference.

STEP 7: **Select DONE on the Group Place menu.**

USER-DEFINED FEATURES

A user-defined feature (UDF) is a feature that is saved to disk and can be used during the construction of future models. It is common practice to create a UDF library of features that are often used in an organization. User-defined features can be positive space (e.g., protrusion) or negative space (e.g., cut and hole). Positive space user-defined features can be placed as the first feature of a part.

In creating a feature that will be saved as a UDF, careful consideration of references and the dimensioning scheme is critical. A UDF functions like a new reference copy. When placing a UDF, Pro/ENGINEER prompts for the selection of new references for the copy.

A UDF can be subordinate or stand-alone. If a UDF is created as subordinate, it depends on the original model for its dimensional values. The original model stores the information defining the UDF. Any changes to the original model and to the original feature will be reflected in any placed UDF. Because of this dependence, the original model must exist. A stand-alone UDF stores the information associated with the feature in the copied-to-model. Once a stand-alone UDF is placed, it loses its associativity with its parent model.

A UDF can be created with one of three possible dimension types.

- **Variable Dimensions** Dimensions that are defined as variable can be modified during the copy process.
- **Invariable Dimensions** Dimensions that are defined as invariable cannot be modified during the copy process.
- **Table-Driven Dimensions** Dimensions that are defined as table-driven receive their values from a family table.

THE UDF MENU

The UDF menu is accessible under the Tools menu from the UDF Library option. This menu is used to create, modify, and manipulate user-defined features. The following options are available:

CREATE

The Create option is used to create a new UDF. The UDF feature will be created in the current working directory. If a specific directory is used as a library to store user-defined features, change the working directory before the creation process, or set the configuration file option *pro_group_dir* to specify the library directory.

MODIFY

The Modify option is used to modify an existing UDF.

LIST

The List option is used to list all available user-defined features in the current working directory.

Dbms

Users of Pro/ENGINEER before release 20 will recognize Dbms as a Data Base Management option. With the UDF menu, Dbms allows for database management options for a UDF. Common file management functions such as Save, Save As, Rename, and Erase are available.

Integrate

The Integrate option is used to define differences that might exist between the original UDF and the copied UDF.

CREATING A USER-DEFINED FEATURE

The following will show the step-by-step process for creating a UDF of the radial hole pattern shown in Figure 7–11. This part (*udf_part.prt*) is available on the book's Web page. The original feature consists of a radial hole that has been patterned as a rotation. The references required to construct the UDF include the placement plane, reference axis, and reference planes. Perform the following steps to create a user-defined feature of this pattern:

Step 1: **Open or create the part with appropriate features.**

Step 2: **Select the working directory where the UDF will be stored.**

Use the File >> Set Working Directory option to select the directory to save the UDF.

Step 3: **Select TOOLS >> UDF LIBRARY on the menu bar.**

Step 4: **Select CREATE on the UDF menu.**

Step 5: **Enter *pattern1* as the UDF name.**

Step 6: **Select STAND ALONE >> DONE.**

A stand-alone UDF will be independent of the model from which the original feature was created. A subordinate UDF will receive its parameters from the original model.

Step 7: **Select YES to include the reference part.**

Step 8: **Select the hole pattern to include as a feature in the UDF.**

On the workscreen or on the model tree, select the hole pattern.

Step 9: **Select DONE on the Select Feature menu.**

Step 10: **Select DONE/RETURN on the UDF FEATS menu.**

The Add option on the UDF Feature menu will allow you to add additional features to the new UDF, while the Remove option will allow you to remove features.

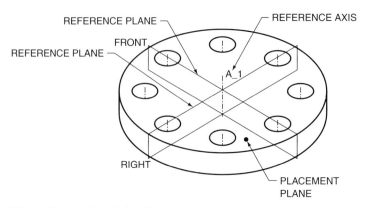

Figure 7-11   User-Defined Feature

STEP 11:   **Enter a prompt for the highlighted axis.**

Pro/ENGINEER will sequentially reveal each reference and request a prompt. The prompt entered by you will be displayed when placing the UDF into a new model. Enter the following prompt: *a Reference Axis to Place Patterned Holes about*

(*Note:* Pro/ENGINEER will add the word "Select" to the front of the prompt)

STEP 12:   **Enter a prompt for the highlighted reference plane.**

Radial holes require the selection of a reference plane. This option will require you to establish a prompt that will request the selection of a new reference plane when the UDF is placed. Enter the following prompt for the highlighted reference plane: *a Reference Plane to Reference the Original Patterned Hole*

STEP 13:   **Select SINGLE >> DONE/RETURN on the Prompts menu.**

When Pro/ENGINEER encounters references that can share the same prompt, it will provide you with the option for selecting a Single prompt for all the references, or an option for selecting individual prompts (Multiple option) for each reference. In this example, the eight holes forming the pattern will each require a placement plane. Pro/ENGINEER recognizes this and gives you the chance to enter one prompt for each hole placement plane.

STEP 14:   **Enter a prompt for the highlighted placement plane.**

This prompt will serve as the placement plane for all three holes. Enter the following prompt: *a Placement Plane for the Holes*

STEP 15:   **Use the NEXT and PREVIOUS options on the Modify Prompts menu to toggle through prompts.**

If a prompt needs to be changed, use the Enter Prompt option.

STEP 16:   **Select DONE/RETURN on the Modify Prompts menu.**

STEP 17:   **Select OK on the UDF Creation dialog box.**

PLACING A USER-DEFINED FEATURE

This guide will step you through the process for placing the UDF created in the "Creating a User-Defined Feature" guide. This UDF (*pattern1.gph*) is also available on the book's Web page. As shown in Figure 7–12, utilizing Pro/ENGINEER's default datum planes, model a Circular protrusion with a diameter of 8.00 in and an extrusion depth of 1.00 in. This part will serve as the base feature to which to attach the base feature.

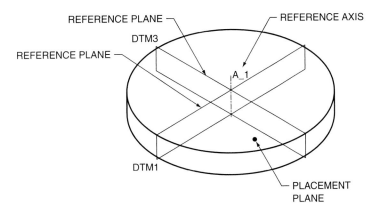

Figure 7–12   Placing a UDF

STEP   1:   Select INSERT >> USER-DEFINED FEATURE.

User-defined features can be placed using the Group option also. User-defined features and groups are closely related. All user-defined features are placed in a model as a group.

STEP   2:   Using the OPEN dialog box, manipulate the directory to locate the *pattern1* UDF created in the previous guide.

User-defined features are saved to disk with an *.gph* file extension. Locate the file *pattern1.gph* that was created in the "Creating a User-Defined Feature" guide. This UDF can also be downloaded from the book's Web page.

STEP   3:   Open the *PATTERN1* user-defined feature.

STEP   4:   Select all three options on the Insert User-Defined Feature dialog box.

Advanced Reference Configuration—This option will allow the placed feature to be independent of the part from which the UDF was created.

View Source Model—When a UDF is created as stand-alone, this option opens in memory the original model from which the UDF was created and displays it in a separate window (Figure 7–13). When new references are selected, the references on the original model will be highlighted.

Make Features Dependent on Dimensions of UDF—This option will keep the dimension values that define the original UDF. The User Scale option will allow a copied UDF to be scaled a user-specified amount.

STEP   5:   Select OK on the Insert dialog box.

STEP   6:   Select the reference axis on the new model (see Figure 7–12).

The User Defined Feature Placement dialog box is used to pick references from your original features for the placement of your UDF. The prompt in the message area is requesting you to Select a Reference Axis to Place Patterned Holes About. This is the prompt entered in Step 11 of the previous section, "Creating a User-Defined Feature." Check the reference window to see if the axis is highlighted. If it is not, your reference colors might need changing.

STEP   7:   Select the second original reference on the dialog box (see Figure 7–13).

STEP   8:   Select a reference plane to reference the original patterned hole.

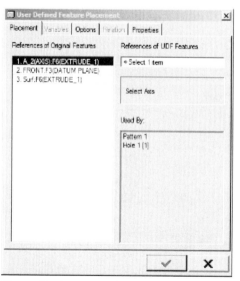

Figure 7–13   User Defined Feature Placement

The prompt in the message area is requesting you to Select a Reference Plane to Reference the Original Patterned Hole. Select the datum as shown in Figure 7–12.

STEP 9: **Select the third original reference on the dialog box.**

STEP 10: **Select a placement plane for the holes.**

The prompt in the message area is requesting you to *Select a Placement Plane for the Holes.* Select the top of the part as shown in Figure 7–12.

STEP 11: **Select the Check Mark on the dialog box.**

STEP 12: **Reselect the placement plane to orient the holes.**

STEP 13: **Select DONE on the Group Placement menu.**

RELATIONS

Mathematical and conditional relationships can be established between dimension values. Chapter 3 introduced the utilization of relations within the sketcher environment. Relations within the sketcher environment are used to establish dimensional relationships between dimensions of a feature. The Relations command that is found under the Part menu is used to establish relationships between any two dimensions of a part. Within Assembly mode, the Relations option can be used to establish relationships between dimensions from different parts.

Dimensions can be shown with numeric values or as symbols. Figure 7–14 shows an example of a part with dimensions shown as symbols. Dimension values are displayed as a *d* followed by the dimension number (e.g., d3). Other parameters that can be displayed symbolically include reference dimensions (e.g., rd3), Plus-Minus Symmetrical Tolerance mode (e.g., tpm4), Positive Plus-Minus Tolerance mode (e.g., tp4), Negative Plus-Minus Tolerance mode (e.g., tm4), and number of instances of a feature (e.g., p5).

Most algebraic operators and functions can be used to define a relation. Additionally, most comparison operators can be used. Table 7–1 provides a list of mathematical operations, functions, and comparisons supported in relation statements. All trigonometric functions use degrees.

Table 7–1 Mathematical Operations

Operator or Function	Meaning and Example
+	Addition d1 = d2 + d3
−	Subtraction d1 = d2 − d3
*	Multiplication d1 = d2 * d3
/	Division d1 = d2/d3
^	Exponentiation d1^2
( )	Group Parentheses d1 = (d2 + d3)/d4
=	Equal to d1 = d2
cos ( )	cosine d1 = d2/cos(d3)
tan ( )	tangent d1 = d3 * tan (d4)
sin ( )	sine d2 = d3/sin (d2)
sqrt ( )	square root d1 = sqrt (d2) + sqrt (d3)
= =	Equal to d1 = = 5.0
>	Greater than d2 > d1
<	Less than d3 < d5
>=	Greater than or equal to d3 >= d4
<=	Less than or equal d5 <= d6
!=	Not equal to d1 != d2 * 5
|	Or (d1 * d2) | (d3 * d4)
&	And (d1 * d2) & (d3 * d4)
~	Not (d3 * d4) ~ (d5 * d6)

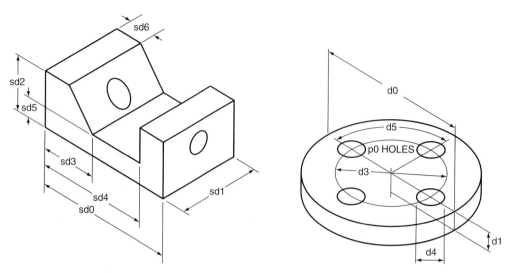

Figure 7–14   Dimension Symbols   **Figure 7–15**   Dimension Symbols

CONDITIONAL STATEMENTS

Pro/ENGINEER's Relations command has the ability to utilize conditional statements for the purpose of capturing design intent. As an example, Figure 7–15 shows a flange that incorporates holes on a bolt circle centerline. In a model such as this, the holes are created typically as a patterned radial hole. Imagine a situation where the number of holes and the diameter of the bolt circle are governed by the diameter of the flange. Suppose, in this example, that the bolt circle diameter is always 2 in less than the diameter of the flange. Also suppose that the design requires 4 holes on the bolt circle if the diameter of the flange is 10 in or less and 6 holes if the flange diameter is greater than 10 in. With the following relational statements, parameters can be established that control this design intent for the bolt circle pattern:

IF d0<=10	(Line 1)
p0=4	(Line 2)
d5=90	(Line 3)
ELSE	(Line 4)
p0=6	(Line 5)
d5=60	(Line 6)
ENDIF	(Line 7)
d3=d0-2	(Line 8)

A conditional relation is defined between an IF statement and an ENDIF statement. The above example utilizes an IF-ELSE statement. With an IF-ELSE statement, if the condition is true the expressions following the IF statement will occur. If the condition is not true, the condition following the ELSE statement will occur. The above example reads as follows:

> If the flange diameter is less than or equal to 10, the number of holes is equal to 4 and the angular spacing between each hole is 90 degrees; else, the number of holes is equal to 6 and the angular spacing is equal to 60 degrees.

In the above example, the condition statement is the diameter of the flange (d0) being less than or equal to 10 (see line 1). If this is true, the number of holes on the bolt circle (p0) will be equal to 4 and the angle between each instance of the hole (d5) will be 90 degrees. If the flange diameter is not less than or equal to 10, the number of holes on the bolt circle will be 6 and the angle between instances will be 60 degrees. In a conditional statement, the

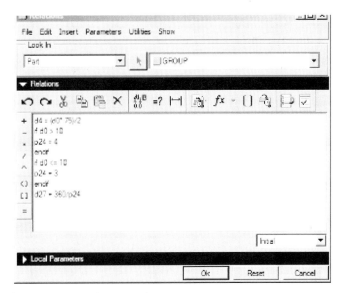

Figure 7–16　Relations Dialog Box

ELSE expression shown in line 4 must occupy a line by itself. The expression ENDIF is used to end the conditional statement. Additionally, in the above example, line 8 is used to make the bolt circle diameter always 2 less than the flange diameter.

ADDING AND EDITING RELATIONS

Relations are added to an object by using the Relations Dialog box (Figure 7–16). Relations are evaluated within a model in the order in which they are defined. In most cases of conflict, the later relation overrides the former.

Perform the following steps to add relations to a model:

STEP 1:　Select TOOLS >> RELATIONS on the menu bar.

STEP 2:　Select the Feature (or Features) applicable to the relation.

Selecting a feature will display the dimensions defining the picked feature. While in the Relations dialog box, dimension will be displayed in the symbolic format. You can use the Dimension Toggle icon on the toolbar to switch between dimension values and dimension symbols.

STEP 3:　Enter the relations equations to capture design intent (Figure 7–16).

STEP 4:　Verify your equations with the Verify Relations icon on the Relations toolbar.

STEP 5:　Select OK on the Relations dialog box.

STEP 6:　Regenerate your model.

Defined relations will not take effect until the model is regenerated.

FAMILY TABLES

A family of parts consists of components that share common geometric features. An example of a family of parts is the hex head bolt. Hex head bolts come in a variety of sizes, but share common characteristics. Bolts can have different lengths and diameters but share

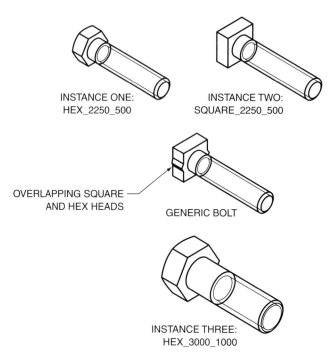

INSTANCE ONE:
HEX_2250_500

INSTANCE TWO:
SQUARE_2250_500

OVERLAPPING SQUARE
AND HEX HEADS

GENERIC BOLT

INSTANCE THREE:
HEX_3000_1000

Figure 7–17   Family Table Example

similar head features and thread parameters (Figure 7–17). Family tables are groups of similar features, parts, or assemblies. Within companies, it is common to find parts and assemblies with common geometric characteristics. Examples can be found in the automotive industry. An automobile manufacturer probably has a large variety of cam shafts. A family table can be created that drives the construction of a company's line of cam shafts. Examples can be found with assemblies also. Imagine the number of alternators that a major automobile company produces. Alternators have similar characteristics but can vary in certain components and features. A family table can be created that controls the design of a company's line of alternators.

Family tables have many advantages. Storing and controlling large numbers of components can be difficult to manage and costly. For a family of parts with large numbers of variations, family tables allow components to be stored to disk, thereby consuming significantly less storage space. Family tables also save time in the modeling process. If a design is known to work effectively, a variation of the design can be created by changing a parameter within the family table. This technique also allows for the standardization of a design.

ADDING ITEMS TO A FAMILY TABLE

No specific option is used to create a family table. A family table is created automatically when an item is selected for addition to the family table. Examples of items that can be added to a family table include dimensions, features, components, and user parameters. Other items that can be added include groups, pattern tables, and reference models. To add an item, select the Add Item option from the Family Table menu then select the item type to add. Some items, such as groups, are better selected by utilizing the model tree.

CREATING A FAMILY TABLE

This section will introduce the creation of family tables by creating a family table of hex head and square head bolts. Three instances will be created: two hex head and one

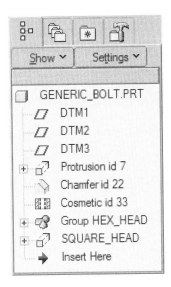

Figure 7-18 Model Tree

square head. Figure 7–17 shows the two instances of the hex head bolt, the instance of the square head bolt, and the generic part. The generic part is a standard Pro/ENGINEER part and is used to create the instances. This part (*generic_bolt.prt*) can be downloaded from the book's Web page. Notice in the figure the overlap of the heads of the hex head bolt and the square head bolt. As shown on the model tree in Figure 7–18, the hex head bolt is actually a grouped feature.

The generic part in this tutorial is a bolt with a major diameter of 0.500 in and a bolt length of 2.25 in. A cosmetic thread feature has been created with a minor diameter of 0.375 in. Additionally, the following dimensional relations have been added to the generic part:

- The height of both bolt heads is set to 0.667 times the major diameter.

- The distance across the flats of both bolt heads is set to 1.5 times the major diameter.

- The cosmetic tread minor diameter is set to 0.75 times the major diameter.

This tutorial will add the bolt's major diameter and length dimensions to the family table. Additionally, the two grouped features that defined the bolt heads will be added. Three instances of the part will be created. The first instance will be a hex head bolt with the same dimensions as the generic part. The second instance will be a square head bolt, also with the same dimensions as the generic part. The third instance will be a hex head bolt with a 1.00-in major diameter and a length of 3.00 in.

Perform the following steps to create a family table of bolts:

Step 1: **Create the bolt with the hex head (or download the *generic_bolt* part from the book's Web page).**

Use appropriate modeling tools to create the bolt. This part, named *generic_bolt,* can be found on the book's Web page too. If you elected to open the existing part, skip Steps 2–4.

Step 2: **Group the features making up the hex head, then suppress the group.**

By grouping the features making up the hex head, these features can be manipulated together. This group will be suppressed to allow for the creation of the square head.

Step 3: **Create the square head feature on the bolt.**

Step 4: **Use the RESUME option to unsuppress the Hex_Head group.**

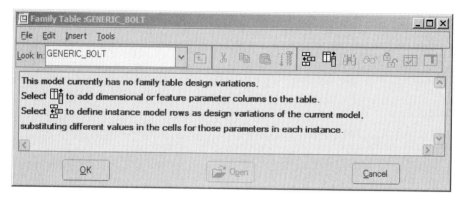

Figure 7-19   Family Table Dialog Box

STEP 5:   Select TOOLS >> FAMILY TABLE on the menu bar.

Each part is capable of one family table. The family table is created when the first item is added to it.

STEP 6:   On the Family Table dialog box select the ADD COLUMN icon (see Figure 7–19).

Examples of items that can be added to a family table include dimensions, features, and components.

STEP 7:   Select the DIMENSION item on the Family Items dialog box (see Figure 7–20), then on the workscreen pick the shaft of the bolt.

The first items that will be added to the family table include the dimensions that define the major diameter of the bolt and the length of the bolt.

STEP 8:   As shown in Figure 7–21, select the dimensions defining the bolt's major diameter and the bolt's length.

The Dimension item type allows selected dimensions to be added to the family table. As shown in Figure 7–20, symbols representing each

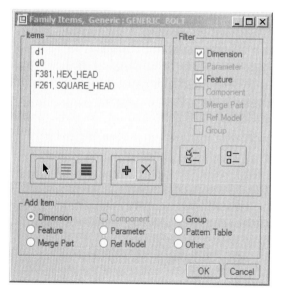

Figure 7-20   Family Items Dialog Box

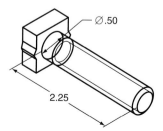

Figure 7-21   Bolt Dimensions

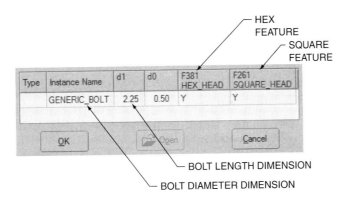

Figure 7-22   Family Table Editor

dimension (e.g., d1 and d0) will be displayed on the dialog box within the
Items area.

Step 9: **Select the FEATURE item on the Family Items dialog box.**

The Feature option allows features, including grouped features, to be added
to the family table. Two features will be added: The Hex_Head group
feature and the Square_Head_Bolt feature.

Step 10: **On the model tree, select the HEX_HEAD group feature and the
SQUARE_HEAD_BOLT feature.**

Step 11: **Select OK on the Family Items dialog box.**

Step 12: On the Family Table dialog box, select the NEW INSTANCE
icon.

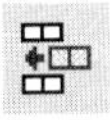

Notice the addition of the ***GENERIC_BOLT*** instance in the table editor of
the Family Table dialog box (Figure 7–22). This name is derived from the
part's name and represents the name of the generic instance of the bolt.
This instance includes the part's default dimensional values and default
features.

Step 13: **As shown in Figure 7–23, enter the information in the table to create
the instance *HEX_2250_500*.**

The new instance's name will be *HEX_2250_500*. This name represents a
hex head bolt that will be 2.250 in in length and 0.500 in in diameter.
The table will not accept a period value or a space in the instance name;
hence, the naming standard shown in this example. Enter 2.25 for the
length of the instance and 0.5 as the bolt diameter. In the table, enter a *Y*
value for the new instance in the column representing the Hex Head

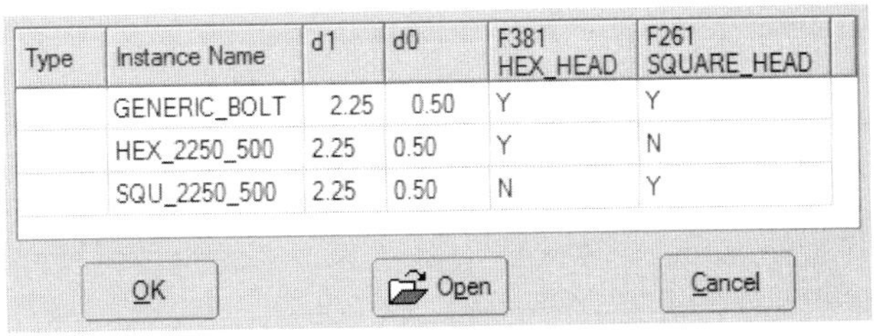

Type	Instance Name	d1	d0	F381 HEX_HEAD	F261 SQUARE_HEAD
	GENERIC_BOLT	2.25	0.50	Y	Y
	HEX_2250_500	2.25	0.50	Y	N
	SQU_2250_500	2.25	0.50	N	Y

OK   Open   Cancel

Figure 7-23   Instance Creation

feature and a *N* value for the column representing the Square Head feature. A *Y* value will reveal the feature in the instance and a *N* value will remove the feature.

STEP 14:   As shown in Figure 7–23, enter parameters in the table to create the instance SQUARE_2250_500.

This instance will consist of a square head bolt with a major diameter of .500 in and a bolt length of 2.250 in.

STEP 15:   On the dialog box, pick the row representing the HEX_2250_500 instance, then select the dialog box's OPEN icon.

Pro/ENGINEER will open a new window with the instance of the part shown. Selecting the Save option from the File menu will save the current instance as a part.

STEP 16:   Close the instance from memory by selecting FILE >> ERASE >> CURRENT.

STEP 17:   Activate the generic part (Window >> Activate) then select the FAMILY TAB menu option.

STEP 18:   Create a HEX_3000_1000 instance.

Create an instance of the generic part that includes a hex head bolt that is 3.0 in in length with a major diameter of 1.0 in.

STEP 19:   Close the Family Table dialog box.

STEP 20:   Save the generic part.

CROSS SECTIONS

Traditional drafting standards utilize cross sections to display the internal details of models. Pro/ENGINEER utilizes cross sections within Drawing, Part, and Assembly modes for the same purpose. Additionally, within Pro/ENGINEER, mass properties of a cross section can be calculated.

Pro/ENGINEER provides the option of creating two types of cross sections: Planar and Offset. Planar Cross Sections are created within a model through a selected planar surface or datum plane. Figure 7–24 shows an example of a revolved part with a section through a datum plane. Offset cross sections are created by extruding a sketched section. Offset cross sections are constructed by normal sketching techniques. Offset sections must be created with an open section. Once created within Part or Assembly mode, cross sections are blanked from the screen. They can be redisplayed with the Show option.

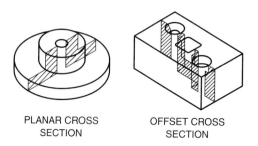

PLANAR CROSS
SECTION

OFFSET CROSS
SECTION

Figure 7–24   Cross Section Examples

MODIFYING CROSS SECTIONS

Cross-section parameters can be modified. The following is a list of parameters that can be modified or manipulated:

NAME

The creation process of a cross section requires the entering of a name. The Name command found under the Cross Section >> Modify menu can be used to rename a cross section.

HATCHING

A cross section creates a hatch pattern along the section or planar line. Parameters defining a hatch pattern such as line style, color, spacing, and offset can be changed by using the Modify >> Hatching option. A commonly used hatching pattern can be stored for later use by using the Hatching >> Save option. The configuration file option *pro_crosshatch_dir* is used to specify the directory where hatch patterns are saved.

DIMENSIONS

Dimension values defining a cross section can be modified with the Dim Values command. A model is automatically regenerated upon finishing the dimension modification process.

CREATING A PLANAR CROSS SECTION

Planar cross sections are created through surface planes or datum planes. Perform the following steps to create a planar cross section:

STEP 1: Open the View Manager dialog box (View >> View Manager) and select the Xsec tab.

STEP 2: Select NEW on the View Manager dialog box.

STEP 3: Enter a name for the cross sectionA.

Cross sections created in Part and Assembly modes can be used for the creation of a section view in Drawing mode. Select a cross-section name that would be appropriate for an engineering drawing.

STEP 4: Select PLANAR >> SINGLE >> DONE on the XSEC CREATE menu.

STEP 5: Select a surface plane or datum plane.

The cross section will be created through the selected plane. The Make Datum option is available for on-the-fly datum planes.

STEP 6: Close the Model Sectioning dialog box.

CREATING OFFSET CROSS SECTIONS

Offset cross sections are created by extruding a sketched cross section. Perform the following steps to create an offset cross section.

STEP 1: Open the View Manager dialog box (View >> View Manager) and select the Xsec tab.

STEP 2: Select NEW on the View Manager dialog box.

STEP 3: Enter a name for the cross section.

Cross sections created in Part and Assembly modes can be used for the creation of a section view in Drawing mode. Select a cross section name that would be appropriate for an engineering drawing.

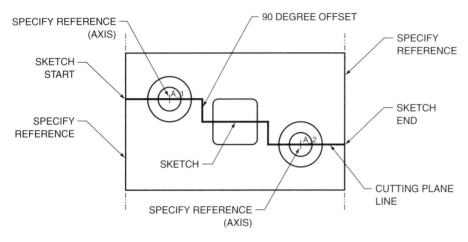

Figure 7–25 Sketching a Cross Section

STEP 4: **Select OFFSET >> BOTH SIDES >> DONE on the XSEC CREATE menu.**

The One Side option will create a cross section on one side of the sketched section, while the Both Sides option will create the section on both sides.

STEP 5: **Select a sketching plane and setup the sketcher environment according to normal sketching requirements.**

STEP 6: **Sketch the cross section using available sketcher tools (Figure 7–25).**

The following rules and techniques apply to a cross-section sketch.

- The section must be open and the first and last entities must be straight line segments.

- Avoid using curved and spline entities for the section sketch. They create a nonmodifiable horizontal hatch pattern.

- When possible, sketch lines that are perpendicular or parallel to each other.

- When creating a cross section through a hole or shaft, use the axis of the feature as a reference.

STEP 7: ✔ **When the section is complete, select the continue icon on the sketcher toolbar.**

STEP 8: **Close the Model Sectioning dialog box.**

Cross sections can be viewed with the Display >> Show X-Section option on the Model Sectioning dialog box.

MODEL TREE

Pro/ENGINEER's model tree (Figure 7–26) is a useful tool for the manipulation of features and parts. The following is a list of some of the uses of the model tree:

- Selecting features and parts that are difficult to select on the workscreen.
- Deleting features and parts from the model.
- Redefining definitions assigned to a feature.
- Inserting features.

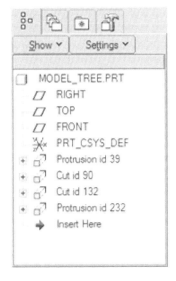

Figure 7-26   The Model Tree

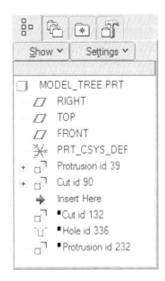

Figure 7-27   Inserting a Feature

- Reordering features.
- Suppressing features.
- Rerouting features.

SUPPRESSING FEATURES

Complicated parts can slow regenerations. Suppressing a feature will remove it from the model tree and from regenerations. Additionally, features can complicate the modeling environment, cluttering the workscreen. An uncluttered workscreen can make selecting and sketching features easier.

The Suppress command is available on the model tree's mouse pop-up menu (right mouse button). When selecting a parent feature for suppression, all child features will be suppressed. Use the Edit >> Resume option to unsuppress features.

INSERTING FEATURES

Pro/ENGINEER adds new features to the end of the model tree. Pro/ENGINEER will allow a new feature to be added before an existing feature. Many times, it is necessary to create a feature that will serve as a reference for an existing feature. This is useful when a new feature must serve as references for an existing part feature. As an example, the sketch plane for a feature can be redefined. If it is the intent of the user to change the sketching plane to a datum plane that does not exist, this new datum plane will have to be inserted into the model tree before the feature under modification.

The Insert Here symbol on the model tree is used to determine the location within the order of regeneration for all new features. This symbol can be dragged to any place after the first feature. Notice in Figure 7–27 how features after the current insertion point are suppressed to avoid improper feature references.

REORDERING FEATURES

Often, it is necessary to change the order of features within the regeneration sequence. As an example, during the modeling process, it might become apparent that features should be grouped and then patterned together. Features to be grouped should be adjacent to each other within the order of regeneration (and on the model tree). If the features to be grouped do not meet this requirement, then one or more of the features can be dragged on the model tree to a new location. Because parent-child relationships might exist between features, care should be taken when planning the reordering of features. Pro/ENGINEER will not let the following situations exist:

- Child features moved before parent features on the model tree.
- Parent features moved after child features on the model tree.

EDITING REFERENCES

When a feature references another feature, a parent-child dependency relationship exists. Rerouting a feature breaks the parent-child relationship that exists between it and another feature. The model tree's Edit References option provides two options for editing a feature's references: Reroute Feature and Replace References.

REROUTE FEATURES OPTION

The Reroute Features option will allow the selection of new or missing references for a feature. This option allows for the replacement of a reference for a child feature that has lost a parent feature. The focus of this option is on the child and its references. This option is available also in the event of a failed regeneration of a feature. Perform the following steps to reroute a feature:

STEP 1:   Select the feature for rerouting on the model tree.

STEP 2:   Select the EDIT REFERENCES option on the model tree's pop-up menu (right mouse button).

STEP 3:   Enter YES as the Roll Back option.

This option is available for rolling back the part to just before the failed feature. This allows younger references that might cause clutter to be removed from the model tree and from the workscreen.

STEP 4:   Select REROUTE FEAT on the Reroute References menu.

STEP 5:   Select a reference option for each feature reference.

Pro/ENGINEER will highlight each feature reference. The following options are available for each reference in turn.

- **Alternate** This option allows for the selection of a new reference.
- **Same Ref** This option will allow the same reference to remain.
- **Ref Info** This option shows information about a reference.
- **Done** Select this option when the rerouting process is finished.
- **Quit Reroute** This option allows for the termination of the reroute process.

STEP 6:   Select DONE on the Reroute menu when required references have been rerouted.

REPLACE REFERENCES OPTION

The Replace References option allows for the replacement of a reference for child features. This option is used when a parent reference must be replaced. An option is available for replacing the reference with one new reference or selecting multiple replacement references. The focus of this option is on replacing the child feature of a parent feature. Perform the following steps to replace references:

STEP 1:   Select the feature for rerouting on the model tree.

STEP 2:   Select the EDIT REFERENCES option on the model tree's pop-up menu (right mouse button).

STEP 3:   **Enter YES as the Roll Back option.**

This option is available for rolling back the part to just before the failed feature. This allows younger references that might cause clutter to be removed from the model tree and from the workscreen.

STEP 4:   **Select REPLACE REF on the Reroute References menu.**

STEP 5:   **Select either FEATURE or INDIV ENTITY on the Select Type menu.**

The Feature option allows for the replacement of all references of a feature. The Individual Entity option allows for the replacement of selected individual entities.

STEP 6:   **Select either SEL FEAT or ALL CHILDREN.**

The Select Feat option allows for the selection of new references for each child feature. The All Children option allows all child features to be referenced to one new feature.

REGENERATING FEATURES

Pro/ENGINEER is a feature-based parametric modeling package. Features are displayed within Pro/ENGINEER on the model tree in the order in which they are created. During part regeneration, the geometry defining each element is recalculated sequentially from the first feature to the last.

Pro/ENGINEER requires a regeneration when dimensions are modified or when dimensional relationships are established. During dimension and feature modification, a part is regenerated from the last modified feature.

REGENERATION FAILURES

Within parametric design, features have explicit relationships with other features. These established feature references cannot conflict. When a conflict is detected, a failed regeneration occurs. There are several reasons why features may conflict. First, section entities cannot overlap. This is a common conflict with swept features. Second, features with bad edges will fail during regeneration. There are two categories of failed regenerations: failed regeneration during feature modeling and failed regeneration during feature modification.

Most feature creation processes utilize either dashboard or a feature definition dialog box. For these features, if a conflict exists during feature creation or redefinition, a repair or resolve button will appear on the dashboard or dialog box. Two options are available for solving the regeneration conflict. The first and most practical solution is to redefine the definitions associated with the feature. As an example, sweep sections often overlap, resulting in a conflict. The sweep's section or trajectory can be redefined to solve the overlap problem. In another example, the Round command utilizes a dialog box. Often, an inappropriate round radius value will create a failed regeneration. A radius value can be redefined through the Feature Definition dialog box.

An alternative to redefinition is to access the resolve environment by selecting the repair or resolve button. The resolve environment provides diagnostic information and provides avenues for modifying other features. The following options are under the Resolve Feature menu:

- **Undo Changes** The undo changes option rolls the modeling process back to the last successful regeneration.
- **Investigate** This option will investigate possible causes of the regeneration failure.
- **Fix Model** The fix model option rolls the modeling process back to before the failure and provides tools for fixing the problem.

- **Quick Fix** The quick fix option provides avenues for redefining, rerouting, suppressing, or deleting the failed feature.

Features that fail and do not utilize a Feature Definition dialog box (e.g., Neck and Rib options) require the use of a Feature Failed menu. The following options are available with the Feature Failed menu:

- **Redefine** This option allows for redefinition of the feature.
- **Show Ref** The show reference option will show references associated with a failed feature. Many failures occur as a result of conflicts with references. Evaluating feature references might highlight the regeneration conflict.
- **Geom Check** The Geometry Check option will check geometry for overlapping and misalignment problems.
- **Feat Info** This option displays information about the failed feature.

SUMMARY

The traditional way within Pro/ENGINEER to create features is to use commands such as Protrusion and Cut. Options exist that will optimize the modeling process by manipulating these features. Like most midrange CAD packages, Pro/ENGINEER has commands for manipulating existing entities and features. Options are available, such as Pattern, that will allow for the copying and arraying of most features. Other options such as Copy also exist for creating instances of a feature. The understanding of these manipulation tools is critical for creating advanced parts within Pro/ENGINEER.

MANIPULATING TUTORIAL 1

This tutorial will create the part shown in Figure 7–28. The following topics will be covered:

- Creating a revolved protrusion.
- Creating an extruded protrusion.
- Creating a coaxial hole.
- Mirroring a feature.
- Rotating a feature.
- Adding dimensional relationships.

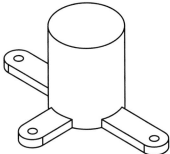

Figure 7–28   Finished Part

CREATING THE BASE PROTRUSION

The first section of this tutorial will require you to create the protrusion feature shown in Figure 7–29. Start Pro/ENGINEER and create a part model named *Rotate*.

Create the base protrusion of the part as a revolve with the following requirements:

- Revolve the section shown in Figure 7–29.
- Use Pro/ENGINEER's default datum planes.
- Sketch the revolved protrusion on datum plane FRONT.
- Use a 360 degree revolution parameter.

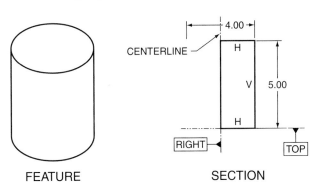

Figure 7–29   Base Protrusion and Section

CREATING AN EXTRUDED PROTRUSION

This segment of the tutorial will create the extruded protrusion shown in Figure 7–30.

STEP   1:   **Set up a solid extruded protrusion for the new feature shown in Figure 7–30 by using the following options.**

- Select datum plane TOP as the sketching plane.
- Use the default sketching orientation.
- Extrude one direction.

STEP   2:   **Once established in the sketcher environment, close the References dialog box.**

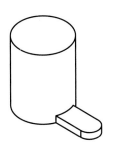

Figure 7–30
Extruded Protrusion

251

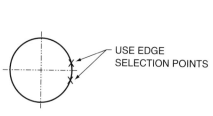

Figure 7–31   Selecting an Edge

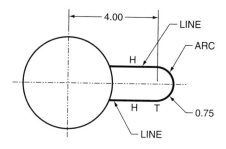

Figure 7–32   The Sketched Section

STEP 3: ☐ **Select the USE EDGE icon on the sketcher toolbar.**

The Use Edge option is used to project existing feature edges as sketch entities within the current sketching environment.

STEP 4: **As shown in Figure 7–31, pick the outside edge of the base revolved feature in the two specified locations. (Only select each location once.)**

The Use Edge option will project the outside edge of the revolved feature onto the sketching plane as a sketcher entity. The selected edges will become references within the sketching environment.

STEP 5: **Sketch the two line entities and the arc entity shown in Figure 7–32.**

Sketch the two new lines and the arc as shown. The ends of each line should be aligned with the edge of the existing revolved protrusion. Since the previous step of this tutorial turned the edges of the revolution into entities and into references, the lines will snap to the edge.

STEP 6: **Modify the dimensions and values to match Figure 7–32.**

STEP 7: ⌐ **Select the Trim Entities icon on the sketcher toolbar.**

Note: ⌐ **The Trim Entities icon is located behind the Dynamic Trim icon.**

INSTRUCTIONAL NOTE Make sure that you select the Trim Entities icon and not the Dynamic Trim icon. The Dynamic Trim option selects entities to trim by requiring the user to dynamically draw a construction spline.

With the Trim option, the two selected entities will be trimmed at their intersection point. You must pick the portion of each entity that should not be deleted.

STEP 8: **Select the four locations shown in Figure 7–33.**

With the Trim option selected, the picked locations identify the entities to trim and the portion of each entity to keep.

STEP 9: **When the section is complete, select the Continue icon.**

If you get an error message on selecting the Continue option, you probably have section entities that overlap. Within any section, you cannot have

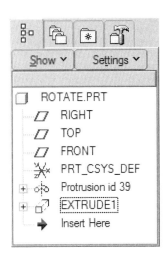

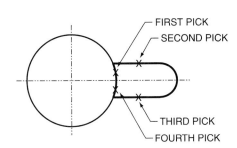

Figure 7–33   Trim Selections

Figure 7–34   Naming a Feature

sections that branch or sketch entities that lie on top of other entities. If you do get an error, observe your sketch to see if there are any obvious problems. If necessary, try to delete entities created with the Use Edge option. You can use the Undo and Redo options to help with the manipulation. A final work-around solution would be to sketch the section's inside arc using the Arc option instead of the Use Edge option.

STEP 10:   Enter a variable depth of 0.500.

STEP 11:   Dynamically rotate your model to observe the direction of extrusion.

STEP 12:   If necessary, reverse the direction of extrusion.

STEP 13:   Build the feature.

STEP 14:   Right-mouse select the new protrusion on the model tree and select the RENAME option.

STEP 15:   Rename the feature *EXTRUDE1* (Figure 7–34).

CREATING A COAXIAL HOLE

This segment of the tutorial will create the coaxial hole shown in Figure 7–35. To create the hole, the datum axis shown will have to be created first. This axis will be created with the Datum Axis command with the Through Cylinder constraint option.

STEP 1:   Select the DATUM AXIS icon on the datum toolbar.

STEP 2:   Pick the *Extrude1* feature in the location shown in Figure 7–36.

STEP 3:   Select THROUGH as the constraint option.

When a cylindrical surface is selected, the Through constraint option will place a datum axis through the center of the existing cylindrical feature. The cylindrical feature can be any solid feature with a rounded surface (e.g., shaft, round, extruded arc).

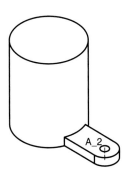

Figure 7–35
Coaxial Hole Feature

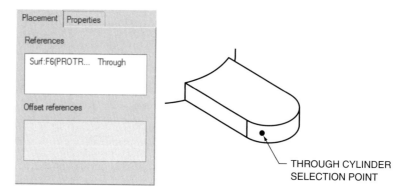

Figure 7-36   Cylinder Selection

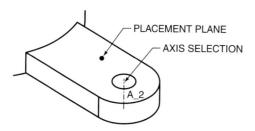

Figure 7-37   Coaxial Hole Creation

STEP 4:   **Select OK on the Datum Axis dialog box.**

STEP 5:   **Create the coaxial hole shown in Figure 7–37.**

With coaxial holes, an axis is the primary reference and the hole's placement plane is a secondary reference. Use the following parameters when creating the hole:

- Create a straight hole.
- Use a hole diameter of 0.500.
- Create the hole with the Through All depth parameter.
- Select the previously created datum axis as the primary reference.
- Use the Placement >> Secondary References option to define the placement plane.
- Use the Coaxial placement option under the Placement menu.

MIRROR THE EXTRUDED FEATURE

This segment of the tutorial will mirror the extruded protrusion, coaxial hole, and datum axis features (Figure 7–38). The Copy >> Mirror option requires the selection of a mirror plane. In this tutorial, one of Pro/ENGINEER's default datum planes will be used as this plane.

STEP 1:   **Select EDIT >> FEATURE OPERATIONS >> COPY.**

STEP 2:   **Select the MIRROR option from the Copy Feature menu.**

STEP 3:   **Select DEPENDENT on the Copy Feature menu, then select DONE.**

With the Dependent option, the copied feature's dimensions are dependent on its parent feature's dimensions. If a dimension on a parent feature changes, the corresponding dimension on its child will change.

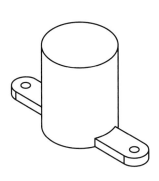

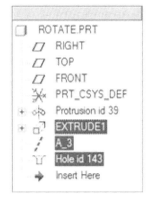

Figure 7–38 Mirrored
Features

Figure 7–39 Feature
Selection

STEP 4:   On the model tree, select the EXTRUDE1 feature, the datum axis feature, and the coaxial hole feature (Figure 7–39).

The last three features on the model tree will be copy-mirrored. Optionally, you can select the features on the workscreen.

STEP 5:   Select DONE on the Select Feature menu.

The Select Feature menu provides an option for selecting additional features to be copied. Select Done to end the selection process.

STEP 6:   As shown in Figure 7–40, select datum plane RIGHT as the plane to mirror the selected features about.

After selecting the plane to mirror about, the mirrored grouped features will be created.

STEP 7:   Select DONE to exit the Feature menu.

STEP 8:   Observe your model tree.

Notice the group feature added to the model tree (Figure 7–41). By selecting the + icon next to the feature on the model tree, you will expand the group to reveal the elements of the group.

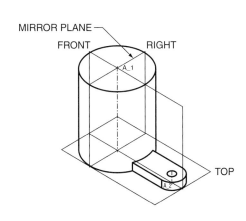

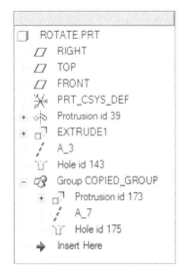

Figure 7–40   Mirror Plane Selection

Figure 7–41   Model Tree

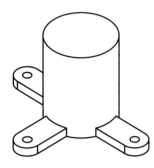

Figure 7–42   Rotated
Features

ROTATING THE EXTRUDED FEATURE

The Copy command has an option for rotating features. Unlike the Pattern command, the Rotate option can array multiple features simultaneously. This segment of the tutorial will rotate the protrusion, hole, and axis features 90 degrees to create the grouped feature shown in Figure 7–42.

STEP 1:   **Select EDIT >> FEATURE OPERATIONS >> COPY.**

STEP 2:   **Select the MOVE option on the Copy Feature menu.**

Move allows for the rotating or translating of selected features.

STEP 3:   **Select INDEPENDENT >> DONE.**

The Independent option will make the copied feature's dimension values independent from its parent feature's values. If a parent feature's dimension value changes, the copied feature's corresponding dimension value will not change.

STEP 4:   **On the model tree, select the EXTRUDE1 feature, the original coaxial hole, and the axis locating the coaxial hole (Figure 7–43).**

Select the features that were mirrored in the previous section of this tutorial.

STEP 5:   **Select DONE on the Select Feature menu.**

STEP 6:   **Select ROTATE on the Move Feature menu.**

The Rotate option will copy a feature about an edge, axis, or curve.

STEP 7:   **Select CRV/EDG/AXIS, then select the axis defining the center of the base revolved protrusion.**

As shown in Figure 7–44, the axis used to define the center of the revolved protrusion feature will be used as the center of the rotation. On the workscreen, select this axis.

STEP 8:   **If necessary, use the FLIP option to match the direction of translation shown in Figure 7–44.**

On the workscreen, an arrow points in the direction of translation. Pro/ENGINEER uses the right-hand rule to determine this direction. Using your right hand, point the thumb in the direction of the arrow; the copy will rotate the direction that the fingers of your hand are pointing.

STEP 9:   **When the direction of translation matches Figure 7–44, select OKAY.**

STEP 10:   **Enter 90 as the rotation angle.**

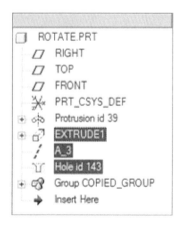

Figure 7–43   Feature Selection

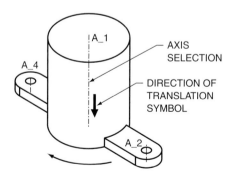

Figure 7–44   Axis Selection

STEP 11:   Select DONE MOVE on the Move Feature menu.

STEP 12:   Select DONE on the GP VAR DIMS menu.

The dimensions from the features that are being copied can be varied during the copy process. The GP VAR DIMS menu provides the option of selecting dimensions to vary during the copy.

STEP 13:   Select OK on the dialog box.

STEP 14:   Select DONE to exit the feature menu.

STEP 15:   Save your mode.

ADDING RELATIONS

This segment of the tutorial will create a dimensional relationship that will make the distance from the center of the base protrusion to the center of the first axial hole equal to the height of the base protrusion. In Figure 7–45, dimension d12 will be set equal to dimension d1.

STEP 1:   Select TOOLS >> RELATIONS on the menu bar.

STEP 2:   **On the workscreen, select the base-revolved protrusion and the EXTRUDE1 feature.**

The dimensions shown in Figure 7–45 should appear similar to your model. Within the Relations menu, dimensions are shown symbolically. The symbols displayed in the illustration may or may not match your symbols. When following this tutorial, use the corresponding symbols from your model.

From Figure 7–45, you will add a relation that will make the dimension displayed with symbol d12 equal to the dimension displayed with the symbol d1.

STEP 3:   **Referring to the dimensions shown in Figure 7–45 (but using your own respective symbols), add an equation that will make the *d12* dimension equal to the *d1* dimension.**

For the symbols shown in Figure 7–45, add the equation d12 = d1.

STEP 4:   ☑ **Verify your equations with the Verify Relations icon on the Relations toolbar.**

STEP 5:   **Select OK on the Relations dialog box.**

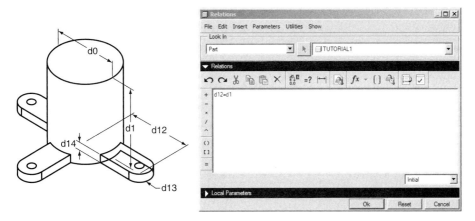

Figure 7–45   Dimension Symbols

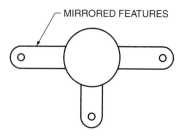

Figure 7-46   Finished Features

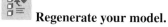

Step 6:   **Regenerate your model.**

After regenerating, notice how the EXTRUDE1 feature and the mirrored feature lengthen (Figure 7–46). Why did the second copy-rotated feature not lengthen with the other two features? The Independent option used in the creation of the rotated copy made its dimensions independent of its parent feature's dimensions.

Step 7:   **Save your model.**

MANIPULATING TUTORIAL 2

This tutorial will create the part shown in Figure 7–47. The primary objective of this tutorial is to demonstrate the Group command and the ability to pattern grouped features. Covered in this tutorial will be the following topics:

- Creating a through axis datum plane.
- Creating an extruded boss feature.
- Creating a coaxial hole and fillet.
- Grouping features.
- Patterning a group.
- Creating a conditional relation.

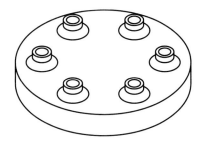

Figure 7–47   Finished Part

CREATING THE BASE PROTRUSION

The base feature for this tutorial is shown in Figure 7–48. Using Pro/ENGINEER's default datum planes and the following requirements, create this feature.

- Revolve the section shown in Figure 7–48.
- Use Pro/ENGINEER's default datum planes.
- Sketch the revolved protrusion on datum plane FRONT.
- Enter a 360 degree angle of revolution.

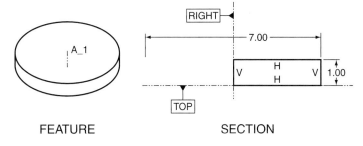

FEATURE   SECTION

Figure 7–48   The Base Protrusion

CREATING A THROUGH AXIS DATUM PLANE

This tutorial will create a datum plane through the center axis of the base feature and at an angle to datum plane FRONT. The boss features shown in Figure 7–47 are patterned groups. To create a revolved patterned group, an angular dimension used within the definition of a feature in the group must be used as the leader dimension. This segment of the tutorial will create the angular datum plane shown in Figure 7–49. The angle defining this datum plane will be the leader dimension for the pattern.

STEP 1:   **Use the Control key and D character combination to set the view's default orientation.**

You can also set the default orientation by selecting View >> Orientation >> Default Orientation on the menu bar. Your view's default orientation is determined by the current setting of the configuration file option *orientation*. This option can be set to trimetric, isometric, or user-defined.

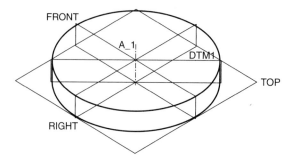

Figure 7–49   Through >> Axis Datum Plane

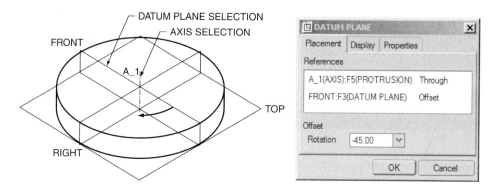

Figure 7–50   Datum Creation

STEP 2:   [icon]   **Select the DATUM PLANE icon on the datum toolbar.**

The Datum Plane tool is used to create new datum plane features. You will create a datum plane through the center axis of the base protrusion that also is defined at an angle to datum plane FRONT.

STEP 3:   **Using the Control key, pick the base protrusion's center axis and datum plane FRONT (see Figure 7–50).**

STEP 4:   **Ensure that the THROUGH and OFFSET constraint options are set on the Datum Plane dialog box (Figure 7–50).**

The Through constraint option will construct a datum plane through an edge, axis, curve, point, vertex, plane, or cylinder. Used in combination with the Offset option, it will create a datum plane through an existing axis and at an angle to an existing plane.

STEP 5:   **Enter °45.00 as the Offset Rotation value.**

On the workscreen, Pro/ENGINEER will graphically display the direction of rotation for the angular datum plane. In this example, entering a value of °45 will create the new datum plane at a 45 degree angle to FRONT.

STEP 6:   **Select OK from the Datum Plane dialog box.**

CREATING THE BOSS FEATURE

This segment of the tutorial will create the extruded protrusion shown in Figure 7–51. This protrusion will serve as the first feature in the group to be patterned.

STEP   1:   [icon]   **Select the Sketch icon on the toolbar.**

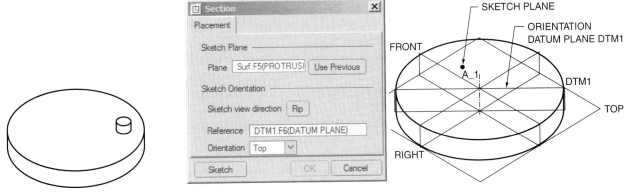

Figure 7-51 Boss Feature   **Figure 7-52** Sketch Orientation

STEP 2: Pick the top surface of the part as the sketching plane then pick datum plane DTM1 as the orientation reference.

STEP 3: On the Sketch dialog box select TOP as the orientation (Figure 7–52), then select SKETCH.

INSTRUCTIONAL POINT The orientation of datum plane DTM1 toward the top of the sketcher environment is an important step of this tutorial. Without this proper orientation, you will get an error message when patterning the group later in this tutorial. As shown in Figure 5–53, datum plane DTM1 should be horizontal on the workscreen. If datum plane DTM1 is not horizontal, restart this segment of the tutorial.

STEP 4: Use the References dialog box to specify datum planes RIGHT, FRONT, and DTM1 as references.

STEP 5: Use the CIRCLE icon to create the circle shown in Figure 7–53.

STEP 6: With the previously created circle still picked, on Pro/ENGINEER's Menu Bar, select EDIT >> TOGGLE CONSTRUCTION.

Elements must be picked before you can turn them into construction entities. If the circle is not selected, use the Pick icon to select it. Elements created as construction entities will not extrude with geometry entities. This construction circle will be used to align and locate the extruded feature.

STEP 7: Modify the construction circle's diameter to equal a value of 4.90.

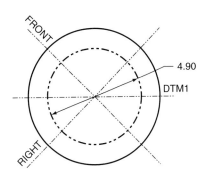

Figure 7-53 Construction Circle

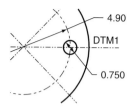

Figure 7–54
Circle Creation

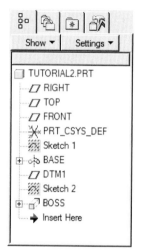

Figure 7–55   Model Tree

STEP 8:   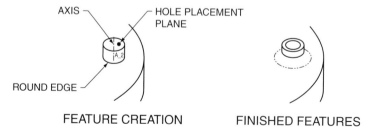 Use the CIRCLE option to create the circle entity shown in Figure 7–54.

Align the center of the new circle at the intersection of the construction circle and datum plane DTM1. Modify the circle's diameter to equal a value of .750.

STEP 9:   Select the Extrude icon on the toolbar.

STEP 10:   Exit the sketcher environment and extrude the protrusion a blind distance of 0.500.

STEP 11:   Build the feature.

The next several steps will give each protrusion feature a more descriptive name.

STEP 14:   On the model tree, right-mouse select the first Protrusion feature (Revolve.1) and rename it BASE with the pop-up menu's RENAME option.

This will allow for the renaming of this feature to make it more descriptive on the model tree (Figure 7–55).

STEP 15:   Use the model tree's pop-up menu to rename the last protrusion feature *BOSS* (see Figure 7–55).

STEP 16:   Save your part.

COAXIAL HOLE AND ROUND

This segment of the tutorial will create the hole and round features shown in Figure 7–56. These two features will be combined with the boss feature and datum plane DTM1 to form a group.

STEP 1:   Create a COAXIAL HOLE as shown in Figure 7–56 by using the following options:

- Use the Hole command.
- Create a straight hole.
- The hole will have a diameter value of 0.500.
- Create the hole using the Through All depth option.

AXIS ─┐   ┌─ HOLE PLACEMENT
                          PLANE

ROUND EDGE ─┘

FEATURE CREATION   FINISHED FEATURES

Figure 7–56   Hole and Round Features

- Use the boss feature's axis as the primary reference.
- Place the hole with the Coaxial option.

STEP 2: **Create a ROUND as shown in Figure 7–56 using the following options:**

- Use the Round command.
- Create a Simple Round.
- Use a constant radius value of 0.250.

GROUPING FEATURES

This section of the tutorial will group datum plane DTM1 with the boss, hole, and round features. When features are grouped, they are turned essentially into one feature. The design of this part requires these features to be arrayed about a bolt circle. Since the normal Pattern command arrays one feature at a time, these features have to be grouped first (Figure 7–57).

STEP 1: **Using your control key, pick the last five elements on the model tree (Figure 7–58).**

Each feature could be selected on the workscreen. Since features have to be adjacent to each other in the order of regeneration, it is often easier to pick features on the model tree.

STEP 2: **Select EDIT >> GROUP on the menu bar.**

STEP 3: **Right-mouse select your new group and rename it BOSS_GROUP.p.**

STEP 4: **Save your model.**

PATTERNING THE BOSS GROUP

The normal Pattern command can pattern only one feature at a time, and does not work with grouped features. The Group menu provides an option for patterning groups. This option will be used to create the rotational pattern shown in Figure 7–59 on the next page.

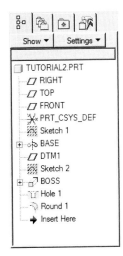

Figure 7–57
Model Tree

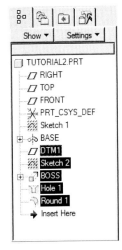

Figure 7–58
Feature Selection

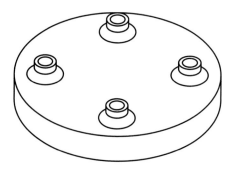

Figure 7-59   A Patterned Group

STEP 1:   **Right-mouse select the *BOSS_GROUP* group on the model tree, then pick the PATTERN command.**

Groups can be patterned in a manner similar to that for patterning individual features. Unlike normal patterning, in patterning groups, only the Identical pattern option is available. As a result, patterned group instances must lie on the same placement plane, instances cannot intersect, and instances cannot intersect an edge.

STEP 2:   **On the workscreen, pick the 45 degree dimension used to define the angle of datum plane DTM1 (Figure 7–60).**

Select the 45 degree dimension shown in Figure 7–60 as the first direction leader dimension. This dimension is the angular reference dimension obtained from the creation of datum plane DTM1. Since DTM1 was included in the group, this angular dimension is available for varying while patterning. Since rotational patterns require an angular dimension, this is the necessary dimension required in the first direction of the pattern.

STEP 3:   **Within the dimension edit box, enter 90 as the leader dimension's increment value.**

Entering 90 as the increment value will create each instance of the pattern 90 degrees apart.

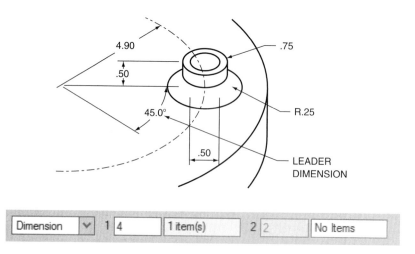

Figure 7-60   Pattern Definition

Step 4: On the dashboard, enter 4 as the number of instances in the first direction.

This group will be patterned in one direction only. Therefore, the selection of a second leader dimension is not necessary.

Step 5: Build the pattern.

Step 6: Save your part.

ESTABLISHING A CONDITIONAL RELATIONSHIP

One of the powerful capabilities of a feature-based/parametric modeling package is its ability to incorporate design intent into a model. One of the ways that intent can be built into a model is through the creation of dimensional relationships. A relation is an explicit mathematical relationship that exists between two dimensions. The flexibility of the Relations command allows conditional statements to be built into a relationship's equation. This tutorial will create a dimensional relationship that utilizes a conditional statement.

The current diameter of the base feature in this tutorial is 7.00 in. This portion of the tutorial will add a conditional relationship that will drive the number of BOSS_GROUP features within the rotational pattern. The following design intent applies:

- A base feature diameter of 5.00 in or less will have four equally spaced boss features.
- A base feature diameter over 5.00 in, but less than or equal to 10 in, will have six equally spaced boss features.
- A base feature diameter over 10 in will have eight equally spaced boss features.
- The centerline diameter of the patterned groups will be 70 percent of the diameter of the base feature's diameter.

As shown in Figure 7–61, the following dimensions with matching symbols will be used. Your symbols may be different.

- **Base Feature Diameter (d0)** This is the diameter value of the base feature. Initially, it is set to a value of 7.00 in.
- **Number of Grouped Boss Features (p0)** This is the number of instances that the grouped boss was patterned. Initially, there are four instances.

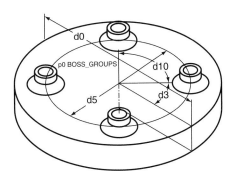

Figure 7–61 Dimension Symbols

- **Diameter of the Pattern (d5)** This is the diameter value that defines the location of the first boss feature. Initially, this value is set to 4.90 in.

- **Increment Value of the Pattern (d10)** This is the number of degrees incremented between each instance of the pattern. Initially, this value is set to 90 degrees.

STEP 1: **Select TOOLS >> RELATIONS on the model tree.**

STEP 2: **On the model tree, expand the pattern feature and the second group feature by selecting the + sign to the left of the required feature's name (Figure 7–62).**

Selecting the + sign next to a feature's name will reveal components, features, and elements defining the feature.

STEP 3: **On the model tree, select the BASE Protrusion feature and each feature defining the second BOSS_GROUP group (Figure 7–62).**

Selecting features of the first grouped feature will not reveal the dimension symbol defining the pattern increment value. Before adding relations, make sure that the dimensions shown in Figure 7–61 are revealed on your workscreen.

STEP 4: **Enter the dimension relations statements shown in Figure 7–63 and Table 7–2.**

Table 7–2 describes the purpose for each statement. Make sure that you use the dimension symbols associated with your model. Your symbols will probably be different from those shown in Figure 7–63 and from those shown in the statements in the table.

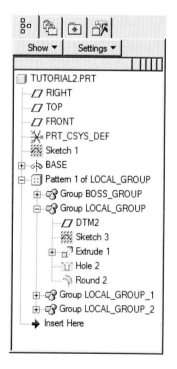

Figure 7–62   Model Tree

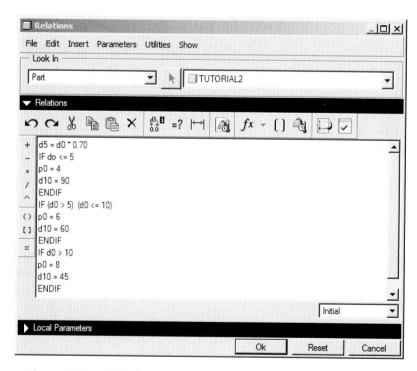

Figure 7–63   The Relations Dialog Box

Table 7-2 Relations Statements

Statement Line	Meaning
d5=d0*.70	Diameter of the bolt-circle is 70% the diameter of the flange
IF d0<=5	If the diameter of the flange is less than or equal to 5.00
p0=4	The number of holes is equal to 4
d10=90	The angular increment value is equal to 90 degrees
ENDIF	End of the definition
IF (d0>5)&(d0<=10)	If the flange diameter is greater than 5.00 and less than or equal to 10.00
p0=6	The number of holes is equal to 6
d10=60	The angular increment value is equal to 60 degrees
ENDIF	End of the definition
IF d0>10	If the flange diameter is greater than 10.00
p0=8	The number of holes is equal to 8
d10=45	The angular increment value is equal to 45 degrees
ENDIF	End of the definition

STEP 5: Select the VERIFY icon on the relations toolbar.

The Verify option ensures that you have no syntax errors within your relations statements. Correct any identified errors.

STEP 6: Select OK to confirm the verification.

STEP 7: If your relations verify properly, select OK to exit the Relations dialog box.

STEP 8: Regenerate your model.

STEP 9: Use the EDIT command (pop-up menu or model tree) to change the BASE feature's diameter dimension to a value of 12.00.

How many boss groups should the part have after regeneration?

INSTRUCTIONAL NOTE For this particular part, changing the flange diameter to a value of 4.00 or less will produce a regeneration error. The Group >> Pattern option defaults to an Identical pattern option. Instances of an identical pattern cannot intersect each other or the placement plane's edge. In this case, making the flange too small can violate both requirements.

STEP 10: Regenerate your model.

STEP 11: Edit the BASE feature's diameter dimension to a value of 9.00, then regenerate your model.

You should have six boss groups.

STEP 12: Edit the BASE feature's diameter dimension to a value of 4.75, then regenerate your model.

STEP 13: Save your part.

PROBLEMS

1. Using Pro/ENGINEER's Part mode, model the part shown in Figure 7–64. Construct the part using the following order of operations:

   a. Create the base geometric feature as a revolved protrusion. Include the hole in the section.

   b. Model one leg feature (including the hole).

   c. Use the Copy >> Move >> Rotate option to create three instances of the leg feature.

   d. Create the remaining three leg instances.

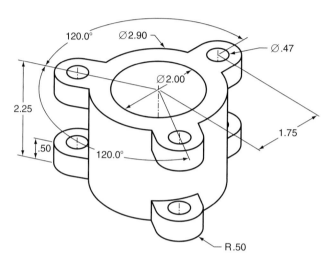

Figure 7–64   Problem 1

2. Model the part shown Figure 7–65. Use the Copy-Rotate option to create the multiple instances of the cut and hole features.

3. Model the part shown in Figure 7–65. Use the Pattern command to create the multiple instances of the cut and hole features.

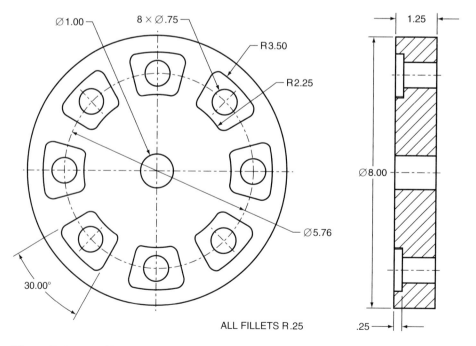

ALL FILLETS R.25

Figure 7–65   •   Problems 2 and 3

4. Model the part shown in Figure 7–66.

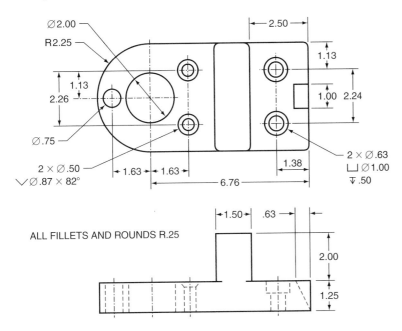

Figure 7–66 Problem 4

5. Model the part shown in Figure 7–67.

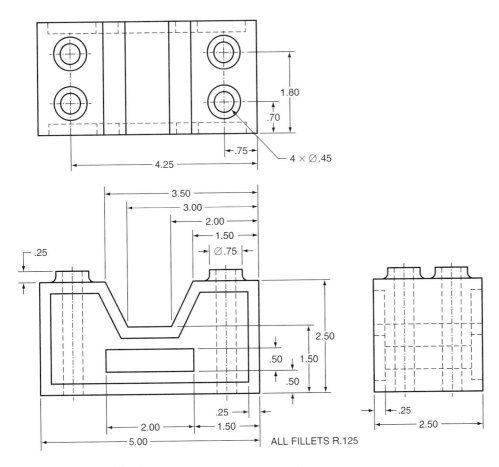

Figure 7-67 Problem 5

6.   Model the part shown in the drawing in Figure 7–68.

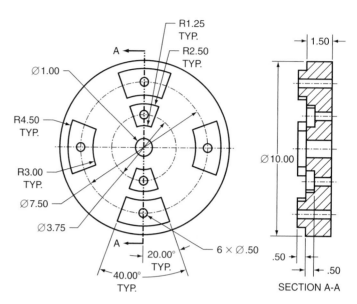

Figure 7–68   Problem 6

QUESTIONS AND DISCUSSION

1.   Describe the difference between the Pattern command and the Copy command. Describe some situations when Copy would be used in place of Pattern.

2.   Describe the difference between the Copy >> Same Refs option and the Copy >> Move >> Translate option. Can you think of a situation when you might use one option over the other?

3.   Describe the difference between copying features independently and copying features dependently.

4.   Describe how a company might utilize user-defined features to streamline the modeling process.

5.   Write relation statements for the following situations:

   •   Dimension *d0* is equal to 2 times dimension *d1*.

   •   Dimension *d0* is equal to the sum of dimension *d1* and *d2* divided by *d3*.

   •   Dimension *d0* is equal to the square root of *d1*.

6.   List items types that can be included in a family table.

7.   How does Pro/ENGINEER utilize cross sections?

8

CREATING A PRO/ENGINEER DRAWING

Introduction

Drawing mode is used to produce detailed engineering drawings of parts and assemblies. Drawing mode has the capability to produce orthographic views and to include section views and auxiliary views. This chapter introduces the fundamentals behind Drawing mode and how to produce an annotated multiview drawing. Upon finishing this chapter, you will be able to:

- Within Drawing mode, create orthographic views of an existing model.
- Setup, retrieve, and create sheet formats.
- Manipulate the settings of a drawing by changing Drawing Setup file options.
- Create a detailed view of a model.
- Apply parametric and nonparametric dimensions to a drawing.
- Show and erase entities such as geometric tolerances, centerlines, and datums.
- Create notes on a drawing, including notes with leaders and balloon notes.
- Create a table on a drawing.

DEFINITIONS

Associative dimension	A parametric dimension that is available for viewing and/or modification in multiple modules of Pro/ENGINEER.
Drawing setup file	A text file used to establish many of Drawing mode's default settings. As an example, the default text height for a drawing is set in the drawing setup file.
Format	A Pro/ENGINEER module used to create drawing borders and title blocks. Formats can be added to a Pro/ENGINEER drawing.
Multiview drawing	The use of multiple orthographic views to graphically display and communicate an engineering design.
Nonparametric dimension	A dimension created in Drawing mode that is not used to construct a part or assembly. Nonparametric dimension values cannot be modified.
Parametric dimension	A dimension that is used to define a part or assembly feature. Parametric dimensions can be modified or redefined.

DRAWING FUNDAMENTALS

Pro/ENGINEER is a fully integrated and associative engineering design package. Since it is an integrated package, an array of design, engineering, and manufacturing tools are

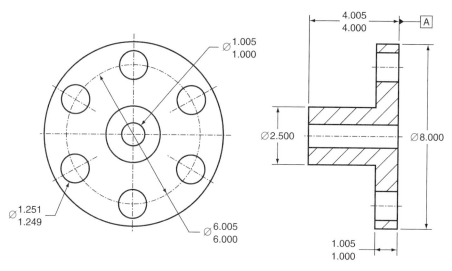

Figure 8–1   A Pro/ENGINEER Drawing

available. Components of a design can be modeled within Part mode, grouped in Assembly mode, and simulated and tested in Pro/MECHANICA, and machining code can be generated in Manufacturing mode (Pro/NC). The two-directional associativity that exists between modes allows changes made in one mode to be reflected in another. Combined within a strong computer network system, a true "paperless manufacturing" environment can be established.

Two-dimensional orthographic drawings were once considered one of the initial steps in the design process. A traditional approach to engineering design might require engineering drawings to be produced first, followed by engineering analysis and computer numerical control (CNC) code production. With Pro/ENGINEER, engineering drawings are considered "downstream" applications that occur after part modeling and analysis.

Pro/ENGINEER's basic package provides a module for creating orthographic drawings (Figure 8–1). Within Drawing mode, detailed multiview drawings can be created from existing models. Dimensions used to create a part (referred to as parametric dimensions) can be revealed in Drawing mode to document a design. Parametric dimensions can be modified in Drawing mode with the changes reflected in other modules of Pro/ENGINEER. Options are available for creating section views, detailed views, and auxiliary views. There are options for creating notes, leaders, nonparametric dimensions, and tables. Additionally, within Drawing mode, there exists a variety of two-dimensional drafting tools.

Drawing mode can be used to create detailed drawings from existing parts and assemblies. When you are creating a new drawing, Pro/ENGINEER provides an option for selecting the model from which to create the drawing. Also, additional models can be added to a drawing from within Drawing mode. Multiple views of a model can be added to a drawing and annotations applied. Pro/ENGINEER provides existing sheet formats (e.g., A, B, C, and D) with detailed title blocks and borders. Format mode can be used to create additional sheet formats.

DRAWING SETUP FILE

Drawing mode uses drawing setup files (DTL) to control the appearance of drawings. Pro/ENGINEER comes with default settings for a variety of drawing parameter options. Examples of parameters include text height, arrowhead size, arrowhead style, tolerance display, and drawing units. Default values can be changed permanently or for individual drawings. Multiple drawing setup files can be created and stored for later use. Pro/ENGINEER's

Table 8–1   Common Drawing Setup File Options

OPTION/ **Description**	**DEFAULT VALUE** **(Optional Value)**
crossec_arrow_length Controls the length of Cutting Plane Line arrowheads.	0.1875
crossec_arrow_width Controls the width of cutting-plane line arrowheads.	.06250
dim_leader_length Controls the length of a dimension line when the dimension line arrowheads fall outside of the extension lines.	0.5000
draw_arrow_length Sets the length of dimension arrowheads.	0.1875
draw_arrow_style Sets the arrowhead style.	Closed (Open or filled)
draw_arrow_width Sets the width of dimension arrowheads.	0.0625
drawing_text_height Sets the height of text in a drawing.	0.15625
drawing_units Sets units for a drawing.	Inch (foot, mm, cm, or m)
gtol_datums Sets the display of geometric tolerance datum symbols.	Std_ansi (std_iso, std_jis, or std_ansi_mm)
leader_elbow_length Sets the length of a leader's elbow.	0.2500
radial_pattern_axis_circle Controls the display of rotational pattern features. Set to *yes* produces a bolt-circle centerline.	No (yes)
text_orientation Sets the orientation of dimension text.	Horizontal (parallel or parallel_diam_horiz)
text_width_factor Sets the width factor for text.	0.8000 (.25 through 8)
tol_display Sets the display of tolerance values.	No (yes)

default drawing setup file (*prodetail.dtl*) can be found in the <pro_engineer load point>\text\ directory.

FILE >> PROPERTIES >> DRAWING OPTIONS

New DTL files are created or current drawing settings are changed with the Properties >> Drawing options command. Pro/ENGINEER utilizes an Options dialog box similar to the configuration file's Options dialog box for drawing setup changes (see Chapter 2). The configuration file option *drawing_setup_file* can be used to establish a specific DTL file. If this option is not set, Pro/ENGINEER uses the default DTL file. The configuration file option *pro_dtl_setup_dir* can be used to set the directory that Pro/ENGINEER searches for DTL files. Table 8–1 provides a list of common DTL file options with default values.

SHEET FORMATS

A format is an overlay for a Pro/ENGINEER drawing. It can include a border, title block, notes, tables, and graphics. A sheet format has the file extension *.frm. Pro/ENGINEER provides a variety of predefined standard formats for use with ANSI and ISO sheet sizes (e.g., A, B, C, and D size sheets). These standard formats can be modified to produce a customized format. Additionally, Pro/ENGINEER's Format mode can be used to create a new sheet format. Figure 8–2 shows an example of an A-size sheet format. A library of standard sheets can be created. The configuration file option *pro_format_dir* can be used to specify the directory path where standard sheets are stored.

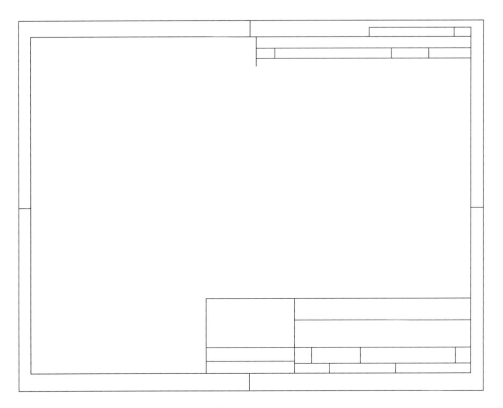

Figure 8–2   ANSI A-Size Format and Title Block

MODIFYING FORMATS

An existing sheet format can be modified to create a customized format. Pro/ENGINEER's existing formats are located in the Format directory under the Pro/ENGINEER's program directory. Perform the following steps to modify an existing format.

STEP 1:   **Start Pro/ENGINEER.**

STEP 2:   **Select FILE >> OPEN.**

A drawing format can be opened from most modes of Pro/ENGINEER.

STEP 3:   **Locate and Open the format to modify.**

Manipulate the Look In directory structure under the File Open dialog box to locate the format to open. On the dialog box, the File Type option can be changed to only display format files.

STEP 4:   **Use available sketch creation and modification tools to modify the format.**

> **INSTRUCTIONAL NOTE**   See the sections "Two-Dimensional Drafting" and "Manipulating Draft Geometry" found later in this chapter for information on creating and modifying a sketch.

Drawing Mode provides a variety of tools for creating two-dimensional drawings. Options are available for creating lines, arcs, circles, splines, and ellipses. Construction options such as copy, mirror, intersect, trim, and offset are also available.

STEP 5:   **Select the SAVE option from the File menu.**

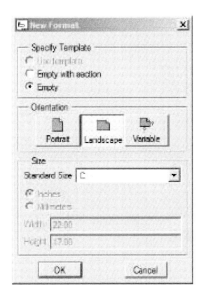

Figure 8-3 New Format Dialog Box

CREATING FORMATS

Formats can be created from scratch. Format mode is the Pro/ENGINEER foundation module used for the creation of standard sheet formats. Sketching tools such as line, circle, arc, and note can be used to create the format section. A new format object can be created by using the File >> New option. When a new format is initially created, Pro/ENGINEER reveals the New Format dialog box (Figure 8–3). The following options exist on this dialog box:

- **Specify Template** Through the Empty suboption, the Specify Template option allows for the selection of a standard sheet size (e.g., A, B, C). The Empty with Section suboption allows an existing section file (*.sec) to be incorporated within the new format.
- **Orientation** The Orientation option is available concurrently with the Empty option. Portrait, landscape, and variable sheet orientations are available.
- **Size** The Size option is available concurrently with the Empty option. This option allows for the selection of a standard sheet size or for the selection of a user-defined sheet size. User-defined sheet sizes can be established in inches or millimeters.

CREATING A NEW DRAWING

The Drawing mode option from the New dialog box is used to create new drawings. When Drawing mode is selected and a file name entered, Pro/ENGINEER introduces the New Drawing dialog box (Figure 8–4). If a part or assembly is currently active in session memory, Pro/ENGINEER defaults to this part or assembly as the model from which to create the drawing. An option is available for browsing to find other existing models. The New Drawing dialog box provides the option for specifying a standard sheet size or for retrieving an existing format.

Three types of items can be added to a drawing: formats, 2D draft entities, and model views. Formats are placed in a drawing by using the Page Setup dialog box (File >> Page Setup). Pro/ENGINEER provides an Open dialog box for browsing to find an appropriate format. Multiple sketching tools allow for the creation of 2D draft entities. Other options are available for adding dimensions and notes. Model views can be added to a drawing with

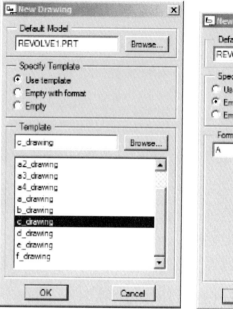
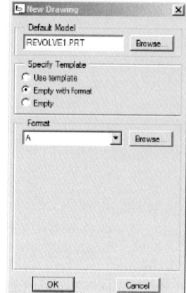
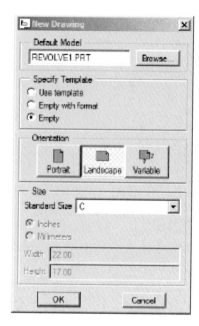

Figure 8–4   New Drawing Dialog Box with Template Options

the Insert View icon. General, section, projected, auxiliary, revolved, and detail views can be added to a model.

The following shows step-by-step how to create a new drawing. Several options are available that can vary the method for creating a new drawing. This process will allow for the creation of a new drawing without specifying a standard sheet size from the New Drawing dialog box.

STEP 1:   Select FILE >> NEW.

STEP 2:   Select Drawing mode from the New dialog box, enter a name for the new drawing, then select OK. (Use the default template file.)

STEP 3:   On the New Drawing dialog box, browse to find the DEFAULT MODEL from which to create the drawing.

Pro/ENGINEER will default to the current active model. You can use the Browse option to search for any existing object.

STEP 4:   Select the EMPTY button from the Specify Sheet option.

The Empty option will require the defining of a sheet size and orientation. Alternatively, the Use Template option is usable for the selection of an existing format, while the Empty with Format option is used to create a drawing without an established size or sheet format.

STEP 5:   Select a sheet orientation and a sheet size, then select OK.

The Empty option requires the selection of a sheet orientation and size. A portrait, landscape, or variable orientation is available. Standard sheet sizes and user-defined sheets sizes are also available.

DRAWING VIEWS

Pro/ENGINEER's Drawing mode provides options for creating a variety of orthographic views to include section views, auxiliary views, detailed views, revolved views, broken views, and partial views. As many views as necessary to fully describe a model can be added to a drawing sheet. Once inserted into a drawing, parameters associated with the model, to include dimension values, can be created. Views are also fully associated. This allows any model changes made in a drawing to be reflected in the part or assembly model. Additionally, changes in one view of a drawing will reflect accordingly in all views of the model.

VIEW MODIFICATION

The Drawing View dialog box used to create views of an existing Pro/ENGINEER model. Drawing mode has several options for manipulating and modifying existing views. The following options are available:

MOVING A VIEW

Views can be dragged about the workscreen with the mouse. When you are placing a view, Pro/ENGINEER requests a center point for the drawing view. The view is set initially at this location. The view can be dragged to any desired location. *Note:* The Unlock View option on the toolbar has to be deselected before moving a view.

MODIFYING VIEWS

The View Modify menu option is available through the Properties option on the mouse's pop-up menu (preselect the view). It provides a variety of tools for modifying a view. The following is a partial list of the available categories and options:

- **View Type** The View Type category is available to change the type of view. As an example, a projection view can be changed to a general or auxiliary view.
- **Scale** This category is used to change the scale of a nonchild view. The view cannot be a child view of another view and the view must have been inserted with a specific, user-defined scale value.
- **View Display** This category is used to change the display of lines on a drawing view. A view's hidden lines can be specified as wireframe, hidden, or no hidden. Additionally, tangent edge lines may be specified to display as a solid line, a centerline, a phantom line, or a dimmed line, or can be set to not be displayed.
- **View Name** Pro/ENGINEER provides each view with a unique name. This option is used to rename a view.
- **View Reorient** When you are inserting a view, Pro/ENGINEER provides the option of orienting the view. As an example, a view can be oriented to allow for a proper front view vantage point. The View Orientation option allows this orientation to be changed.
- **Section** This category allows a cross section to be created or replaced.
- **Z-Clipping** The Z-Clipping option allows for the exclusion of all graphics on a view behind a selected plane. This option is advantageous for views with background graphics that can clutter the drawing.

DELETE VIEW

The Delete View option permanently removes a view from the drawing. Views that are parents of other views cannot be deleted.

RELATE VIEW

The Relate View option (Edit >> Group >> Relate to View) is used to assign draft entities to a selected view. As an example, a note might be placed into a drawing using the Insert >> Note option. This note can be assigned to a specific view with the Relate View option.

DRAWING MODELS

Multiple models can be accessed within one drawing. The Drawing Models (File >> Properties) option allows additional models to be added to the current drawing. The Set option allows a specific model to be set as the active model in the drawing.

VIEW TYPES

Pro/ENGINEER provides a variety of view types to serve the documentation needs of a model. The following is a list of the available types:

GENERAL VIEWS

The general view is the basic view type available. It is required as the first view placed into a drawing and is used by other types as a parent view. General views require a user-defined orientation. The Drawing View dialog box (Figure 8–5) is used to place and orient the view to match orthographic view projection requirements. In Figure 8–6, the front view of the object was inserted as a general view.

PROJECTION VIEWS

The Projection view is an orthographic projection from a General view or from an existing view. Projection views follow normal lines of projection according to conventional drafting standards. As an example, a Projection view would be used to create a right-side view off of an existing front view. Projection views become child views of the view from which they are projected. In Figure 8–6, the top and right-side views were created as projection views.

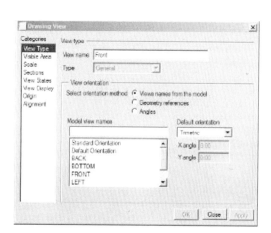

Figure 8–5   Drawing View Dialog Box

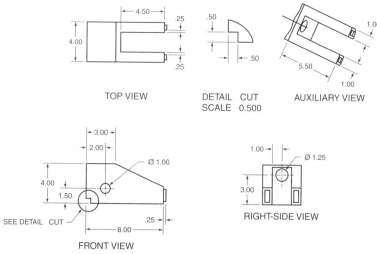

Figure 8–6   View Examples

AUXILIARY VIEWS

Auxiliary views are used to project a view when normal lines of projection will not work. They are used to show the true size of a surface that cannot be shown from one of the six primary views. Auxiliary views are projected from a selected edge or axis. The auxiliary view shown in Figure 8–6 is projected off the front view.

DETAILED VIEWS

Often, features are too small to describe fully with a standard projection view. In such a case, it is common practice to enlarge portions of a drawing to allow for more accurate detailing. Figure 8–6 shows an example of a detailed view.

REVOLVED VIEWS

Revolved views are used to show the cross section of a part or feature. A cross section is required and can be retrieved or constructed from within the Revolve option. Once selected, the cross section is revolved 90 degrees from the cutting plane. A revolved view can be either a full view or a partial view.

VIEW VISIBILITIES

In addition to view types such as general, projection, and auxiliary, Drawing mode provides the option of controlling the visibility of selected portions of a view. The following View Visibility options are available:

HALF VIEWS

Often, it is not necessary to show an entire model in a view. A good example would be a symmetrical object. Half views remove the portion of a model on one side of a selected datum plane or planar surface. Figure 8–7 shows an example of a half view. Half views can be used with General, Projection, and Auxiliary views only.

PARTIAL VIEWS

It is not necessary to document an entire view when only a small portion of a feature needs to be detailed. The Partial View option allows a selected small portion of a view to be created. The area to be revealed is enclosed in a sketched spline. Figure 8–7 shows an example of a partial section view. Unlike Detailed views, Partial views must follow normal lines of projection. Partial views can be used with general, projection, and auxiliary views only.

BROKEN VIEWS

As with partial and half-views, in many views, it is not necessary to show the entire object. As shown in Figure 8–7, broken views are used often with long, consistent

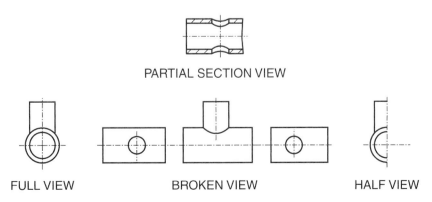

PARTIAL SECTION VIEW

FULL VIEW   BROKEN VIEW   HALF VIEW

Figure 8–7   View Types

cross sections. With a broken view, multiple horizontal or vertical breaks may be used if necessary. The Drawing Setup file option *broken_view_offset* is used to set the offset distance between portions of a view. The Move View option can be used to move a portion of a broken view.

SINGLE-SURFACE VIEWS

A single-surface view is a projection of an individual part surface. It may be used with any view type except detailed and revolved. When a single-surface view is created, only the selected surface is revealed. The Of Surface option is used to create a single-surface view.

MULTIPLE MODELS

When starting a new drawing, Pro/ENGINEER provides the option of selecting a model (part or assembly) from which to create the drawing. From an active model, drawing views can be created from this model. Pro/ENGINEER also provides an option for adding models to a drawing. By having multiple models associated with a drawing, you can detail multiple models on the same drawing sheet.

A model has to be added to a drawing before a view of the model can be created. To add a new model to a drawing, use

File >> Properties >> Drawing Models >> Add Model

When a model is added to a drawing, it becomes the current model. To set another model current, use the Set Model option. To delete a model from a drawing, use the Delete Model option.

CREATING GENERAL VIEW

Views are created with the Drawing View dialog box. This dialog box is accessible through the selection of the *Create a General View* icon or through the Insert >> Drawing View >> General option on the menu. Categories are available on this dialog for selecting a view type, a view visibility, and a scale. Additionally, a section view of a model can be specified. By default, when no view exists in a drawing, general is the only view type available. After a general view has been added to the drawing, projection becomes the default view type. Perform the following steps to add a full nonsectioned general view to a drawing:

STEP 1: Select the CREATE A GENERAL VIEW icon on the toolbar.

STEP 2: On the workscreen, pick a location for the view.

After you select the location, the view will be placed with the default view orientation.

STEP 3: Under the View Type category, select GENERAL as the view type to create.

Other options include Projection, Auxiliary, Detailed, and Revolved.

STEP 4: Enter a name for the view.

STEP 5: Orient the model using the View Orientation grouping on the dialog box.

The View Orientation grouping under the View Type category (see Figure 8–5) requires the selection of two perpendicular, planar surfaces or datum planes. The first reference is set by default to Front while the second reference is set to Top. Additional reference options available include Back, Bottom, Left, Right, Horizontal Axis, and Vertical Axis. A selected plane will face in the direction of the set reference option. As an example, if Top is selected as the reference, a selected plane will orient toward the Top of the workscreen.

STEP 6: **Under the Visible Area category, select FULL VIEW as the view visibility.**

Alternatively to a standard full view, a half, partial, or broken view can be created for a part drawing. For assembly drawings, an additional option for creating an exploded view is available.

STEP 7: **Under the Scale category, enter a scale for the view.**

This option is available for general views only. The Scale option allows for a user-specified scale for the view. The Default Scale option fits the view to the size of the model. The configuration file option *default_draw_scale* is used to set a default initial scale factor.

STEP 8: **If necessary, under the Sections category, define a section for the view.**

The following section options are available: No Section, 2D Cross Section, and 3D Cross Section. The Section option is used to create a section view of a model, while the Of Surface option is used to create a single-surface view. For views without a section, use the NoXsec option.

STEP 9: **If necessary, define other parameters under the remaining categories.**

STEP 10: **Select OK on the dialog box.**

SETTING A DISPLAY MODE

Views may be displayed as Wireframe, Hidden, No Hidden, or Shading. Views set with the Hidden option will print with dashed hidden lines. By default, the display mode of a view is dependent on the display mode set for the environment. An object's display mode can be set on the toolbar or in the Environment dialog box. To promote drawing clarity, individual display modes can be set for each view of a drawing by using the View Display category on the Drawing View dialog box.

Tangent edges can be displayed in a variety of styles on a model. The Environment dialog box setting *Tangent Edges* is used to set a default style. The configuration file option *tangent_edge_display* can be used to set a default tangent display style. The following tangent edge styles are available within Pro/ENGINEER:

- **Solid** Tangent edges are displayed as a solid line.
- **None** Tangent edges are not displayed.
- **Phantom** Tangent edges are displayed with a phantom line.
- **Centerline** Tangent edges are displayed with a centerline.
- **Dimmed** Tangent edges are displayed in the color set for dimmed entities, menus, and commands.
- **Default** Tangent edges will be displayed as set in the Environment dialog box.

> **MODELING POINT** Pro/ENGINEER has a default color for command lines or entities that are dimmed. As an example, if a command is inappropriate for a specific operation, it will appear dimmed on the menu. The configuration file option system_dimmed_menu_color is used to set the color of dimmed entities and commands.

To change the display mode of an individual view, perform the following steps:

STEP 1: **Using the Control key, preselect views for display modification.**

STEP 2: Select EDIT >> PROPERTIES.

The Properties option is also available through the mouse pop-up menu.

STEP 3: Select the View Display category on the Drawing View dialog box.

Select either Follow Environment, Wireframe, Hidden Line, No Hidden, or Shading from the Display Style option.

STEP 4: Select a Tangent Edge display style.

From the View Display menu, select a Tangent Edge display style.

STEP 5: Select OK on the Drawing View Dialog box.

DETAILED VIEWS

Detailed views are used on drawings to highlight portions of a component. A component might have a section with small and complicated features that would be hard to detail within the realm of the drawing's scale. Other sections of a drawing might require a detailed view because of a specific importance factor. Figure 8–8 shows an example of a drawing with a detailed view.

Within Pro/ENGINEER, a detailed view can be created any time after a general view has been placed. The scale of a detailed view is independent of its parent view. Additionally, detailed views do not lie along normal lines of projection. By default, a detailed view reflects its parent view. The display mode of a detailed view (line and tangent display) is the same as its parent view. Any cross sections shown in a parent view will be reflected in its detailed views. This view relationship can be broken with

Edit >> Properties >> View Disp >> Det Indep

Perform the following steps to create a detailed view:

STEP 1: Select INSERT >> DRAWING VIEW >> DETAILED on the menu bar.

STEP 2: Select a reference point on the edge of an entity in the parent view.

Select the entity in the view that includes the portion of the drawing used to create the detailed view. You have to select the edge of an entity. Pro/ENGINEER uses this reference point to regenerate the detailed view.

STEP 3: Draw a spline around the geometry to include in the detailed view.

Use the left mouse button to select points on the spline. Use the middle mouse button to close the spline. Objects included inside the spline boundary will be included in the detailed view.

STEP 4: Select the location for the view.

On the workscreen, select the location for the detailed view.

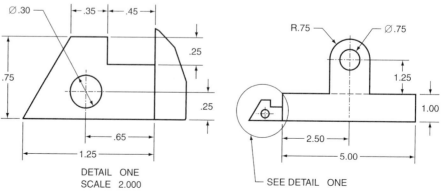

Figure 8–8   Detailed View

STEP 5: If necessary, use the Drawing View dialog box (Edit >> Properties) to change parameters such as view name and scale.

SHOWING AND ERASING ITEMS

Drawing mode is used to create annotated presentations of models created in Part and Assembly modes. One of the uses of Drawing mode is to create orthographic views of a model. Orthographic drawings frequently consist of items such as dimensions, centerlines, notes, and geometric tolerances. Within Part mode, features are fully defined through the use of dimensions, constraints, and references. Dimensions used to define a part are considered parametric. When a feature is created, the dimensioning scheme defining the feature should match the intent for the design of the part. These parametric dimensions can be used in Drawing mode to annotate the drawing of the model.

Other items created in Part or Assembly mode can be used in Drawing mode. Drawing mode's Show/Erase option is used to show an item created in Part or Assembly mode or used to erase an item created in Part or Assembly mode. The Erase selection within the Show/Erase option can be misleading. Since the Show/Erase option is used to manipulate the display of items created in Part or Assembly mode, these items cannot be deleted in Drawing mode. A more descriptive name for the Show/Erase option would be Show/No-Show. Items that are erased are actually hidden from display.

The manipulation of the display of items is accomplished through the Show/Erase dialog box (Figure 8–9). Listed are the items that can be shown or erased. To show or erase an item type, the button associated with the item type has to be selected. Multiple item types can be selected at one time. Several options are available for controlling the items that are displayed:

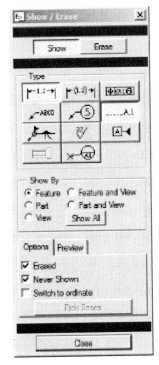

Figure 8-9
Show/Erase Dialog Box

- **Show All** The Show All option will display all items of a particular type. As an example, if Show All is selected with the Dimension item type, all parametric dimensions defining a model will be displayed.
- **Feature** The Feature option will display items on a selected feature. The user must select the feature on the workscreen.
- **Part** The Part option will display items on a selected part. The user must select the part on the workscreen or on the model tree. This option is useful for assembly drawings where items on individual parts must be displayed.
- **View** The View option will display items on a selected view. The user must select the view on the workscreen.
- **Feature and View** The Feature and View option will display items on a selected feature within the view from which the feature was selected.
- **Part and View** The Part and View option will display items on a selected part within the view from which the part was selected.
- **Erased** The Erased option will show only items that have previously been erased.
- **Never Shown** The Never Shown option will show only items that have never been previously shown.
- **Preview** Located under the Preview tab is the Preview option. The Preview option is used to preview items before they are displayed on the workscreen. Options are available for accepting or not accepting an item.

SHOWING ALL ITEM TYPES

The Show All option is used to show all items of a selected type. As an example, you can show all Geometric Tolerances or you can show all Centerlines. Perform the following steps to Show All of a selected item:

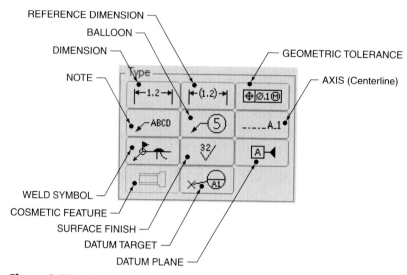

Figure 8–10   Item Types

Step 1: 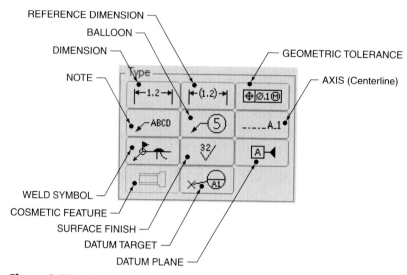 Select the Show/Erase icon on the toolbar.

Step 2: Select the SHOW option on the Show/Erase dialog box.

> The Show option will show items of a select type. The Show and Erase options cannot be selected at the same time.

Step 3: Select an item type to show (Figure 8–10).

> Examples of item types to show include dimensions, centerlines (Axis option), datums, and geometric tolerances.

Step 4: Select the SHOW ALL option.

> The Show All option will display all items of a selected type. This is a useful option for displaying item types that will not overwhelm a drawing, such as centerlines. When this option is used for dimensions, a drawing can become cluttered.

Step 5: Select OK to confirm the use of Show All.

> After you select OK, the items will be displayed on the workscreen.

Step 6: Select Preview option (if required), then Close the dialog box.

SHOWING/ERASING LIMITED ITEM TYPES

While the Show All option is used to show all items of a selected type, options are available for you to show or erase limited numbers of an item type. Perform the following steps to show or erase a limited number of items:

Step 1: Select the Show/Erase icon on the toolbar.

Step 2: Select either the SHOW or the ERASE option on the Show/Erase dialog box.

> The Show option will show items of a selected type while the Erase option will hide items of a selected type. The Show and Erase options cannot be selected at the same time.

Step 3: Select an item type to show (Figure 8–10).

> Examples of item types to show include dimensions, axes, datum planes, and geometric tolerances.

Step 4:   **Select an option for showing items.**

You can select either Feature, Part, View, Feat & View, or Part & View. Feature will show items on a selected feature, the Part option will show items on a selected part, and View will show items in a selected view.

Step 5:   **Select OK on Select menu or select the middle mouse button.**

Pop-up Menu

Drawing mode provides a pop-up menu for the modification of drawing mode entities to include dimension text and dimension properties. Using the pop-up menu, an item can be selected for modification during the drawing process. To modify an item with the pop-up menu, select the item with the left mouse button, then with the right mouse button reselect the item (use the Control key for multiple selections). The reselection of the item with the right mouse button will reveal the pop-up menu.

Dimensions

Dimensions can be modified in multiple ways with the pop-up menu. The following dimension modification options are available:

- Dimensions can be switched to another view.
- Arrows can be flipped.
- Witness lines may be shown or erased (Properties option).
- Arrow styles may be changed (Properties option).
- Dimension nominal values may be changed.
- Dimensions can be erased.
- Dimension text style can be changed (Properties option).

Geometric Tolerances

Geometric tolerances created can be modified or manipulated by using the pop-up menu. The following options are available:

- Geometric tolerance attachments can be modified.
- Geometric tolerances can be switched to another view.
- The leader type of a geometric tolerance can be changed.
- Geometric tolerances can be edited.
- Geometric tolerances can be erased.
- Geometric tolerances can be deleted.
- The properties of a geometric tolerance can be modified.

Notes

Notes can be modified and/or manipulated in the following ways by using the pop-up menu:

- Notes may be switched to another view.
- Note text styles can be modified.

Views

Views can be modified with the pop-up menu in the following ways:

- Views can be deleted.
- Cross-section arrows can be added to a view.
- The scale of a view can be changed (Properties option).

- A view's cross section can be replaced.
- The view text can be modified.
- The view type can be changed (Properties option).
- The display of hidden lines can be changed (Properties option).
- The display of tangent edges can be changed (Properties option).

DIMENSIONING AND TOLERANCES

Pro/ENGINEER's Drawing mode has the capability to display dimensions created in Part or Assembly mode and the capability to create dimensions within Drawing mode itself. Dimensions created in Part mode can be displayed by using the Show/Erase option. These dimensions are associative and can be modified. Because of the associativity between modes, any parametric dimension modified in Drawing mode is also modified in other modes, such as Part and Assembly. While parametric dimensions can be hidden with the Show/Erase option, they cannot be deleted.

Driven Dimensions can be created within Drawing mode by using the Insert >> Dimension option. Dimensions created in this manner are not associative and are not modifiable. The model geometry drives the value of each dimension. This dimension option provides the following suboptions:

- **On Entity** Attaches the dimension's witness line to the point where the entity was picked.

- **Midpoint** Attaches the dimension's witness line to the midpoint of a selected entity.

- **Intersect** Attaches the dimension's witness line to the closest intersection point of two entities.

- **Center** Attaches the dimension's witness line to the center of a circle, arc, or ellipse.

- **Make Line\Horizontal Line** Creates a horizontal reference line through a vertex that can be used as a reference for vertical dimensions.

- **Make Line\Vertical Line** Creates a vertical reference line through a vertex that can be used as a reference for horizontal dimensions.

- **Make Line\2 Points** Creates a dimension that measures the distance between two points.

MANIPULATING DIMENSIONS

Drawing mode provides a variety of options for manipulating dimensions. These tools are used to create a clear and readable engineering drawing.

MOVING DIMENSIONS

Drawing mode provides two tools for moving dimension entities. For the manipulation of dimension text placement, the Move Special dialog box is available (Edit >> Move Special). For quicker dimension manipulation, dimension text and dimension line placement can be dragged with the mouse. When a dimension is selected, grip points are provided that allow an entire dimension to be repositioned to include the dimension's text, dimension line, and witness lines. Additional grip points are provided that allow for the moving of only the dimensions number value.

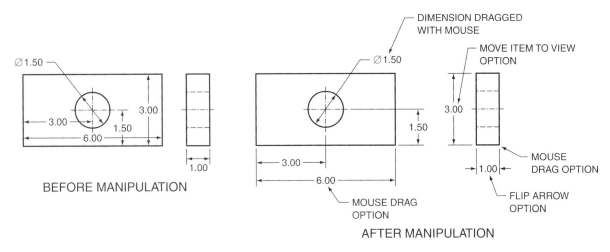

Figure 8–11   Manipulation Tools

MOVE ITEM TO VIEW

The Move Item to View option is used to switch the view in which a dimension is located (Figure 8–11). It is available under the Edit menu or the mouse pop-up menu. You must preselect each dimension before using this option.

FLIP ARROW

The Flip Arrow option is used to change the direction that an arrowhead points. This option can be used with linear and radial dimensions (Figure 8–11). The Flip Arrow option is accessible through the pop-up menu by preselecting the dimension text with the left mouse button, followed by reselecting the dimension text with the right mouse button.

WITNESS LINE JOG

Dimensions can become too confined when you are annotating small features. On a normal dimension, the distance between two witness lines is the same as the nominal size of the dimension. The Jog command, available by selecting Insert >> Jog on the menu, can be used to create a jog in a witness line (Figure 8–12).

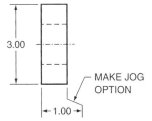

Figure 8–12
Make Jog Option

DIMENSION TOLERANCES AND MODIFICATION

Pro/ENGINEER allows for the creation of dimensions that adhere to ANSI or ISO tolerance standards. See Chapter 2 for more information on setting a tolerance standard. Within Drawing mode, tolerances may be displayed in a variety of formats. Available formats include Limits, PlusMinus, and PlusMinusSymmetric. In Drawing mode, before a dimension can be displayed as a tolerance, the Drawing Setup file option *Tol_Display* has to be set to Yes. Use the drawing Options dialog box (File >> Properties) and set the *Tol_Display* option to modify the tolerance display for an individual drawing.

When ANSI is the tolerance standard, tolerance values and formats can be set with the Dimension Properties dialog box (Figure 8–13). This dialog box is accessible through the mouse pop-up menu's Properties option. Some of the dimension properties that can be set include:

- **Tolerance Mode** Nominal, PlusMinus, and PlusMinusSymmetric may be selected.
- **Tolerance Values** Tolerance values associated with a particular format may be entered.

- **Decimal Places** The number of decimal places of a dimension may be selected.
- **Dimension Display** Dimensions can be set as Basic, Inspection, or Neither.
- **Witness Line Display** Dimension witness lines can be erased or displayed.
- **Dimension Text** The Dimension Text tab (Figure 8–13) is used to add text and symbols around a dimension value. When you are adding text, the dimension value cannot be deleted. The symbol pallet (Figure 8–14) is available for the selection of symbols. Figure 8–15 shows an example of how a typical dimension text note can be changed into a counterbored hole note.
- **Text Style** The style of dimension text can be modified.

> **MODELING POINT**   Notice in Figure 8–15 the 1.50 diameter hole as displayed in the drawing before modification. This value is a parametric dimension used to define the size of the hole. This dimension can be modified, which after regeneration will modify the part's hole diameter. By entering the hole's assigned dimension symbol in the Dim Text tab (e.g., &d6) instead of its dimension value, you can use the "after modification" hole note to modify its corresponding parametric dimension.

GEOMETRIC TOLERANCES

Geometric tolerances are used to control geometric form, orientation, and location. An example of a geometric tolerance would be specifying that a planar surface is flat to within a tolerance value. The Geometric Tolerance dialog box is used to create a geometric tolerance characteristic (Figure 8–16). This dialog box can be accessed through the Insert >> Geometric Tolerance menu option. See Chapter 2 for more information on establishing a geometric tolerance. Before a datum can be utilized in a drawing, it has to be first set through the Edit >> Properties option on Pro/ENGINEER's menu bar.

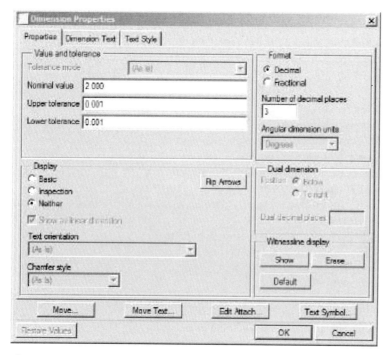

Figure 8–13   Dimension Properties Dialog Box

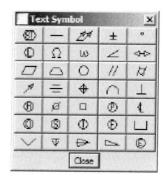

Figure 8–14   Symbol Pallet

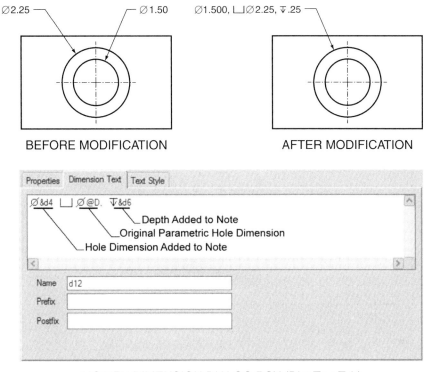

Figure 8–15   Dimension Text Modification

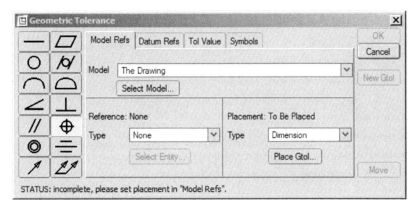

Figure 8–16   Geometric Tolerance Dialog Box

CREATING NOTES

Drawing mode allows for the creation of notes. Notes may be stand-alone or they can be attached to a leader. Additionally, notes can be entered from the keyboard or they can be input from a text file. Pro/ENGINEER provides the symbol palette (Figure 8–14) for adding symbols to a note.

NOTE WITHOUT LEADER

Perform the following steps to create a note without a leader by entering the note through the keyboard:

STEP 1:   Select **INSERT >> NOTE** on the menu bar.

STEP 2: Select NO LEADER from the Note Types menu.

STEP 3: Select a format/placement type.

Available Format/Placement types include Horizontal, Vertical, Angled, Left-Justify, Right-Justify, Center-Justify, and Related to Dimension Text.

STEP 4: Select MAKE NOTE.

STEP 5: On the workscreen, select the location for the note.

Using the mouse cursor, select the location for the text. After creating the text, you can reposition it with the Move option.

STEP 6: Enter note in Pro/ENGINEER's textbox.

When creating a note through the keyboard, Pro/ENGINEER will provide a textbox for the creation of the note. Select Enter on the keyboard to end the creation of the note. For the creation of symbols within the note, Pro/ENGINEER provides a symbol palette.

NOTE WITH A STANDARD LEADER

Perform the following steps to create a note with a standard leader by entering the note through the keyboard:

STEP 1: Select INSERT >> NOTE on the menu bar.

STEP 2: Select LEADER from the Note Types menu.

STEP 3: Select a leader type.

Pro/ENGINEER provides the option of creating either a standard leader, a normal leader, or a tangent leader. The Normal Leader (Normal Ldr) option will create the leader perpendicular to the selected entity while Tangent Leader (Tangent Ldr) will create the leader tangent to the selected arc, circle, or ellipse. Standard is the default and allows for multiple leader attachment points.

STEP 4: Select a format/placement type.

Available format/placement types include Horizontal, Vertical, Angled, Left-Justify, Right-Justify, Center-Justify, and Related to Dimension Text.

STEP 5: Select the MAKE NOTE option.

STEP 6: Select an attachment type (for standard leaders only), then select an appropriate attachment point.

When you are placing a standard leader, Pro/ENGINEER provides the following leader attachment types:

- **On Entity** Attaches a leader to a model or to a draft entity.
- **On Surface** Attaches a leader to a surface. The surface may be a model surface, datum plane, or cosmetic thread.
- **Free Point** Attaches a leader anywhere on the drawing. With the Free Point option, the leader does not have to be attached to an entity or surface.
- **Midpoint** Attaches a leader at the midpoint of a model edge or draft entity.
- **Intersect** Attaches a leader at the intersection of two entities.

STEP 7: Select OK on the Select menu to end attachment point locations.

STEP 8: Enter note in Pro/ENGINEER's textbox.

When you are creating a note through the keyboard, Pro/ENGINEER will provide a textbox for the creation of the note. Select Enter on the keyboard to end the creation of the note. For the creation input of symbols, Pro/ENGINEER provides the symbol palette.

CREATING DRAWING TABLES

A drawing table is similar to a table that might be created in a popular word processing application (Figure 8–17). A drawing table is composed of cells arranged in columns and rows. Standard text can be entered into a text cell. Perform the following steps to create a table:

STEP 1: Select the Insert Table icon on the toolbar.

STEP 2: Select a table direction creation method.

The following creation methods are available:

- **Descending** Creates the table from the top down.
- **Ascending** Creates the table from the bottom up.
- **Rightward** Creates the table from the left toward the right.
- **Leftward** Creates the table from the right toward the left.

Select either Descending or Ascending then select either Rightward or Leftward.

STEP 3: Select BY NUM CHARS as the cell creation method.

The following cell creation methods are available:

- **By Num Chars** You create a table by graphically picking the number of characters to include in each cell (Figure 8–18).
- **By Length** You create a table by specifying the size of each in drawing units.

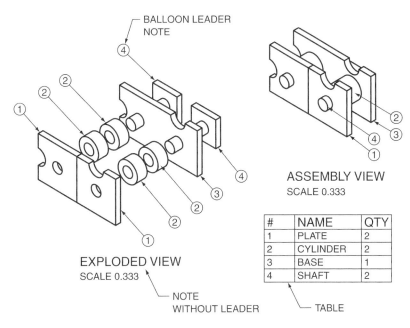

Figure 8-17 Notes, Leaders, and Tables

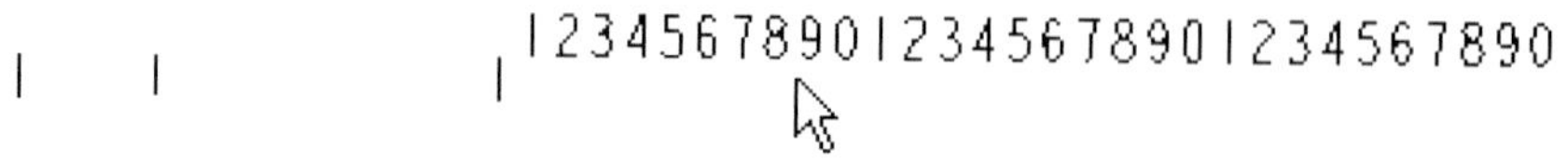

123456789012345678901234567890

Figure 8–18   *Creating a Table by the Number of Characters*

STEP 4: **On the workscreen, select the location for the cell.**

This option will locate the table according to the table direction creation method selected previously. As an example, when creating a table with the Descending and Leftward options, you will select the upper-right corner of where the table is to be located. When creating a table with the Ascending and Rightward options, you will select the lower-left corner of where the table is to be located.

STEP 5: **On the workscreen, mark off the number of characters to include in each column of the table (Figure 8–18).**

STEP 6: **Select DONE to finish the number of columns.**

STEP 7: **On the workscreen, mark off the number of characters to include in each row of the table.**

MODELING POINT   Tables can be modified with the Table menu on the menu bar or with the mouse pop-up menu. Options are available for inserting rows and columns, for merging cells, and for modifying text justification. Other useful manipulation tools include rotating a table and resizing rows and columns.

STEP 8: **Select DONE to finish the number of rows.**

Selecting DONE will create the table.

TWO-DIMENSIONAL DRAFTING

While Pro/ENGINEER is considered primarily a three-dimensional design application, with Drawing mode being used to create annotated detailed drawings of 3D models, Drawing mode can be used to create strictly 2D drawings. Any two-dimensional geometry created in Drawing mode is nonparametric and cannot be associated with any other Pro/ENGINEER mode.

Pro/ENGINEER's drafting capabilities are similar to those of two-dimensional modeling tools found in many popular midrange computer-aided drafting applications. The following options are available:

- **Draft Geometry** Used to create lines, circles, arcs, splines, ellipses, points, and chamfers.

- **Construction Geometry** Lines and circles are used to create draft geometry. They are displayed in phantom font.

- **Draft Dimensions** Used to dimension two-dimensional geometry.

- **Draft Cross Sections** Used to create section lines.

DRAFT GEOMETRY

Pro/ENGINEER provides a variety of tools for creating two-dimensional draft geometry. By default, Drawing mode allows for the creation of individual entities only. As an example, when sketching a line between two points, only one line will be created between the two points. To continue the line creation process, the start and end of the next line will have to be selected. To connect entities during the sketching process, Drawing mode provides the

Chain option. On the Drawing toolbar, selecting the Chain icon will allow drafted entities to be connected. With this option, the end of one entity becomes the start of the next.

The pop-up menu (right mouse button) provides several options for creating precise draft entities. As an example, for lines, polar or relative coordinates can be entered. For a circle entity, the pop-up menu provides an option for entering the circle's radius. The pop-up menu provides the Select References option for constraining an entity to a selected existing entity. One example of this would be constraining a new line entity perpendicular to an existing line. Another example would be constraining a new circle tangent to an existing line. An existing entity has to be specified as a reference before a new entity can be constrained to it.

The following options are available for the creation of draft entities. Drafting in Drawing mode is similar to sketching in a sketcher environment. Refer to Chapter 3 for specific information on creating sketch entities. Each of the following options is available on either the Drawing toolbar or the Sketch menu.

 ### LINES

Lines can be created through the Line icon or through the Sketch Line option. For a chain of lines, the Chain icon or the Sketch >> Chain option is available. The pop-up menu provided by the right mouse button provides options for creating a line at a specific angle and with either polar coordinates or relative coordinates.

 ### CIRCLES

Circles can be created through the Circle icon or the Sketch >> Circle option. The pop-up menu provides an option for specifying a precise radius for the circle.

 ### ARCS

Arcs can be created through the use of one of two arc icons or through the Sketch >> Arc suboptions. Two types of arcs are available: a three point arc and a center/endpoints arc.

 ### SPLINES

The Spline option is available on the drawing toolbar or the Sketch menu. To create a spline, locate the spline's start point first, intermediate points along the spline second, then the spline's endpoint last. The pop-up menu provides an option for specifying a spline's initial start angle.

 ### ELLIPSES

Pro/ENGINEER provides two options for creating ellipses. The first option creates a spline through the selection of the endpoints of the spline's axes. The second option creates a spline through the definition of the spline's center vertex, followed by the spline's major and minor axes definitions.

 ### POINTS

The Point option is used to create a point on the workscreen.

 ### CHAMFERS

The Chamfer option is used to create a line that intersects and trims two nonparallel lines. Chamfer options available in Drawing mode are similar to options available under the Chamfer command in Part mode.

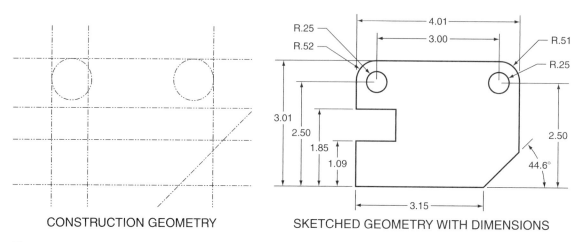

CONSTRUCTION GEOMETRY   SKETCHED GEOMETRY WITH DIMENSIONS

Figure 8-19   Construction Geometry

CONSTRUCTION GEOMETRY

Construction geometry is used within Pro/ENGINEER for the construction of two-dimensional draft geometry. The options are provided for creating construction lines and construction circles. Figure 8–19 shows an example of a two-dimensional drawing after the creation of construction geometry and after the creation of draft geometry. Construction geometry appears as phantom line font on the workscreen, but will not print.

LINE STYLES AND FONTS

The default line style for sketching draft geometry is a continuous line. Within Pro/ENGINEER, a continuous object line does not have a line style but has a line font set as *Solidfont*. There is a slight difference between line styles and line fonts. A line style is a defined group consisting of a line font, color, and weight. As an example, a Phantom line style has a Phantomfont line font with a color of blue. The line styles available for draft geometry include hidden, geometry, leader, cut plane, phantom, centerline, and none (Figure 8–20). Use the mouse pop-up menu's Line-Style option to set a default line style for a specific line type (e.g., geometry, phantom, hidden, leader).

How a line style appears on the workscreen may be different from how it appears when printed. As an example, a line defined as a hidden line style will appear on the workscreen as a blue continuous line; it will print as a dashed hidden line, though. This is because the color of a line, in the absence of a set line font, will dictate the printed line font and the printed line weight. Figure 8–21 shows the color dialog box with available colors that can

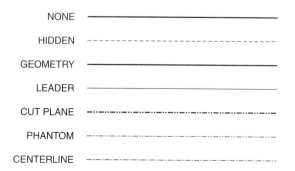

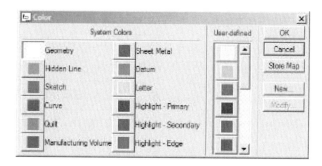

Figure 8-20   Pro/ENGINEER Line Styles   **Figure 8-21**   Color Dialog Box

be assigned to a line. Each color within the dialog box corresponds to a particular line font, line weight, and pen number. As an example, the Geometry color (pen 1) will print a thick, continuous line. The Letter color (pen 2) will print a continuous, medium thick line while the Hidden color (pen 3) will print a dashed, thin line.

A line font is the geometric definition of a line. The available fonts include Solidfont, Dotfont, Ctrlfont, Phantomfont, Dashfont, Ctrlfont_S_L, Ctrlfont_L_L, Ctrlfont_S_S, Dashfont_S_S, Phantomfont_S_S, and Ctrlfont_MID_L.

MODELING POINT   The weight of a printed line is governed by its pen number. Correspondingly, the pen number of a line is governed by its color. As an example, the Hidden color (blue) is assigned to pen 1.

The line weight produced from a pen number can be modified with the configuration file option pen#_line_weight (where # is equal to the pen number). Each option can be set equal to a value ranging from 1–16. Each increment value equals a value of 0.005 inches. As an example, if pen1_line_weight is set to 2, the line weight will be equal to 0.010 in (2 x 0.005 = 0.010).

DRAFT DIMENSIONS

Draft dimensions (Insert >> Dimension) are used to annotate draft geometry created in drawing mode. The New References dimension option is similar to the standard dimensioning option found in a sketcher environment. See Chapter 3 for more information on creating draft dimensions.

DRAFT CROSS SECTIONS

Drawing mode's Edit >> Fill >> Hatched option is used to create section lining or hatch within an enclosed area. The Fill >> Solid can also be used to completely fill an area. Figure 8–22 shows an example of several types of cross sections that can be created. As shown in the first example, cross sections can include islands of unfilled areas.

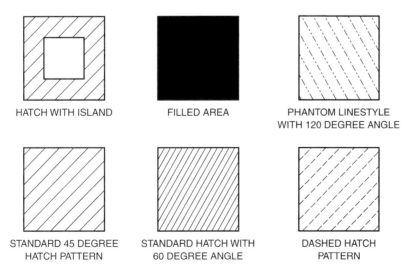

| HATCH WITH ISLAND | FILLED AREA | PHANTOM LINESTYLE WITH 120 DEGREE ANGLE |

| STANDARD 45 DEGREE HATCH PATTERN | STANDARD HATCH WITH 60 DEGREE ANGLE | DASHED HATCH PATTERN |

Figure 8–22   Hatch Examples

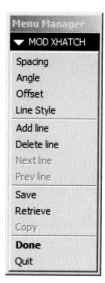

Figure 8-23
Modify Cross Hatch Menu

Draft cross sections are created with either the Hatched or Solid option by the pre-selecting of drafting entities. An area to incorporate section lining must be completely closed. The Solid option will fill the selected area with the color associated with the current line style. The Hatched option requires the entering of a cross-section name. When you complete the selection of a cross section's boundary, Drawing mode displays the Modify Cross Hatch menu (Figure 8–23). This menu allows for the modification of hatch spacing, angle, offset, and line style. This menu can also be accessed through the Edit >> Properties option.

MANIPULATING DRAFT GEOMETRY

Most new users of two-dimensional computer-aided drafting applications use draft geometry techniques such as the line, arc, and circle options to create drawings. While it is obvious that these are central commands for the creation of draft geometry, other powerful options are available that allow for the manipulation of existing draft entities. These options can be used in combination with basic geometry creation techniques to construct a drawing. Within Drawing mode, many of these tools are identical or similar to options available within a typical Pro/ENGINEER sketching environment. Within drawing mode, the following manipulation options are available:

COPY

The Copy option can be found under the Edit menu. It is used to create identical instances of selected entities (Figure 8–24). It functions in a similar manner to a Windows application's Copy command. Items are copied to Pro/ENGINEER's clip board and can be later pasted back into the drawing.

GROUP

The Group option can be found under the Edit menu. It is used to combine entities. One advantage of a group is that entities in a group can be manipulated together. As an example, the entities in a group can be moved all at once, or all entities can be deleted together. Use the Group >> Create command to build a group and use the Group >> Explode command to break a group into its individual entities.

ORIGINAL ENTITIES   TRANSLATED ENTITIES

Figure 8–24   Horizontal Translation

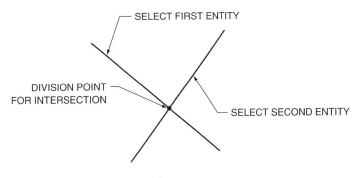

Figure 8–25   Intersection Tool

DIVIDE AT INTERSECTION

The Divide at Intersection option can be found under the Edit >> Trim menu. It is used to break entities at their selected intersection point. As shown in Figure 8–25, selecting the intersection of two lines will create four entities. This option is identical to the Intersect option found within Part mode's sketching environment.

 ### MIRROR

The Mirror option can be found on the sketch toolbar. It is used to create an identical reflected image of selected entities (Figure 8–26). The user is required to select the entities to be mirrored and to select a draft line to mirror the draft geometry about.

 ### OFFSET

The Offset option can be found on the sketcher toolbar. It is used to create new draft geometry by offsetting from a selected entity at a user-defined distance (Figure 8–27). Single entities or chained entities may be offset. The Single Entity (Single Ent) option will offset only one entity at a time. Suboptions are available for tapering the offset and trimming the offset. The Entity Chain (Ent Chain) option allows a connected chain of entities to be offset at once.

ROTATE

The Rotate option can be found under the Edit >> Transform menu. It is used to rotate one or more draft entities.

TRANSLATE

The Translate option can be found under the Edit >> Transform menu. Selected items are moved a user-defined distance. A Translate and Copy option is available for creating a copy of a selected entity.

CORNER

The Corner option can be found under the Edit >> Trim menu. It is used to trim two entities back forming a corner.

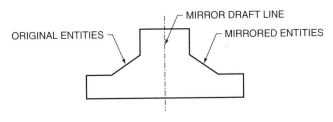

Figure 8–26   Mirrored Entities

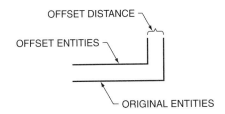

Figure 8–27   The Offset Tool

SUMMARY

Drawing mode is used to create drawings of Pro/ENGINEER parts or assemblies. Multiple views of a model can be displayed to include projection views, section views, and partial views. Dimensions and other items such as cosmetic threads and geometric tolerance notes can be displayed in a drawing by using the Show/Erase option. Drawing mode has the capability for the creation of nonassociative two-dimensional drawings. Notes, leaders, and nonparametric dimensions can be added to a drawing.

DRAWING TUTORIAL 1

This tutorial will demonstrate the creation of the drawing shown in Figure 8–28. The part used in this tutorial is shown in Figure 8–29. The start of this tutorial will require you to model this part. As shown in Figure 8–28, four different view types will be created: general, projection, auxiliary, and detailed. This tutorial will cover:

• Starting a drawing.

• Adding a drawing format.

• Creating a general view.

• Creating projection views.

• Creating a detail view.

• Creating notes.

• Modifying the drawing setup file.

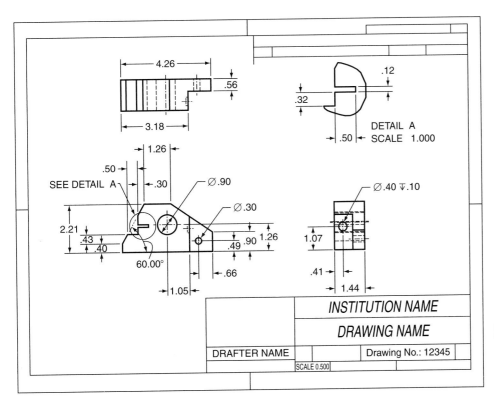

Figure 8–28 Multiview Drawing

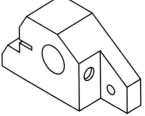

Figure 8–29
Model for Drawing

CREATING THE PART

Model the part shown in Figures 8–28 and 8–29, naming it *view1*. The dimensioning scheme shown in Figure 8–28 matches the design intent for the part. When modeling this part, make sure that these dimensions are incorporated into your design.

STARTING A DRAWING

This section of the tutorial will create the object file for the drawing to be completed. Before starting this tutorial, ensure that you have completed the part model shown in Figure 8–29.

STEP 1:   Start Pro/ENGINEER.

If Pro/ENGINEER is not open, start the application.

STEP 2:   Set an appropriate working directory.

STEP 3:   Select FILE >> NEW.

STEP 4:   In the New dialog box, select DRAWING mode then enter *view1* as the name of the drawing file (Figure 8–30).

STEP 5:   Select OK on the New dialog box.

After you select OK on the dialog box, Pro/ENGINEER will reveal the New Drawing dialog box. This dialog box is used to select a model, a sheet size, and a format.

STEP 6:   Select the BROWSE option on the New Drawing dialog box and locate the *view1* part (Figure 8–31).

Use the Browse option to locate the *view1* part created in the first section of this tutorial. The part will serve as the default model for the creation of drawing views. If *view1* is the active part during the creation of this drawing file, it will be displayed by default in the model selection box.

STEP 7:   Select the EMPTY option (Figure 8–31) under the Specify Template option.

The Specify Sheet option allows for the selecting of a standard or user-defined sheet size. The Retrieve Format option allows for the retrieval of a predefined sheet format. When you retrieve a format, the size of the sheet defining the format will define the sheet size for the new drawing.

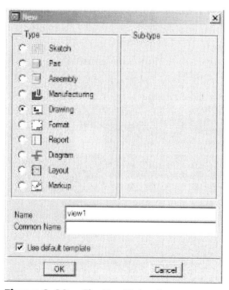

Figure 8–30   The New Dialog Box

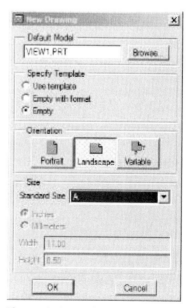

Figure 8–31   The New Drawing
Dialog Box

Step 8: **Select LANDSCAPE as the Orientation option.**

Step 9: **Select *A* as the standard sheet size (Figure 8–31).**

Pro/ENGINEER provides a variety of standard sheet sizes (e.g., A, B, C, A1, A2). A unique sheet size can be entered in either inch or millimeter units.

Step 10: **Select OK on the New Drawing dialog box.**

After you select OK, Pro/ENGINEER will launch its Drawing mode.

ADDING A DRAWING FORMAT

This section of the tutorial will provide instruction on how to add a format to a drawing sheet. Formats can be retrieved from a variety of sources. Pro/ENGINEER provides sheet formats in the *format* subdirectory of the Pro/ENGINEER load point directory. Formats can also be imported through IGES and DXF. The configuration file option *pro_format_dir* is used to specify a default format directory.

Step 1: **Select FILE >> PAGE SETUP on the menu bar.**

The Page Setup dialog box is used to set the size and format of specific sheets. Within this tutorial, you will change the format size from A size to an existing A format (a.frm).

Step 2: **On the Page Setup dialog box, select the existing A Size format option and pick Browse on the drop-down menu (Figure 8–32).**

Step 3: **On the Open dialog box, select and open the a.frm file.**

A format added to a drawing is associated with its original format file. Any changes made to the format through Format mode will be reflected in all drawings using that specific format.

Step 4: **Select OK to exit Page Setup dialog box.**

CREATING THE GENERAL VIEW

General views serve as the parent view for all views projected off of it. This section of the tutorial will create a general view as the front view of the drawing (Figure 8–33).

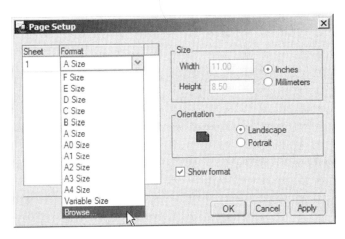

Figure 8–32   Opening an A Size Format

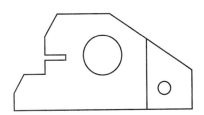

Figure 8–33   The Front View of the Part

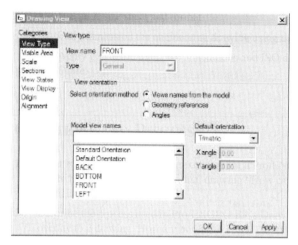

Figure 8–34   Drawing View Dialog Box

STEP 1:   Select the Create a General View icon on the drawing toolbar.

STEP 2:   On the workscreen, select the location for the drawing view.

The drawing under creation in this tutorial will consist of a front, right side, top, and detailed view. The general view currently being defined will serve as the front view. On the workscreen, pick approximately where the front view will be located. Later, the mouse can be used to reposition each view's position.

STEP 3:   On the Drawing View dialog box, enter FRONT as the view's name (Figure 8–34).

Notice that the dialog box has a view type of General. Since no views currently exist within the drawing, General is the only view type available.

STEP 4:   Under the View Type category, select the Geometry References orientation option. (Figure 8–35).

The Geometric References option will allow you to define the orientation of the model within the drawing sheet. You will accomplish this through the selection of orientation references.

STEP 5:   On the Orientation dialog box, select FRONT as the Reference 1 option.

A planar surface selected with the Front reference option will orient the surface toward the front of the workscreen. Other available reference options include Back, Top, Bottom, Left, Right, Vertical Axis, and Horizontal Axis.

STEP 6:   Pick the front of the model (Figure 8–35).

STEP 7:   On the Orientation Dialog box, select TOP as the Reference 2 option.

MODELING POINT   When you are selecting references to orient a model, it is often helpful to turn off all datum planes and to display the model as No Hidden. This technique provides clarity when you are selecting a reference surface.

A planar surface selected with the Top reference option will orient the surface toward the top of the workscreen.

STEP 8:   Select the top of the model (Figure 8–35).

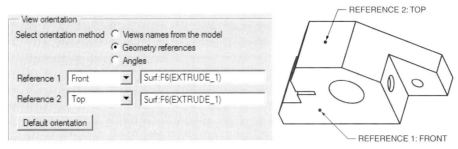

Figure 8-35   Orienting the Model

STEP   9:   Under the dialog box's Scale category, enter 0.500 as the custom scale value for the view.

A scale value of 0.500 will create a view at half scale.

STEP 10:   Explore other category options on the dialog box but don't make any additional changes.

STEP 11:   Select OK on the Drawing View dialog box.

Your front view should appear as shown in Figure 8–33.

INSTRUCTIONAL POINT   Within Part and Assembly modes, rotating is the most commonly used dynamic viewing capability, so the middle mouse button defaults to this option. Within Drawing mode, drawing sheets cannot be rotated, so panning becomes the more common dynamic viewing choice. Within Drawing mode, the middle mouse button is used to pan, while the Control key and middle mouse button combination is used to dynamically zoom.

CREATING PROJECTION VIEWS

Once a general view has been created, views can be projected off of it. Options exist for creating projection, auxiliary, and/or section views. This segment of the tutorial will create a right-side view and a top view, both projected from the front view (Figure 8–36). The front view will be the parent view of these projected views.

STEP   1:   Select INSERT >> DRAWING VIEW >> PROJECTION on the menu bar.

The Projection option will project a view from an existing view. In this step of the tutorial, the right-side view will be projected from the front view.

STEP   2:   On the workscreen, pick the location for the Right-Side view.

Pro/ENGINEER will create a projected view based on its parent view. Your view should appear as shown in Figure 8–37. If necessary, you can use your mouse to reposition this view.

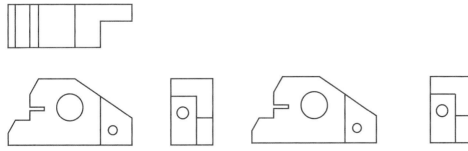

Figure 8-36   Front, Top, and Right-Side Views   **Figure 8-37**   Front and Right-Side Views

STEP 3: **Select INSERT >> DRAWING VIEW >> PROJECTION on the menu bar.**

Next, you will add the Top view.

STEP 4: **Pick the front view as the parent view for your new view.**

STEP 5: **On the workscreen, select the location for the top view.**

Your views should appear as shown in Figure 8–36.

STEP 6: Ensure that the Lock View Movement icon is not selected.

When selected, this option disallows the movement of views with the mouse. This option is also available with the mouse's pop-up menu.

STEP 7: **With the mouse, select and drag each view to better locations on the workscreen.**

When a general view is moved, all views projected from it will be repositioned to keep normal lines of projection. When moving a projected view, it will remain aligned with its parent view.

STEP 8: **Save your drawing.**

CREATING A DETAILED VIEW

Pro/ENGINEER's Drawing mode has an option for creating a detailed view. Detailed views are used to highlight and expand an area of an object that requires special attention. An example would be a small, complex feature that would be hard to dimension on a normal projection view. This segment of the tutorial will create the detailed view shown in Figure 8–38.

STEP 1: **Select INSERT >> DRAWING VIEW >> DETAILED on the menu bar.**

STEP 2: **On the front view, pick an entity centered within the detail area (Figure 8–39).**

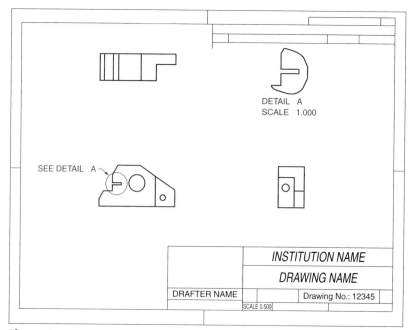

Figure 8–38   Detailed View of Part

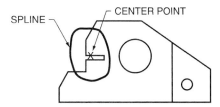

Figure 8–39   Sketching the Spline

Pro/ENGINEER requires the selection of an entity in an existing view. This selection point is used to calculate the regeneration of the detailed view.

STEP 3: **On the workscreen, select the location for the detailed view (Figure 8–38).**

Using the mouse, select on the workscreen where the detailed view will be located.

STEP 4: **As shown in Figure 8–39, sketch a spline around the area to detail.**

Using the mouse, sketch a spline that will include the area to be detailed. Use the left mouse button to select spline points and the middle mouse button to close the spline.

STEP 5: **Select a location on the workscreen for the detail note.**

As shown in Figure 8–38, this will locate the note SEE DETAIL A. This note can be moved later if necessary.

Note: The Drawing View dialog box (Edit >> Properties) can be used to change the scale and name of a detailed view.

STEP 6: **Reposition the detailed view by dragging with the mouse.**

If necessary, reposition the detailed view or any other view. The Lock View Movement option has to be unselected.

MODELING POINT   Don't forget that you can use dynamic zooming (Control key and middle mouse button) and dynamic panning (middle mouse button) when performing these operations.

STEP 7: **If necessary, select the detail view's note and drag it to an appropriate location.**

Views, notes, dimensions, and other annotations can be moved by selecting the entity with the left mouse button. Your drawing should appear as shown in Figure 8–38.

STEP 8: **Save your drawing.**

ESTABLISHING DRAWING SETUP VALUES

Pro/ENGINEER's drawing setup file is used to set option values associated with a drawing. Examples of options that can be set include text height, arrowhead size, and geometric datum plane symbol. Multiple drawing setup files can be created. This segment of the tutorial will modify the values of the current drawing setup file.

STEP 1: **Select FILE >> PROPERTIES on the menu bar.**

STEP 2: **Select DRAWING OPTIONS on the File Properties menu.**

After you select Drawing Options, Pro/ENGINEER will open the active drawing settings within an Options dialog box. This dialog box is similar to the Preferences dialog box used to make changes to the configuration file option.

STEP 3:   **Change the text and arrowhead values for the current drawing.**

Within the Options dialog box, make changes to the options shown in the following table. To change an option's value, perform the following steps:

1. Type the option's name in the Option field.
2. Type or select the option's value in the Value field.
3. Select the Add/Change icon.

Drawing Setup Option	New Value
drawing_text_height	0.125
text_width_factor	0.750
dim_leader_length	0.175
dim_text_gap	0.125
draw_arrow_length	0.125
draw_arrow_style	FILLED
draw_arrow_width	0.0416

STEP 4:   **Apply the modified values for the active drawing setup file, then Exit the text editor.**

CREATING DIMENSIONS

Pro/ENGINEER provides options for creating two types of dimensions in Drawing mode: parametric and nonparametric. The Show/Erase dialog box can be used to show parametric dimensions that were created in Part and Assembly modes. During part modeling, a sketched feature has to be fully defined by utilizing dimensions, constraints, and references. The dimensions defining a feature can be revealed in Drawing mode through the Show/Erase option. The second option, Insert >> Dimension, can be used to create nonparametric dimensions.

This segment of the tutorial will create the dimension annotations as shown in Figure 8–40.

STEP 1:   Select the Show/Erase option on the toolbar.

The Show/Erase option is used to show and not show items that were defined in Part and/or Assembly modes. Figure 8–41 shows the Show/Erase dialog box. Items that can be shown include parametric dimensions, reference dimensions, geometric tolerances, notes, balloon notes, axes (centerlines), symbols, surface finish, datum planes, and cosmetic features.

STEP 2:   **On the Show/Erase dialog box, select the Dimension item type (Figure 8–42).**

The dimension item type option will show parametric dimensions that were created in Part and Assembly modes.

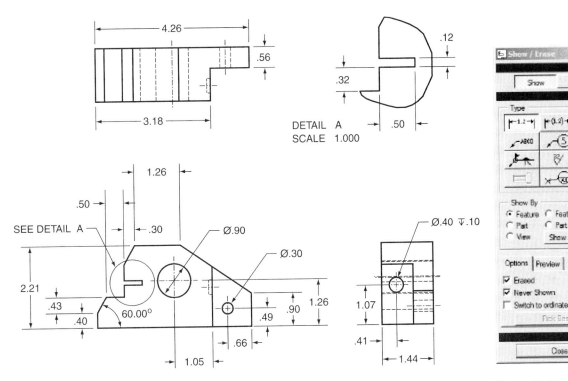

Figure 8-40 Dimensioned Drawing

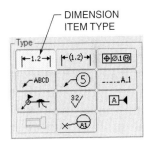

Figure 8-41
The Show/Erase Dialog Box

STEP 3: On the Show/Erase dialog box, under the Options tab, check ERASED and NEVER SHOWN options (Figure 8–41).

The Erased option will show items that have previously been erased and the Never Shown option will show items that have never been shown.

STEP 4: On the Show/Erase dialog box, under the Preview tab, check the WITH PREVIEW option (Figure 8–43).

STEP 5: Select SHOW ALL (Figure 8–41) on the Show/Erase dialog box and confirm the selection.

The Show All option will show all available item types. In this step of the tutorial, all dimensions available from the referenced model will be shown. As revealed in Figure 8–44, this can create a confusing and cluttered drawing. You will use the Move, Switch View, Flip Arrows, and Erase options to clean the drawing.

STEP 6: Select ACCEPT ALL under the Preview option (Figure 8–43).

The previously selected With Preview option allows for the previewing of shown item types. Use the Accept All option to accept the shown dimensions.

STEP 7: CLOSE the Show/Erase dialog box.

DIMENSION
ITEM TYPE

Figure 8-42
The Dimension Item Type

Figure 8-43
With Preview Option

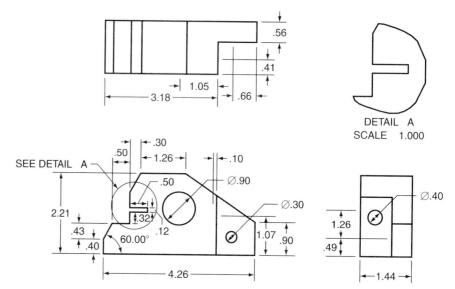

Figure 8-44   Dimensions Shown on a Drawing

Step   8:   **Use the mouse to drag dimensions and the mouse's pop-up menu's MOVE ITEM TO VIEW and FLIP ARROWS options to reposition the dimensions to match Figure 8–45.**

The mouse pop-up menu is available by first selecting the text of a dimension with your left mouse button, followed by reselecting the same dimension text with your right mouse button. (*Note*: A slight delay in releasing the right-mouse button is necessary.) A variety of modification and manipulation tools are available through the revealed pop-up menu.

Use the following options to reposition your dimensions on the workscreen:

- **Mouse Move and Drag**   Once selected, dimensions, including text and witness lines, can be moved and dragged on the screen with your mouse.

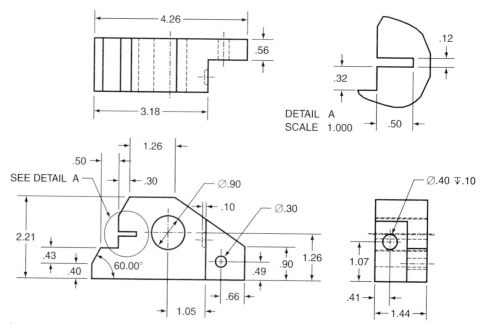

Figure 8-45   Repositioned Dimensions

- **Move Item to View** The Move Item to View option is used to switch a dimension from one view to another. It is available under the pop-up menu. Multiple dimensions can be selected by a combination of the Control key and the left mouse button. Once dimensions are preselected, the pop-up menu is available through the right mouse button.

- **Flip Arrows** The Flip Arrows option is used to flip dimension arrowheads. It is available under the pop-up menu.

STEP 9:   **Use the SHOW/ERASE option to show all centerlines.**

The Axis option (Figure 8–46) on the Show/Erase dialog box is used to show centerlines. On the Show/Erase dialog box, select the Axis option then select Show All. Accept all available centerlines, then close the dialog box.

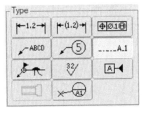

Figure 8-46
The Axis Item Type

STEP 10: **Use the left mouse button's move and drag capability to create witness line gaps.**

Where a dimension's witness line is linked to an entity (Figure 8–47), a visible gap (approximately 1/16 in wide) is required by drafting standards. After selecting the dimension's text with the left mouse button, drag the witness line's grip point to create the gap.

STEP 11: **Select INFO >> SWITCH DIMENSIONS on the menu bar.**

STEP 12: **Make a mental note or write down the dimension symbol for the hole depth shown in Figure 8–48.**

Make sure that you record the symbol for your hole depth. Your symbol value will probably be different from the figure's value. You will use this dimension symbol within a dimension note to be created in the next few steps.

STEP 13: **Select INFO >> SWITCH DIMENSIONS to display the numerical values of your dimensions.**

STEP 14: **Pick the .40-in diameter hole dimension, then select the pop-up menu's PROPERTIES option.**

You can also access the Properties dialog box by double picking the dimension on the work screen.

STEP 15: **Use the Dimension Text tab on the Dimension Properties dialog box to modify the hole's note (Figure 8–49).**

You will add the .10-in depth value to the .40-in diameter hole note by including the dimension's symbol in the hole note. Under the Dimension Text tab, use the Text Symbol option and the keyboard to add the hole depth attributes shown in the illustration.

STEP 16:   **Select the SHOW/ERASE icon on the drawing toolbar.**

The .10 dimension shown in Figure 8–50 is not needed in the drawing. The Show/Erase option can be used to remove it from the drawing.

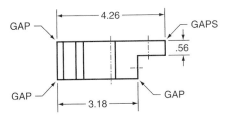

Figure 8-47 Witness Line Gaps

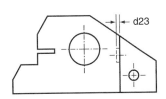

Figure 8-48
Dimension Symbol Display

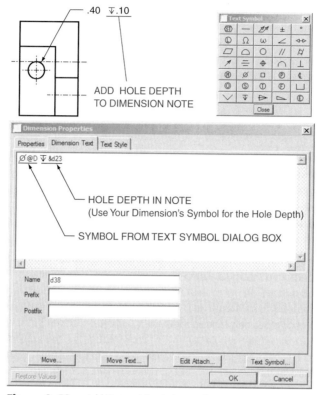

Figure 8-49   Add Text and Symbols to a Dimension Note

STEP 17:   On the Show/Erase dialog box, select the ERASE option then select the DIMENSION item type (Figure 8–51).

STEP 18:   On the workscreen, select the .10 dimension text then select the middle-mouse button.

The Erase option does not delete a selected item. Instead, it removes it from the drawing screen. Any erased item can be redisplayed with the Show option.

STEP 19:   Close the Show/Erase dialog box.

STEP 20:   Select the CLEAN DIMENSIONS icon on the Drawing toolbar.

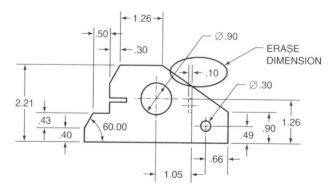

Figure 8–50   Erase Dimension

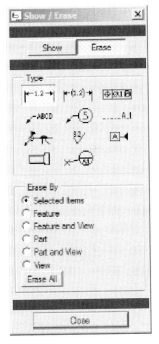

Figure 8–51
The Dimension Item Type

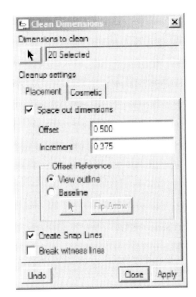

Figure 8–52 The Clean
Dimensions
Dialog Box

The Clean Dimensions option will consistently space dimensions on a drawing view. The minimum distance between a dimension line and the object is 3/8 in, while the minimum spacing between two dimension lines is 1/4 in. These values can be increased when appropriate. After you select the Clean Dims option, the Clean Dimensions dialog box will appear (Figure 8–52).

STEP 21: **Using your control key, select the front, top, right-side, and detailed views, then select OK on the Select menu.**

Pro/ENGINEER's default values for the Clean Dimensions dialog box are 0.500 for the object Offset setting and 0.375 for the Dimension Increment setting. You will keep these default settings.

STEP 22: **Select APPLY, then CLOSE the dialog box.**

After you apply the clean dimension settings, Pro/ENGINEER will create snap lines that will evenly space any dimension line. Depending on the complexity of the drawing, the applied settings might not create an ideal drawing. *Snap lines will not print when the drawing is plotted.*

STEP 23: **Use drawing modification and manipulation tools to refine the placement of dimensions.**

Use available tools such as mouse move and drag, move to view, and flip arrows to tweak the placement of dimensions (See Step 8). Your final dimensioning scheme should appear as shown in Figure 8–53.

STEP 24: **Save your drawing.**

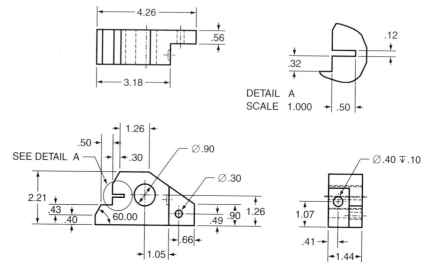

Figure 8–53   The Final Dimensioning Scheme

CREATING NOTES

Annotations can be added to a drawing by using the Insert >> Note option. Within the Note Types menu, options are available for creating notes with or without leader lines. Also, options are available for justifying text and for entering text from a file. This segment of the tutorial will create text for the title block. The finished title block will appear as shown in Figure 8–54.

STEP 1:   **Using the pop-up menu's TEXT STYLE option, modify the SCALE 0.500 note to have a text height of 0.100 in.**

For modifying the properties of a scale note, the Enter Text dialog box is utilized.

When a general view is added to a drawing, Pro/ENGINEER inserts a scale value. In this tutorial, the first general view has a scale value equal to 0.500. When selecting the SCALE 0.500 note, you have to select both the word SCALE and the number 0.500.

STEP 2:   **Move the SCALE 0.500 note to the title block (Figure 8–54).**

STEP 3:   **Select INSERT >> NOTE and create the additional title block shown in the figure.**

The Insert >> Note option is used to add text to a drawing. Use the following options to create each note:

• **No Leader** Creates a note without a leader line.

• **Enter** Allows for the creation of a note through the keyboard.

• **Horizontal** Creates notes horizontal on the work screen.

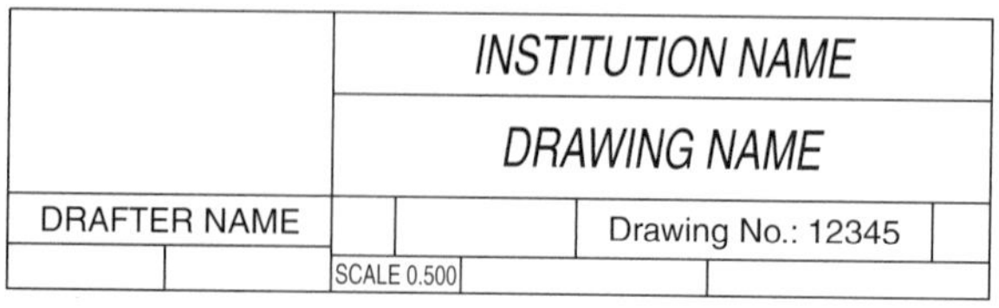

	INSTITUTION NAME		
	DRAWING NAME		
DRAFTER NAME		Drawing No.: 12345	
	SCALE 0.500		

Figure 8–54   Title Block Information

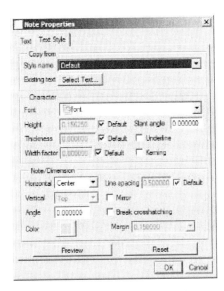

Figure 8–55   Modifying the Style of a Text

STEP 4:   Select CENTER on the Note Types menu.

The Center option will center-justify the note text. Other justification options available include Left, Right, and Default.

STEP 5:   Select MAKE NOTE to enter the text for the note.

STEP 6:   Within the workscreen, select the location for the *INSTITUTION NAME* note.

STEP 7:   In the textbox, enter your institution's name as the note text.

In Pro/ENGINEER's textbox, enter the text then select the Enter key on the keyboard.

STEP 8:   Select ENTER on the keyboard to end the Make Note option.

STEP 9:   Repeat the note-making steps to create the remaining notes.

STEP 10:   Select DONE/RETURN from the Note Types menu to exit the note creation menu.

STEP 11:   On the workscreen, using a combination of the Control key and the left mouse button, preselect the *INSTITUTION NAME* and the *DRAWING NAME* notes.

STEP 12:   Access the pop-up menu's Text Style option.

You will modify the text style and the text height of the Institution Name and Drawing Name notes.

STEP 13:   On the Text Style dialog box, select FILLED as the Font (Figure 8–55).

STEP 14:   On the Text Style dialog box, enter a text Height of 0.250.

STEP 15:   Select OK on the Text Style dialog box.

STEP 16:   Save your drawing.

SETTING DISPLAY MODES

The display of hidden lines on a drawing can be controlled with the display model options found on the toolbar (e.g., Wireframe, Hidden Line, and No Hidden). Additionally, the Tangent Edge display mode can be set with the Environment dialog box. The problem with using these two methods to set the display mode of a drawing is that a selected setting, such as hidden line display, will affect the entire drawing. Often, it is necessary to set a different display mode for a view or for an individual entity. The following steps will set the front, top, and right-side views with a Hidden Line display. Additionally, the detailed view will be set with a No Hidden display.

Step   1:   **Using the Control key, on the workscreen, select the front, top, and right-side views.**

Step   2:   **Using the pop-up menu, select the PROPERTIES option.**

Step   3:   **Observe how View Display is the only category available on the Drawing View Dialog box.**

Step   4:   **Select HIDDEN LINE as the Display style.**

The default display style is Follow Environment. Other options available include Wireframe, No Hidden, and Shading.

Step   5:   **Select NONE as the Tangent edges display style.**

Step   6:   On the drawing sketcher toolbar, pick the Select Item icon.

Step   7:   **On the workscreen, select the detail view.**

You will next set a No Hidden display mode for the detail view.

Step   8:   **Using the pop-up menu, select the PROPERTIES option.**

Step   9:   **Select VIEW DISPLAY as the category to modify.**

Step 10:   **Deselect the USE PARENT VIEW STYLE option.**

The Use Parent View Style option makes the Detail's display style dependent upon its parent's display style.

Step 11:   **Select NO HIDDEN as the display style.**

Step 12:   **Select OK on the dialog box.**

Step 13:   **SAVE your drawing and purge old versions of the drawing by using the FILE >> DELETE >> OLD VERSIONS option.**

DRAWING TUTORIAL 2

This tutorial will create the drawing shown in Figure 8–56. As with the first tutorial in this chapter, the start of this tutorial will require you to model the part. As shown in the figure, two different view types will be created: a general view and a projection view.

As shown in Figure 8–56, this tutorial will demonstrate the creation of tolerances on a Pro/ENGINEER drawing. This tutorial will cover:

- Starting a drawing with a drawing format.
- Creating a general view.
- Creating a projection view.
- Adding geometric tolerances to a drawing.
- Setting dimensional tolerances.
- Modifying the drawing setup file.

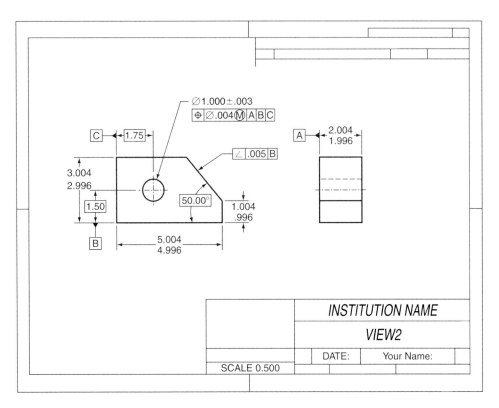

Figure 8–56   Pro/ENGINEER Drawing with Geometric Tolerances

CREATING THE PART

Model the part shown in Figure 8–57, naming the part *VIEW2*. When modeling the part, make sure that the locations of your datum planes match the locations of the datum planes shown in the drawing. As a reference, later in this tutorial, the default datum planes RIGHT,

315

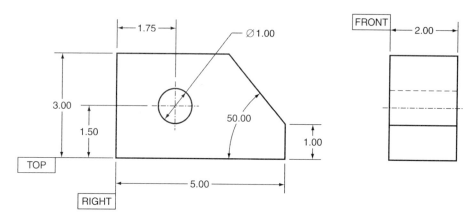

Figure 8–57   Part and Dimensioning Scheme

TOP, and FRONT will be renamed A, B, and C respectively. The required dimensioning design intent is shown. When modeling this part, make sure that these dimensions are incorporated into your design.

STARTING A DRAWING WITH A TEMPLATE

This section of the tutorial will create the object file for the drawing to be completed. You will utilize an existing drawing template file to set views, formats, and other drawing settings. Before starting this segment of the tutorial, ensure that the part shown in Figure 8–57 has been completed.

STEP 1:   **Start Pro/ENGINEER.**

If Pro/ENGINEER is not open, start the application.

STEP 2:   **Set an appropriate working directory.**

STEP 3:   **Select FILE >> NEW (Use the Default Template option).**

STEP 4:   **In the New dialog box, select Drawing mode, then enter *VIEW2* as the name of the drawing file.**

STEP 5:   **Select OK on the New dialog box.**

After you select OK on the dialog box, Pro/ENGINEER will reveal the New Drawing dialog box. This dialog box is used to select a model to create the drawing from and to set a sheet size.

STEP 6:   **On the New Drawing dialog box, select the BROWSE option and locate the *VIEW2* part (Figure 8–58).**

Use the Browse option to locate the *VIEW2* part created in the first section of this tutorial. The part will serve as the Default Model for the creation of the drawing views.

STEP 7:   **On the New Drawing dialog box, under the Specify Template option, select the EMPTY WITH FORMAT option (Figure 8–58).**

The Specify Sheet option allows for the selection of a standard or user-defined sheet size. The Empty-with-Format suboption allows for the retrieval of a predefined sheet format. When you retrieve a format, the size of the sheet defining the format will define the sheet size for the new drawing.

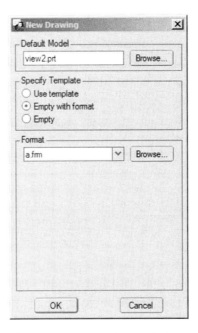

Figure 8–58 New Drawing Dialog Box

MODELING POINT   Template files under the Use Template option come with preestablished settings such as views and model display properties. As an example, the a_drawing template comes complete with preexisting front, top, and right-side views. Other information that can preexist in a template file includes drawing notes not derived from drawing's model (e.g., notes and symbols) and parameter notes defined within the model.

STEP 8: **Select the BROWSE option under the Format section of the New Drawing dialog box.**

The second Browse option on the dialog box allows you to browse the directory structure to locate existing sheet formats.

STEP 9: **Open a.frm as the standard sheet format.**

Pro/ENGINEER provides a variety of standard sheet formats (e.g., A, B, C, D, and E). A user-defined format can also be selected. When picking an existing format, the size of the format defines the size of the drawing sheet.

STEP 10: **Select OK on the New Drawing dialog box.**

ESTABLISHING DRAWING SETUP VALUES

This section of the tutorial will temporarily set Pro/ENGINEER's drawing settings. The Advanced >> Draw Setup option will be used to change the default settings for this specific drawing.

STEP 1: **Select FILE >> PROPERTIES on the menu bar.**

STEP 2: **Select DRAWING OPTIONS on the File Properties menu.**

After selecting Draw Setup, Pro/ENGINEER will open the active drawing settings within an Options dialog box. This dialog box is similar to the Options dialog box used to make changes to configuration file options.

STEP 3:   **Change the text, arrowhead, and datum values for the current drawing.**

Within the Options dialog box, make changes to the options shown in the following table. To change an option's value, perform the following steps:

1. Type the option's name in the Option field.

2. Type or select the option's value in the Value field.

3. Select the Add/Change icon.

The item *tol_display* will display dimensions in tolerance mode. The item *gtol_datums* when set to a value of *STD_ISO* will display datums in the ISO format.

Drawing Setup Option	New Value
drawing_text_height	0.125
text_width_factor	0.750
dim_leader_length	0.175
dim_text_gap	0.125
tol_display	YES
draw_arrow_length	0.125
draw_arrow_style	FILLED
draw_arrow_width	0.0416
gtol_datums	STD_ISO

STEP 4:   **Apply and Close the modified values for the active drawing setup file.**

Save your drawing setup file values then exit the text editor.

STEP 5:   **Select DONE/RETURN to close the Advanced Drawing Options menu.**

CREATING THE GENERAL VIEW

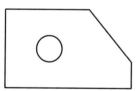

Figure 8–59
The Front View

General views serve as the parent view for all views projected off of it. This section of the tutorial will create a General view as the Front view of the drawing, as shown in Figure 8–59.

STEP 1:   ⬚ **Select the Create a General View icon on the drawing toolbar.**

STEP 2:   **On the workscreen, select the location for the drawing view.**

The drawing under creation in this tutorial will consist of a Front view and a Right-side view. The general view currently being defined will serve as the Front view. On the workscreen, pick approximately where the Front view will be located.

STEP 3:   **On the Drawing View dialog box, enter *FRONT* as the view's name.**

Notice on the dialog box has a view type of General. Since no views currently exist within the drawing, General is the only view type available.

STEP 4:   **Under the View Type category, select the Geometry References orientation option (Figure 8–60).**

The Geometric References option will allow you to define the orientation of the model within the drawing sheet. You will accomplish this through the selection of orientation references.

STEP 5:   **On the Orientation dialog box, select FRONT as the Reference 1 option.**

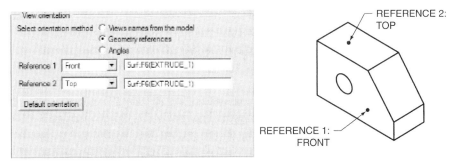

Figure 8–60   Orienting the Model

A planar surface selected with the Front reference option will orient the surface toward the front of the workscreen. Other available reference options include Back, Top, Bottom, Left, Right, Vertical Axis, and Horizontal Axis.

> **INSTRUCTIONAL NOTE**   Depending on your part construction method, your default orientation might be different from the orientation shown in Figure 8–60. Adjust your view orientation reference selections accordingly. Your final view should match the front view shown in Figure 8–59.

STEP 6: On the Orientation dialog box, select FRONT as the Reference 1 option.

STEP 7: Pick the front of the model (Figure 8–60).

STEP 8: On the Orientation Dialog box, select TOP as the Reference 2 option.

A planar surface selected with the Top reference option will orient the surface toward the top of the workscreen.

STEP 9: Select the top of the model (Figure 8–60).

STEP 10: Under the dialog box's Scale category, enter 0.500 as the custom scale value for the view.

STEP 11: Select OK on the Drawing View dialog box.

Your front view should appear as shown in Figure 8–33.

STEP 12: Save your drawing.

Drawings are saved with an *.drw file extension.

CREATING THE RIGHT-SIDE VIEW

This segment of the tutorial will create the right-side view of the part (Figure 8–61).

STEP 1: Select INSERT >> DRAWING VIEW >> PROJECTION on the menu bar.

STEP 2: On the workscreen, select the location for the right-side view.

Pro/ENGINEER will create a projected view based on its parent view. Your view should appear as shown in Figure 8–61.

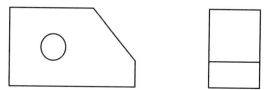

Figure 8-61   The Front and Right-Side Views

STEP 3:    **Unselect the Lock View Movement icon.**

The Lock View Movement option will lock the dragging of drawing views. It is selected by default.

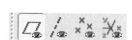

Figure 8-62
Model Display Toolbar

STEP 4:   **Use your mouse to drag the front and right-side views to an appropriate location.**

STEP 5:   **On the Toolbar, select the Hidden Line display icon (Figure 8–62).**

Selecting Hidden Line as the model display mode will produce hidden lines when the drawing is plotted.

SETTING AND RENAMING DATUM PLANES

This section of the tutorial will set and rename your datum planes. To utilize datum planes within a drawing, and with geometric tolerances, they have to first be set. The Set Up >> Geom Tolerance option is used to set datums in Part mode. The Edit >> Properties option in Drawing mode will be used in this tutorial to set datums.

STEP   1:   **On the toolbar, turn ON the display of datum planes (Figure 8–63).**

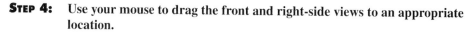

Figure 8-63
Datum Display Toolbar

STEP   2:   **On the workscreen or on your model tree, pick datum plane FRONT.**

Select datum plane FRONT. If during the construction process the location of your datum plane FRONT does not match the location of FRONT as shown previously in Figure 8–57, adjust your datum selection to match this figure.

STEP   3:   **On the menu bar, select EDIT >> PROPERTIES.**

After you select this datum plane, the Datum dialog box will appear (Figure 8–64). The Properties option is also available on the pop-up menu.

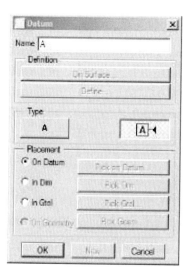

Figure 8-64   Datum Dialog Box

STEP 4: Rename this datum plane A.

As shown in Figure 8–64, in the Name textbox, rename FRONT to a value of A.

STEP 5: On the Datum dialog box, set the datum by selecting the datum plane symbol button (Figure 8–64).

If your datum plane symbol does not appear as shown in the figure, you did not properly set the drawing setup file option *gtol_datums* to *STD_ISO* as required earlier in this tutorial.

STEP 6: Select OK on the Datum dialog box to accept the values.

After selecting OK, notice on the workscreen how the datum symbol is now displayed in the ISO format (Figure 8–65).

STEP 7: Drag datum plane A to the location shown in Figure 8–65.

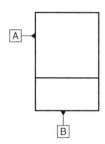

Figure 8–65
ISO Datum Plane
Symbol

INSTRUCTIONAL NOTE   If your location of datum plane A does not match the location shown in Figure 8–65, you probably extruded the base protrusion of the part in the wrong direction. This can be fixed in Part mode by redefining the direction of extrusion.

STEP 8:  Repaint your workscreen by selecting the Repaint icon on your toolbar.

STEP 9: Use the EDIT >> PROPERTIES option to set datum plane TOP and rename it B (Figure 8–66).

STEP 10: Repaint your workscreen.

STEP 11: Use the EDIT >> PROPERTIES option to set datum plane RIGHT and rename it C (Figure 8–66).

STEP 12: Repaint your workscreen.

STEP 13: Select SHOW/ERASE option on your toolbar.

As shown in Figure 8–67, two datum plane B symbols are available in the drawing. Unless more are needed for the clarity of a drawing, only one datum plane symbol is required. You will use the Show/Erase dialog box to erase the second datum plane B symbol.

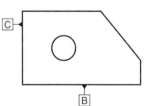

Figure 8–66
Setting Datum Planes

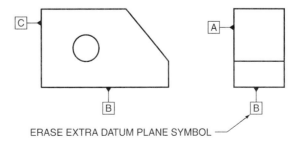

ERASE EXTRA DATUM PLANE SYMBOL

Figure 8–67   Erase the Extra Datum Plane Symbol

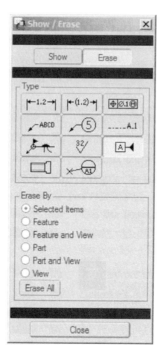

Figure 8-68   Datum Plane
Item

STEP 14:   On the Show/Erase dialog box, select the ERASE option and then select the DATUM type (Figure 8–68).

STEP 15:   On the Show/Erase dialog box, make sure that SELECTED ITEMS is the selected Erase By option.

The Selected Items option will erase only items that are selected on the workscreen.

STEP 16:   On the workscreen, select the extra datum plane symbol B (Figure 8–67) then select OK on the Select menu.

STEP 17:   Close the Show/Erase dialog box.

CREATING DIMENSIONS

This segment of the tutorial will show the parametric dimensions that were created when the model was constructed in Part mode. You will use the Show/Erase option to show all dimensions and centerlines.

STEP 1:   Select the SHOW/ERASE option.

The Show/Erase option is used to show or not show items that were defined in Part and/or Assembly modes.

STEP 2:   If necessary, on the Show/Erase dialog box, select the SHOW option and deselect the DATUM option (Figure 8–69).

From the previous segment of this tutorial, the Erase option and the Datum item type should still be selected. If they are, select the Show option and deselect the Datum option.

STEP 3:   On the Show/Erase dialog box, select the DIMENSION and AXIS item types (Figure 8–69).

The Dimension item type option will show parametric dimensions that were created in Part mode. The Axis item type will create centerlines.

STEP 4: **On the Show/Erase dialog box, under the Options tab, check the ERASED and NEVER SHOWN options.**

The Erased option will show items that have previously been erased and the Never Shown option will show items that have never been shown.

STEP 5: **On the Show/Erase dialog box, under the Preview tab, uncheck the WITH PREVIEW option (Figure 8–70).**

Because of the limited number of dimensions in this tutorial, you will not preview your dimensions.

STEP 6: **Select the SHOW ALL option then select YES to confirm the show all.**

The Show All option will show all available item types. In this step of the tutorial, all dimensions available from the referenced model will be shown.

STEP 7: **Close the Show/Erase dialog box.**

Your drawing should appear similar to Figure 8–71. Previously, you set the Drawing Setup File option *tol_display* equal to a value of YES. This created the tolerance display shown on your drawing.

STEP 8: **Use the mouse to drag dimensions and datum symbols and the mouse's pop-up menu's MOVE ITEM TO VIEW and FLIP ARROWS options to reposition the dimensions to match Figure 8–71.**

The mouse pop-up menu is available by first selecting the text of a dimension with your left mouse button, followed by reselecting the same dimension text with your right mouse button. (*Note:* A slight delay in releasing the right mouse button is necessary.) A variety of modification and manipulation tools are available through the revealed pop-up menu.

Use the following options to reposition your dimensions on the workscreen:

- **Mouse Move and Drag** Once selected, dimensions including text and witness lines and datum plane symbols can be moved and dragged on the screen with your mouse.

- **Move Item to View** The Move Item to View option is used to switch a dimension from one view to another. It is available under the pop-up menu. Multiple dimensions can be selected by a combination of the Control key and the left mouse button. Once dimensions are preselected, the pop-up menu is available through the right mouse button.

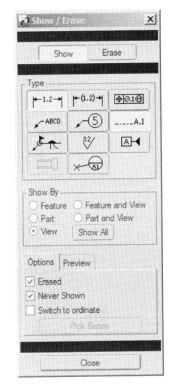

Figure 8–69
The Show/Erase Dialog Box

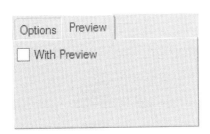

Figure 8–70 Showing without a Preview

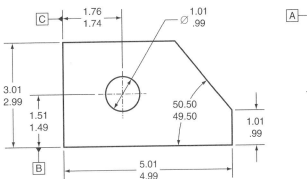

Figure 8–71 Drawing with Dimensions and Centerlines

- **Flip Arrow** The Flip Arrows option is used to flip dimension arrowheads. It is available under the pop-up menu.

STEP 9: **Use Drawing mode's move and drag capabilities to create dimension witness line gaps.**

Where a dimension's witness line is linked to an entity, a visible gap (approximately 1/16 in wide) is required. Drag the end of each witness line to create these gaps in each view.

STEP 10: **Select the CLEAN DIMENSIONS icon.**

The Clean Dimensions option will consistently space dimensions on a drawing view.

STEP 11: **On the workscreen, use the Control key to pick the front and right-side views, then select OK on the Select menu.**

Pro/ENGINEER's default values for the Clean Dimensions dialog box are 0.500 for the object Offset setting and 0.375 for the dimension Increment setting. You will keep these default settings.

STEP 12: **Select APPLY then CLOSE the dialog box.**

After applying the clean dimension settings, Pro/ENGINEER will create snap lines that will evenly space linear dimension lines. Depending on the complexity of the drawing, the applied settings might not create an ideal drawing.

STEP 13: **Use the mouse move and drag capabilities to refine the placement of dimensions and datum plane symbols.**

Your final dimensioning scheme should appear as shown in Figure 8–71.

STEP 14: **Save your drawing.**

SETTING GEOMETRIC TOLERANCES

Geometric tolerances are used to control the form and/or location of geometric features. Within Pro/ENGINEER, geometric tolerances can be incorporated into a model through Part, Assembly, or Drawing mode. This tutorial will first establish a position tolerance (±0.004 at MMC) for the 1.00-in nominal diameter hole. Second, an Angular tolerance (±0.005) will be provided for the 50 degree angled surface. Your final geometric tolerances should appear as shown in Figure 8–72.

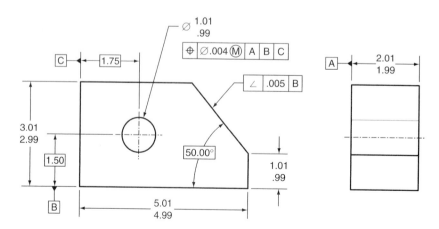

Figure 8–72   Drawing with Geometric Tolerances

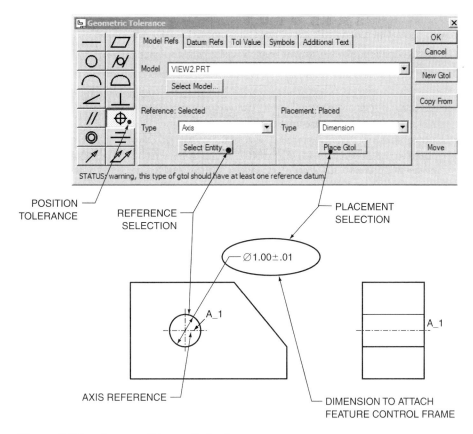

Figure 8–73 Geometric Tolerance Dialog Box

STEP 1: **On the menu bar, select INSERT >> GEOMETRIC TOLERANCE.**

After selecting the Geometric Tolerance option, the Geometric Tolerance dialog box will be displayed. This is the same Geometric Tolerance dialog box that that is utilized in Part mode. From this step forward, the steps for creating geometric tolerances are the same in Drawing mode as they are in Part mode.

STEP 2: **On the Geometric Tolerance dialog box, select the POSITION tolerance symbol (Figure 8–73).**

STEP 3: **On the Toolbar, turn on the display of axes (Figure 8–74).**

This portion of the tutorial will create a position tolerance for the 1.00-in nominal size hole. A position tolerance affects the axis of a hole. Because of this, in the next step of this tutorial, you will be required to select the hole's axis.

Figure 8–74
Datum Axis Display

STEP 4: **On the Geometric Tolerance dialog box, change the reference type to AXIS (Figure 8–73).**

STEP 5: **If necessary, select the SELECT ENTITY option on the dialog box.**

STEP 6: **On the workscreen select the axis at the center of the hole (Figure 8–73).**

STEP 7: **On the Geometric Tolerance dialog box, set DIMENSION as the placement type (Figure 8–73).**

STEP 8: **If necessary, select the PLACE GTOL option on the dialog box.**

STEP   9:   **On the workscreen, select the 1.00-in diameter dimension value (Figure 8–73).**

When selecting the dimension value, you will have to select the dimension text. The nominal size of the hole is 1.00 in. The dimension should be currently displayed with a limit tolerance format. This will be modified in a later step.

After performing this step, notice on the workscreen how the Feature Control frame is now visible.

STEP 10:   **On the Geometric Tolerance dialog box, select the DATUM REFS tab.**

STEP 11:   **On the Geometric Tolerance dialog box, under the PRIMARY Datum Reference tab, set the Basic datum to A (Figure 8–75).**

Notice how the Feature Control frame updates dynamically.

STEP 12:   **On the Geometric Tolerance dialog box, under the SECONDARY Datum Reference tab, set the Basic datum to B (Figure 8–75).**

STEP 13:   **On the Geometric Tolerance dialog box, under the TERTIARY Datum Reference tab, set the Basic datum to C.**

STEP 14:   **On the Geometric Tolerance dialog box, under the TOL VALUE tab, set the Overall Tolerance to a value of 0.004 and set the MATERIAL CONDITION to MMC (Figure 8–76).**

STEP 15:   **On the Geometric Tolerance dialog box, under the SYMBOLS tab, check the Diameter Symbol option (Figure 8–77).**

Since a Position tolerance for a hole creates a cylindrical tolerance zone, it requires a diameter symbol with the tolerance value. Notice how the Feature Control frame updates dynamically.

STEP 16:   **Select OK on the Geometric Tolerance dialog box.**

STEP 17:   **If prompted, select YES to confirm the setting of basic dimensions (if available).**

Your drawing should appear as shown in Figure 8–78. Notice how the value of each location dimension is enclosed in a box. This box represents a basic

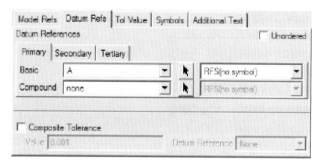

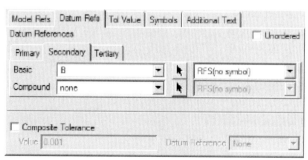

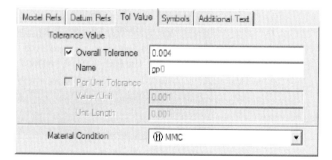

Figure 8–75   Primary and Secondary Datum Selection   **Figure 8–76**   Tolerance Value

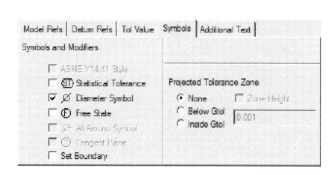

Figure 8–77   Diameter Symbols Selection

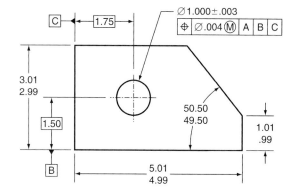

Figure 8–78   Position Tolerance

dimension. The confirmation required in this step sets the hole's location dimensions to Basic. If you do not get this message, use the Dimension Properties dialog box (**pop-up menu >> Properties**) to set each location dimension as basic.

STEP 18:   **Select INSERT >> GEOMETRIC TOLERANCE.**

The remaining steps of this segment of the tutorial will create the angular geometric tolerance.

STEP 19:   **On the Geometric Tolerance dialog box, select the ANGULAR tolerance characteristic (Figure 8–79).**

STEP 20:   **On the Geometric Tolerance dialog box, change the reference type to SURFACE, then on the workscreen select the edge of the angled surface.**

STEP 21:   **Change the Placement Type to LEADERS (Figure 8–79).**

The Feature Control frame will be attached to the surface with a leader.

STEP 22:   **Select ON ENTITY >> ARROW HEAD on the Attachment Type menu.**

STEP 23:   **On the workscreen, pick the edge of the angled surface (select only once).**

The selected location on the edge will be the attachment point for the leader. Pick the edge only once.

STEP 24:   **Select the DONE option on the Attachment Type menu.**

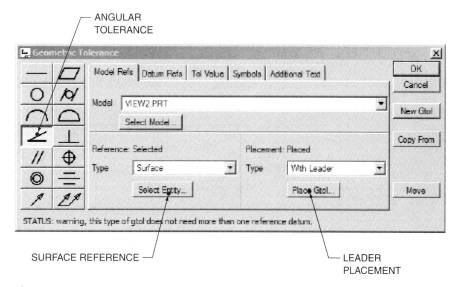

Figure 8–79   Geometric Tolerance Dialog Box

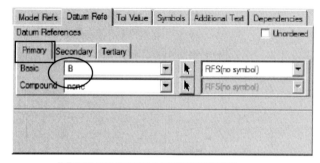

PRIMARY DATUM REFERENCE

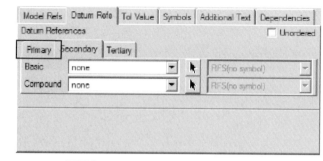

SECONDARY DATUM REFERENCE

Figure 8–80   Datum Reference Selection

STEP 25: On the workscreen, select a location for the Feature Control frame.

STEP 26: On the Geometric Tolerance dialog box, under the Datum Refs tab, set the PRIMARY datum reference to B (Figure 8–80).

An Angular Geometric tolerance requires one datum reference. In this example, datum plane B will be used as the datum reference.

STEP 27: On the Geometric Tolerance dialog box, under the Datum Refs tab, set the SECONDARY and TERTIARY datum references to NONE (Figure 8–80).

STEP 28: On the Geometric Tolerance dialog box, under the Tol Value tab, set the Overall Tolerance to a value of 0.005 and set the Material Condition to RFS (no symbol) (Figure 8–81).

STEP 29: On the Geometric Tolerance dialog box, under the Symbols tab, uncheck the DIAMETER SYMBOL button (Figure 8–81).

Angular Geometric tolerances do not create cylindrical tolerance zones. Hence, they do not require diameter symbols.

STEP 30: Select OK on the Geometric Tolerance dialog box.

Because of the Angular Geometric tolerance, the dimension defining the size of the angle (50 degree nominal size) should be set as basic.

INSTRUCTIONAL NOTE   If you make a mistake with a geometric tolerance's feature control frame, use the pop-up menu's Properties option to make necessary changes.

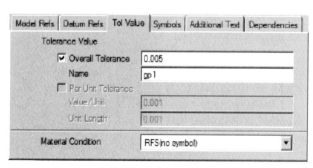

TOLERANCE VALUE

TOLERANCE SYMBOLS

Figure 8–81   Tolerance Value and Diameter Symbol

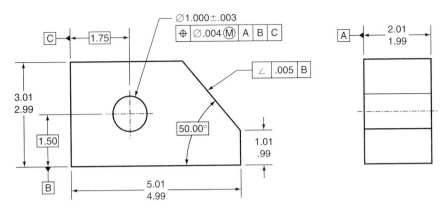

Figure 8–82 Geometric Tolerances

Step 31: Select the 50 angular dimension value with your left mouse button, then with your right mouse button, access the pop-up menu.

Step 32: Select the PROPERTIES option on the pop-up menu to access the Dimension Properties dialog box.

Step 33: On the Properties tab of the Dimension Properties dialog box, select BASIC as the dimension display type.

Step 34: Select OK to exit the dialog box.

Your drawing should appear as shown in Figure 8–82.

SETTING DIMENSIONAL TOLERANCES

This segment of the tutorial will create the dimensional tolerances shown in Figure 8–83. Previously in this exercise, you set the Drawing Setup File option *display_tol* to YES. This option is used to display dimensions in a tolerance format.

Step 1: On the workscreen, select the hole's diameter dimension value, then access the pop-up menu with the right mouse button.

This portion of the tutorial will create a tolerance dimension defining the size of the hole.

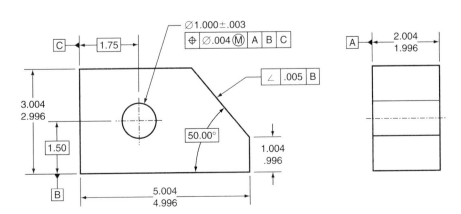

Figure 8–83 Dimensional Tolerances

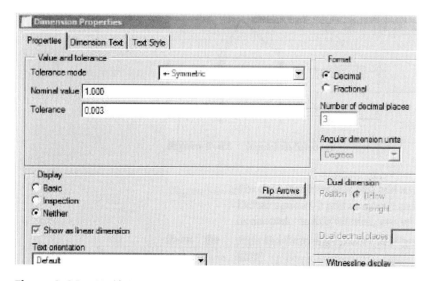

Figure 8–84 Modify Dimension Dialog Box

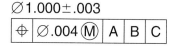

Figure 8–85 Hole Dimension Note

STEP **2:** Select the PROPERTIES option on the pop-up menu.

STEP **3:** On the Dimension Properties dialog box, change the Tolerance Mode to =-SYMMETRIC (Figure 8–84).

STEP **4:** On the Dimension Properties dialog box, change the Number of Decimal Places to a value of 3 (Figure 8–84).

STEP **5:** On the dialog box, change the Tolerance value to 0.003 (Figure 8–84).

This will change the tolerance of the hole dimension to ±0.003 in.

STEP **6:** Select OK to exit the Dimension Properties dialog box.

Your dimension note should appear as shown in Figure 8–85.

STEP **7:** Using a combination of your Control key and left mouse button, pick the remaining four dimensions that are currently displayed with a limit format (see Figure 8–86).

This portion of the tutorial will modify the remaining dimension to have a tolerance value of ±0.004 with a limit format.

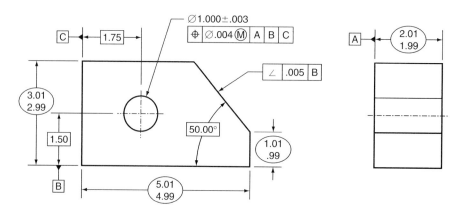

Figure 8–86 Dimensions to Change Format

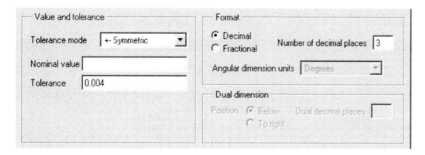

Figure 8–87 Modify Dimension Dialog Box

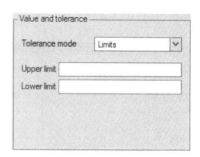

Figure 8–88 Limits Tolerance Format

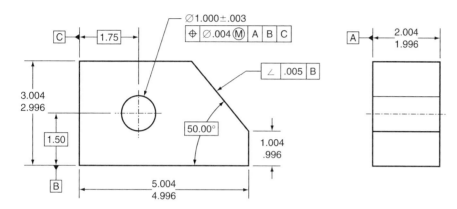

Figure 8–89 Final Dimensioning and Tolerance Scheme

STEP 8: Select EDIT >> PROPERTIES or select the Properties option on the pop-up menu.

STEP 9: On the Dimension Properties dialog box, change the Number of Decimal Places to a value of 3 (Figure 8–87).

STEP 10: On the Dimension Properties dialog box, change the Tolerance Mode to +–SYMMETRIC then change the Tolerance value to 0.004 (Figure 8–87).

Changing the Tolerance Mode to +–*Symmetric* will allow you to change the Tolerance value for all four dimensions at once. In the next step, you will change the Tolerance Mode back to Limits.

STEP 11: Change the Tolerance Mode to LIMITS (Figure 8–88) then select OK on the Modify Dimension dialog box.

Your drawing should appear as shown in Figure 8–89.

CREATING THE TITLE BLOCK

Use the Insert >> Note option to create the title block information shown in Figure 8–90. Move the scale note (SCALE 0.5000) to the title block. The pop-up menu's Properties option can be used to modify the height and style of any text.

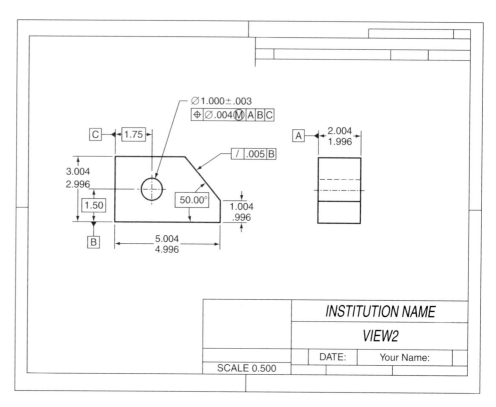

Figure 8–90   Finished Drawing

PROBLEMS

1. Model the part shown in Figure 8–91. For this problem, meet the following requirements:

    - The dimensions shown in the figure meet design intent. During part modeling, incorporate these dimensions.

    - Create an engineering drawing with front, top, and right-side views.

    - Use an A size sheet and format.

    - Fully dimension the engineering drawing using the part's parametric dimensions.

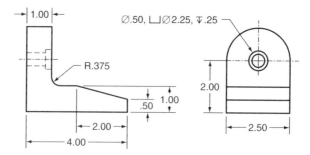

Figure 8–91   Problem 1

2. Model the part shown in Figure 8–92. For this problem, meet the following requirements:

    - The dimensions shown in the figure meet design intent. During part modeling, incorporate these dimensions.

    - Create an engineering drawing with front, top, and right-side views.

- Use an A size sheet and format.
- Fully dimension the engineering drawing using the part's parametric dimensions.

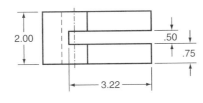

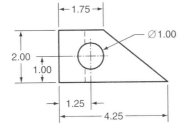

Figure 8–92   Problem 2

3. Model the part shown in Figure 8–93. For this problem, meet the following requirements:

- The dimensions shown in the figure meet design intent. During part modeling, incorporate these dimensions.
- Create an engineering drawing with front, top, and right-side views.

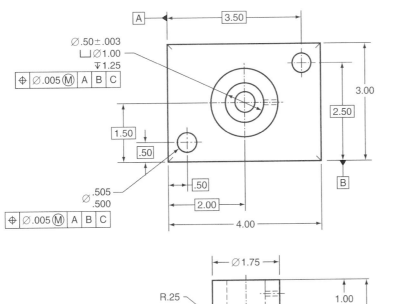

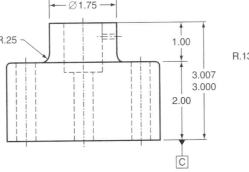

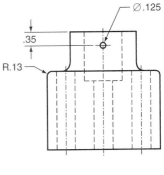

Figure 8–93   Problem 3

- Use an A size sheet and format.

- Fully dimension the engineering drawing using the part's parametric dimensions.

- Apply geometric tolerance annotations and ISO datum plane symbols as shown in the figure.

QUESTIONS AND DISCUSSION

1. What is the default dimension text height of a Pro/ENGINEER drawing and how can this text height be changed for an individual drawing? How can the default text height be changed permanently?

2. Describe the process used in Pro/ENGINEER's Drawing mode to add a border/title block overlay to a drawing. How are border/title block overlays created in Pro/ENGINEER?

3. Describe the difference between erasing a drawing view and deleting a drawing view.

4. Describe the difference between a general view and a projection view.

5. Describe the difference between the Drawing mode views half, partial, and broken.

6. How can a hidden line display be set permanently in a specific drawing view?

7. How can tangent edge display be turned off permanently in a specific drawing view?

8. Describe the process used in Drawing mode to display parametric dimensions and centerlines.

9. Describe the process for adding symbols to a dimension note.

CHAPTER

SECTIONS AND ADVANCED DRAWING VIEWS

Introduction

Pro/ENGINEER's Drawing mode provides a variety of options for creating orthographic and detailed drawings. This chapter will explore some of the advanced views available in this mode. Highlighted will be section views and auxiliary views. Upon finishing this chapter you will be able to:

- Create a full section view in Drawing mode.
- Create a half section.
- Create an offset section.
- Create a broken out section.
- Create an aligned view.
- Create an auxiliary view.

DEFINITIONS

Auxiliary view	Any orthographic view that is not one of the six principle views.
Cutting plane line	A thick line used to show the cutting pattern of a section view.
Drawing setup file	A file used to establish environmental settings for a drawing. Examples of possible settings include text height and arrowhead size.
Offset section	A section view whose cutting plane is offset to include more features within the section.
Section lining	A pattern used on a section view to show where a model is cut.

SECTION VIEW FUNDAMENTALS

Section views are utilized within a working drawing to show details of a design that would be difficult to view in a traditional orthographic projection. Figure 9–1 shows an example of a full section view. A section view simulates what a model would look like if it were actually cut apart. On the section view, section lining is used to show where the part is cut. Industry standards provide section lining patterns for a variety of materials. The default section lining used in Pro/ENGINEER represents iron (see Figure 9–1). A section lining's line style, weight, spacing, and angle can be changed with the Edit >> Properties option.

Another important line type associated with a section view is the cutting plane line (Figure 9–1). A cutting plane line typically lies in a view adjacent to the section view and

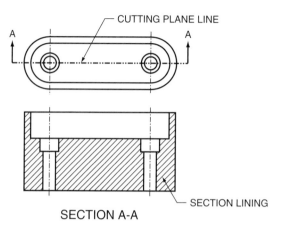

Figure 9–1   Full Section View

represents the cutting path of the cross section (a removed section is an example of a view where the cutting plane line does not lie in a view adjacent to its section view). Arrowheads terminate the ends of a cutting plane line and point in the viewing direction of the section. The Drawing Setup File options *crossec_arrow_length* and *crossec_arrow_width* control the size of a cutting plane line's arrowheads.

Within Pro/ENGINEER, section views are created from defined cross sections. Pro/ENGINEER provides options for creating cross sections in Part, Assembly, and Drawing modes. Chapter 7 provides details on how to create a cross section in Part mode. Cross Sections created in Part and Assembly modes can be retrieved and used to create section views in Drawing mode. While you are creating a section view, Drawing mode also provides an option for creating a cross section.

SECTION VIEW TYPES

The most common section view used is the full section. Other types of section views are available to serve a variety of documentation needs.

FULL SECTION

The full section view is the traditional type of section view used on most engineering drawings. As shown in Figure 9–2, a full section passes completely through a model. A full section is available for general, projection, and auxiliary views.

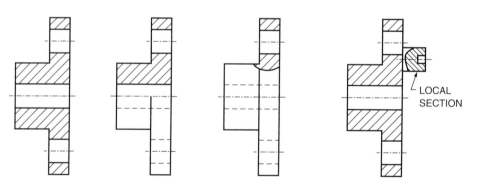

FULL SECTION   HALF SECTION   LOCAL SECTION   FULL & LOCAL SECTION

Figure 9–2   Section View Types

HALF SECTION

The half section view is similar to the full section, except only half of the view is sectioned. As shown in Figure 9–2, for a symmetrical model, half sections provide the advantage of a section on half the view, while also presenting the other half with traditional projection. A half section is available for general, projection, and auxiliary views. It is not available with the half, broken, and partial view types.

LOCAL

The Local Section option is used to create a broken-out section view. As shown in Figure 9–2, a local section creates a section in a specific, user-defined area. Local sections are available for general, projection, and auxiliary views. It is not available with the half and broken view types.

FULL & LOCAL

The Full & Local Section option is a section view with both a full section and a local section (Figure 9–2). The full section is placed first.

FULL SECTIONS

A full section is a section view that runs completely through a model. Figures 9–1, 9–2, and 9–3 show examples of full section views. Cross Sections used to construct a Full Section are created along a planar surface. Often, a datum plane is used as this surface. Perform the following steps to construct a full section view:

STEP 1: Select INSERT >> DRAWING VIEW.

STEP 2: Select a view type.

Full section views can be created as a projection, auxiliary, general, detailed, or revolved view.

STEP 3: Place the view.

STEP 4: Prepare to edit the view with the Drawing View dialog box (Edit >> Properties).

STEP 5: On the Drawing View dialog box, select the SECTION category.

The Section category is used to select or create a cross section for your drawing.

STEP 6: Select the 2D CROSS SECTION option.

STEP 7: ✛ Select the Add Cross Section icon.

A full section is a section that runs completely through a model (Figure 9–2).

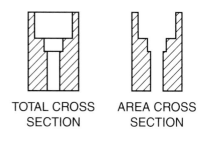

TOTAL CROSS   AREA CROSS
SECTION   SECTION

Figure 9–3 Total and Area Section Views

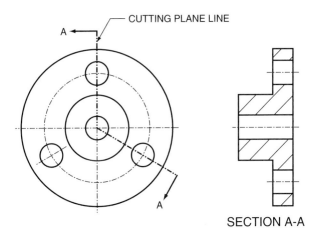

CUTTING PLANE LINE

SECTION A-A

Figure 9-4 Cutting Plane Line

STEP 8: **Select PLANAR >> DONE as the cross section creation method.**

Planar will create a cross section through a part at the location of a planar surface. While Planar is used to create a straight cross section, the Offset option will create a cross section that does not lie along a straight line.

STEP 9: **Enter a name for the section view.**

In Pro/ENGINEER's textbox, enter a name for the section view. Within the area of mechanical drafting, sections are often named with an alphabetic character.

STEP 10: **Select either a planar surface or a datum plane.**

This plane will create the cross section that will define the section view. The plane has to lie parallel to the section view location.

STEP 11: **Define FULL as the Sectioned Area.**

STEP 12: **In the Arrow Display option, select a view to locate the cutting plane line (Figure 9–4).**

STEP 13: **Select a view to locate the cutting plane line (Figure 9–4).**

Pro/ENGINEER's message area is requesting a "view for arrows where the section is perpendicular." The cutting plane line defining the cross section cut will be located in this view. If a cutting plane line is not necessary, select the middle mouse button.

STEP 14: **Select OK on the Drawing View dialog box.**

HALF SECTIONS

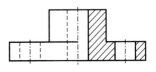

Figure 9–5
Half Section View

Section views are used to improve the clarity of an engineering design by providing an avenue where the interior details of a model can be viewed. Often, it is not necessary to create a full section. Full sections show the entire view of a model as a section. As a result, some details outside the section might lose their clarity. Half sections are used to create a section view through only half of a model. Figure 9–5 shows an example of a half section. Perform the following steps to construct a half section view:

STEP 1: **Select INSERT >> DRAWING VIEW.**

STEP 2: **Select a view type.**

Half section views can be created as a projection, auxiliary, general, detailed, or revolved view.

Step 3: Place the view.

Step 4: Prepare to edit the view with the Drawing View dialog box (Edit >> Properties).

Step 5: On the Drawing View dialog box, select the SECTION category.

The Section category is used to select or create a cross section for your drawing.

Step 6: Select the 2D CROSS SECTION option.

Step 7: ✛ Select the Add Cross Section icon.

A half section is a section that runs only halfway through a model (Figure 9–5).

Step 8: Select PLANAR >> DONE as the cross section creation method.

Planar will create a cross section through a part at the location of a planar surface. While Planar is used to create a straight cross section, the Offset option will create a cross section that does not lie along a straight line.

Step 9: Enter a name for the section view.

In Pro/ENGINEER's textbox, enter a name for the section view. Within the area of mechanical drafting, sections are often named with an alphabetic character.

Step 10: Select either a planar surface or a datum plane.

This plane will create the cross section that will define the section view. The plane has to lie parallel to the section view location.

Step 11: Define LOCAL as the Sectioned Area.

Step 12: For the reference box under the Section category, pick a datum plane that will define the boundary between the sectioned half of the view and the non-sectioned half (Figure 9–6).

You may need to scroll the cross section definition area horizontally to see options available for displaying a section.

Step 13: In the Arrow Display option, select a view to locate the cutting plane line (Figure 9–4).

Step 14: Select a view to locate the cutting plane line (Figure 9–4).

Pro/ENGINEER's message area is requesting a "view for arrows where the section is perpendicular." The cutting plane line defining the cross section cut will be located in this view. If a cutting plane line is not necessary, select the middle mouse button.

Step 15: Select OK on the Drawing View dialog box.

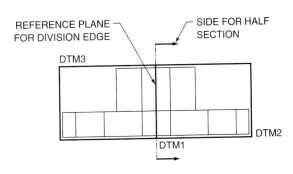

Figure 9–6   Reference Plane Selection

OFFSET SECTIONS

Most section views are situated along straight cutting planes. Often, features cannot be fully described through a straight cut. Drafting standards allow for an offset cutting plane. Figure 9–7 shows an example of an offset section view with its corresponding offset cutting plane. With an offset section, the cutting plane is offset by the use of 90 degree bends that allow the cutting plane to pass through features that require sectioning. While normal section views are created within Pro/ENGINEER by using a planar surface or datum plane, the cutting plane line for an offset section is sketched. The part file (*offset.prt*) and drawing file (*offset.drw*) for practicing this guide are available on the book's web page. Perform the following steps to construct a projected offset section view:

STEP 1:   Select INSERT >> DRAWING VIEW.

STEP 2:   Select a view type.

Section views can be created as a projection, auxiliary, general, detailed, or revolved view.

STEP 3:   Place the view.

STEP 4:   Prepare to edit the view with the Drawing View dialog box (Edit >> Properties).

STEP 5:   On the Drawing View dialog box, select the SECTION category.

The Section category is used to select or create a cross section for your drawing.

STEP 6:   Select the 2D CROSS SECTION option.

STEP 7:   ✚ Select the Add Cross Section icon.

A half section is a section that runs only halfway through a model.

STEP 8:   Select OFFSET from the Cross Section Create menu.

STEP 9:   Select BOTH SIDES >> DONE.

STEP 10:   In Pro/ENGINEER's textbox, enter a name for the section view.

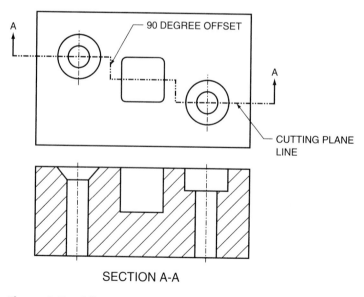

Figure 9–7   Offset Section

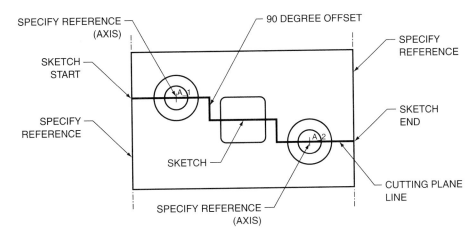

Figure 9-8   Sketching an Offset Section

STEP 11:   **Switch to the model's window.**

If necessary, use the Windows taskbar or Pro/ENGINEER's Application Manager to switch to the part or assembly model from which the drawing is being created. When constructing an offset section, for projection views, you are required to select a sketching plane on the actual model. When selecting a sketching plane from a projection view, ensure that the model from which the drawing is being produced is in the active window. You must switch to this window at this step in the view creation process.

STEP 12:   **Select or create a sketching plane then orient the sketching environment.**

Pro/ENGINEER creates offset sections by sketching a cutting plane line. This step requires you to select or create a planar surface that will be suitable for sketching the cutting plane line.

STEP 13:   **Sketch the cutting plane line (Figure 9–8).**

Utilizing appropriate sketching tools, sketch the cutting plane line. As shown in the figure, when sketching the cutting plane line, ensure that you differentiate between the lines that form the cut and the lines that form the offset.

MODELING POINT   When sketching a cutting plane line, select the axis of each hole to include in the section as a reference.

STEP 14:   ✔ **Select the Continue icon to exit the sketching environment.**

STEP 15:   **Select or create a sketching plane, then orient the sketching environment.**

Offset section views can be created as a full, half, or local cross section type. Additionally, a total cross section (TotalXsec), area cross section (Area Xsec), aligned crosssection (Align Xsec), or total align cross section type can be used.

STEP 16:   **In the Arrow Display option, select a view to locate the cutting plane line (Figure 9–4).**

STEP 17:   **Select OK to exit the Drawing View dialog box.**

BROKEN OUT SECTION

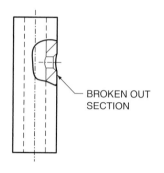

└─ BROKEN OUT
       SECTION

Figure 9–9
Broken Out Section

Broken out sections allow for the display of internal details of a model without the creation of a full section. Figure 9–9 shows an example of a typical broken out section. Broken out sections do not utilize a cutting plane line. Within Pro/ENGINEER, a broken out section can be created with a general, projection, or detail view. The part file (*broken_out.prt*) and drawing file (*broken_out.drw*) for practicing this guide are available on the book's web page. Perform the following steps to create a broken out section:

STEP 1:   Select INSERT >> DRAWING VIEW.

STEP 2:   Select a view type.

Broken out section views can be created as a projection, auxiliary, general, detailed, or revolved view.

STEP 3:   Place the view.

STEP 4:   Prepare to edit the view with the Drawing View dialog box (Edit >> Properties).

STEP 5:   On the Drawing View dialog box, select the SECTION category.

The Section category is used to select or create a cross section for your drawing.

STEP 6:   Select the 2D CROSS SECTION option.

STEP 7:   ✛ Select the Add Cross Section icon.

A half section is a section that runs only halfway through a model (Figure 9–5).

STEP 8:   Select PLANAR >> DONE as the cross section creation method.

STEP 9:   Enter a name for the section view.

In Pro/ENGINEER's textbox, enter a name for the section view. Within the area of mechanical drafting, sections are often named with an alphabetic character.

STEP 10:   Select either a planar surface or a datum plane.

This plane will create the cross section that will define the section view. The plane has to lie parallel to the section view location.

STEP 11:   Define LOCAL as the Sectioned Area.

The Local option is the key selection for creating a broken out section. This option creates a section view within a sketched spline boundary. This boundary will be sketched in a later step.

STEP 12:   Define the reference by picking an entity approximately at the center of where the broken out view will be created (Figure 9–11).

This selection is required for the normal regeneration of the broken out portion of the sectioned view.

STEP 13:   On the workscreen, sketch a spline to create the boundary of the broken out section (Figure 9–11).

Use the left mouse button to locate spline points and the middle mouse button to close the spline.

OPTIONAL: Within the guide, the following steps are optional for defining a Broken View. (Figure 9–10)

STEP 14:   Select VISIBLE AREA on the Drawing View dialog box.

STEP 15:   Select BROKEN VIEW as the View Visibility.

STEP 16:   ✛ Select the Add Break Line icon.

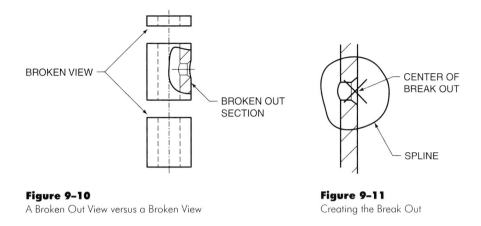

Figure 9–10
A Broken Out View versus a Broken View

Figure 9–11
Creating the Break Out

STEP 17:   Pick two locations to define the break in the view.

STEP 18:   Select OK on the Drawing View dialog box.

ALIGNED SECTION VIEWS

Engineering graphics is a language used to communicate design intent. Standards and conventions exist that govern the way designs are displayed on engineering drawings. Within the realm of engineering graphics, designs are often displayed in multiview projection. Multiview projection does not always present the best display of a design. Figure 9–12 shows a design displayed using normal lines of projection. Normal lines of projection for a multiview drawing project at a 90 degree angle. With the drawing shown in Figure 9–12, the part does not form a 90 degree angle. This presents a projection problem. Clarity for this design can be improved with the use of an aligned view. As shown in Figure 9–13, the angled feature of the model can be aligned with normal lines of projection to create a drawing with more clarity.

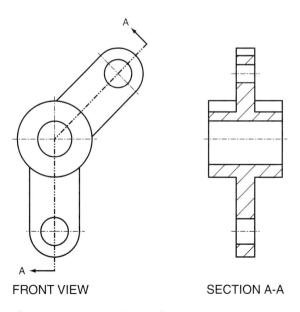

FRONT VIEW   SECTION A-A

Figure 9–12   Normal Lines of Projection

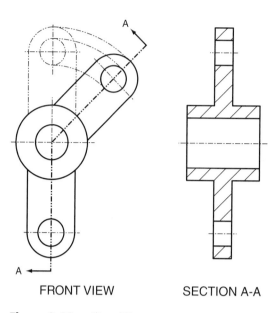

FRONT VIEW   SECTION A-A

Figure 9–13   Aligned View

Pro/ENGINEER's Drawing mode has the capability of producing an aligned section view. The following guide shows how to produce the Aligned Section view shown in Figure 9–13. The part (*align.prt*) and drawing (*align.drw*) files for this guide are available on the book's Web page.

INSTRUCTIONAL NOTE   Cross sections can be created in Part, Assembly, or Drawing mode. During the creation of a section view, the cross section can be created within the inserting of the view. In this guide, the cross section was created in Part mode as an offset section. This cross section will be retrieved during the view creation process.

STEP 1: Select INSERT >> DRAWING VIEW.

STEP 2: Select PROJECTION as the view type.

STEP 3: Place the view.

STEP 4: Prepare to edit the view with the Drawing View dialog box (Edit >> Properties).

STEP 5: On the Drawing View dialog box, select the SECTION category.

The Section category is used to select or create a cross section for your drawing.

STEP 6: Select the 2D CROSS SECTION option.

STEP 7: ✛ Select the Add Cross Section icon.

Note: Optionally, if you choose to retrieve an existing cross section, skip to step 15.

STEP 8: Select OFFSET from the Cross Section Create menu.

STEP 9: Select BOTH SIDES >> DONE.

STEP 10: In Pro/ENGINEER's textbox, enter a name for the section view.

STEP 11: Switch to the model's window.

If necessary, use the Windows taskbar or Pro/ENGINEER's Application Manager to switch to the part or assembly model from which the drawing is being created. When constructing an offset section, for projection views, you are required to select a sketching plane on the actual model. When selecting a sketching plane from a projection view, ensure that the model from which the drawing is being produced is in the active window. You must switch to this window at this step in the view creation process.

STEP 12: Select or create a sketching plane, then orient the sketching environment.

Pro/ENGINEER creates offset sections by sketching a cutting plane line. This step requires you to select or create a planar surface that will be suitable for sketching the cutting plane line.

STEP 13: Sketch the cutting plane line.

Utilizing appropriate sketching tools, sketch the cutting plane line.

STEP 14: Select the Continue icon to exit the sketching environment.

STEP 15: In the Sectioned Area box on the Drawing View dialog box, select FULL(ALIGNED) as the section type.

This option will create a full section aligned about a selected axis.

STEP 16: To define the reference element for the aligned section, pick an axis of revolution (Figure 9–14).

STEP 17: In the Arrow Display option, select a view to locate the cutting plane line (Figure 9–15).

STEP 18: Select OK to exit the Drawing View dialog box.

Figure 9–14
Aligned Cross Section

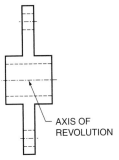

Figure 9–15
Axis of Revolution

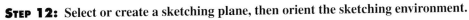

REVOLVED SECTIONS

Revolved sections are used to show the cross section of a spoke, rail, or rib type feature. Additionally, it is used with features that are extruded, such as wide flange beams. Revolved sections are useful for representing the cross section of a feature without having to create a separate orthographic view. They are displayed by revolving the cross section 90 degrees. Figure 9–16 shows three different ways to locate the cross section in relation to its parent view. The following guide will demonstrate how to superimpose a revolved section onto a view.

STEP 1: Create or identify the view from which to obtain the revolved section.

A revolved section can be created from a projection, auxiliary, or general view. Additionally, one can even be created from an existing revolved view.

STEP 2: Select INSERT >> DRAWING VIEW.

STEP 3: On the workscreen, select a parent for the new view.

STEP 4: Place the view.

The Revolved option allows for either a full view or a partial view only. A revolved view by default has to be a section view.

STEP 5: On the workscreen, pick the location for the revolved view.

For a superimposed revolved section view, select on the view used to create the revolved section.

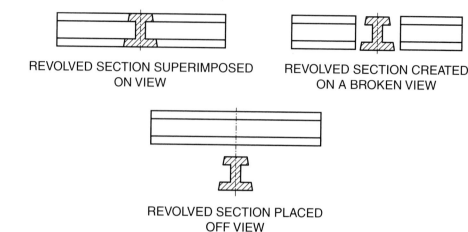

REVOLVED SECTION SUPERIMPOSED
ON VIEW

REVOLVED SECTION CREATED
ON A BROKEN VIEW

REVOLVED SECTION PLACED
OFF VIEW

Figure 9–16   Revolved Sections

STEP 6:   **Select an existing cross section or Create a new one.**

The cross section created or retrieved in this step will be used as the revolved section.

STEP 7:   **Select a symmetry axis for the revolved section or select the middle mouse button to accept the default.**

The symmetry axis is the location about which the revolved section view will be centered.

STEP 8:   **Select OK to exit the Drawing View dialog box.**

STEP 9:   **Use the mouse drag capability to refine the placement of the revolved section.**

AUXILIARY VIEWS

Within the language of engineering graphics, any object has six principle views. An auxiliary view is any orthographic projected view that is not one of the six principle views. Auxiliary views are used frequently to show the true size of an inclined surface. Figure 9–17 shows an example of an auxiliary view and how it helps to better represent the inclined surface. This guide will demonstrate the creation of an auxiliary view. The part file (*auxiliary.prt*) and drawing file (*auxiliary.drw*) for practicing this guide is available on the book's Web page.

STEP 1:   **Select INSERT >> DRAWING VIEW on the menu bar.**

STEP 2:   **Select AUXILIARY as a view type.**

Auxiliary is one of the five principle view types available in Drawing mode. An existing view is required before an auxiliary view can be created.

STEP 3:   **On the workscreen, pick an edge or an axis to project the auxiliary view from Figure 9–19.**

STEP 4:   **On the workscreen, select a location for the auxiliary view.**

STEP 5:   **If necessary, use the Drawing View dialog box (Edit >> Properties) to modify additional properties for the auxiliary view.**

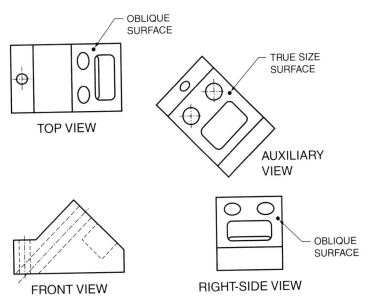

Figure 9-17   An Auxiliary View

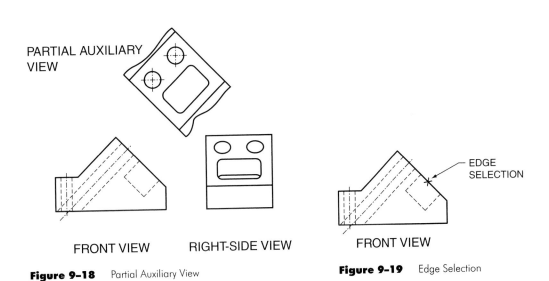

Figure 9-18   Partial Auxiliary View

Figure 9-19   Edge Selection

An auxiliary view can be created as a half view or as a partial view. Figure 9–18 shows an example of a partial auxiliary view. You can use the View Visibility category on the Drawing View dialog to create a partial view.

SUMMARY

Pro/ENGINEER is an integrated engineering design tool that allows for a full range of applications to include modeling, assembly, manufacturing, and analysis. Because of these tools, Pro/ENGINEER is a design package, not a drafting application. Despite this, there is still a need to document a design. Pro/ENGINEER's Drawing mode provides multiple tools for creating a high-quality engineering drawing. Included in these tools is the capability to create a variety of view types, such as section, detail, and auxiliary. Since Pro/ENGINEER is a fully associative computer-aided design application, models created in Part and Assembly modes can be used to create views in Drawing mode. Additionally, parametric dimensions that define a design can be used within Drawing mode to document the design.

ADVANCED DRAWING TUTORIAL 1

This tutorial exercise will provide instruction on how to create the drawing shown in Figure 9–20. The first step in this tutorial will require you to model the part from which the drawing will be created.

Within this tutorial, the following topics will be covered:

- Starting a drawing.
- Establishing drawing setup values.
- Creating a general view.
- Creating an aligned section view.
- Creating a partial broken out section view.
- Annotating a drawing.

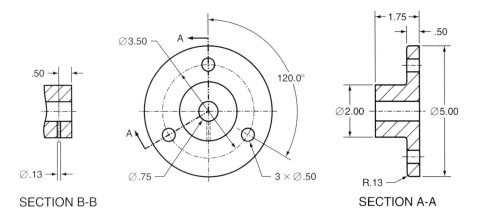

SECTION B-B   SECTION A-A

Figure 9–20   Completed Views

CREATING THE PART

Using Part mode, model the part shown in the drawing in Figure 9–20. Name the drawing file *section1*. When defining features, the dimensioning scheme shown in the figure matches the design intent for the part. Incorporate this intent into your model. Create the base feature as a revolved protrusion. Create the sketch for this feature on datum plane *RIGHT*. Create the bolt circle hole pattern using the Radial Hole option and the Pattern command. The cross section for the drawing will be created in Drawing mode.

STARTING A DRAWING

This section of the tutorial will create the object file for the drawing to be completed. Do not start this section until you have modeled the part portrayed in Figure 9–20.

STEP 1:   **Start Pro/ENGINEER.**

If Pro/ENGINEER is not open, start the application.

STEP 2:   **Set an appropriate working directory.**

Figure 9-21   New Drawing Dialog Box

STEP 3:   Select FILE >> NEW.

STEP 4:   In the New dialog box, select DRAWING mode, then enter *SECTION1* as the name of the drawing file.

STEP 5:   Select OK on the New dialog box.

After you select OK on the dialog box, Pro/ENGINEER will reveal the New Drawing dialog box. The New Drawing dialog box is used to select a model to create the drawing from and to set a sheet size and format.

STEP 6:   Under the Default Model textbox, select the BROWSE option and locate the *SECTION1* part (Figure 9-21).

Use the Browse option to locate the *section1* part created in the first section of this tutorial. The part will serve as the default model for the creation of drawing views.

STEP 7:   Select the EMPTY WITH FORMAT option under the Specify Template section.

You will use the Specify Sheet option to select an A size format.

STEP 8:   Select the BROWSE option under the Format option on the New Drawing dialog box, then open *a.frm* as the Standard Sheet format (see Figure 9-21).

The Browse option will allow you to browse the directory structure to locate an existing sheet format.

STEP 9:   Select OK on the New Drawing dialog box.

After you select OK, Pro/ENGINEER will launch its Drawing mode.

ESTABLISHING DRAWING SETUP VALUES

This section of the tutorial will temporarily set Pro/ENGINEER's drawing settings. The drawing Options dialog box will be used to change the default settings for this specific drawing.

STEP 1: Select FILE >> PROPERTIES >> DRAWING OPTIONS.

After selecting Drawing Options, Pro/ENGINEER will open the active drawing setup file with the Options dialog box. This dialog box is similar to the Properties dialog box used to make permanent changes to your modeling environment.

STEP 2: Change the text and arrowhead values for the current drawing.

Within the Options dialog box, make the changes shown in the following table. The *radial_pattern_axis_circle* item is used to create a bolt circle centerline around the hole pattern.

Drawing Setup Item	New Value
drawing_text_height	0.125
text_width_factor	0.750
dim_leader_length	0.175
dim_text_gap	0.125
draw_arrow_length	0.125
draw_arrow_style	FILLED
draw_arrow_width	.0416
radial_pattern_axis_circle	YES

STEP 3: Apply the new options and close the dialog box.

CREATING THE GENERAL VIEW

General views serve as the parent view for all views projected off of it. This section of the tutorial will create a general view as the front view of the drawing (see Figures 9–20 and 9–22).

STEP 1: Select the CREATE GENERAL VIEW tool on the drawing toolbar.

STEP 2: On the workscreen, select the location for the front view.

The drawing under creation in this tutorial will consist of a front view and a full section right-side view. The general view currently being defined will serve as the front view. On the workscreen, pick approximately where the front view will be located. Once the location is placed, you can drag the view with the mouse to a more suitable location. Before attempting to drag a view, ensure that you deselect the Lock View option on the drawing toolbar.

STEP 3: Use the Geometry References option to orient the view to match Figure 9–22.

On the Drawing View dialog box, use the appropriate geometric references to set the view shown in Figure 9–22. Because of the nature of the part, you will have to select a datum plane as a reference. Pay careful attention that your hole locations match those shown in the illustration.

STEP 4: Enter .500 as the scale value for the view.

STEP 5: Select OK on the dialog box.

STEP 6: Save your drawing.

Drawings are saved with an *.drw file extension.

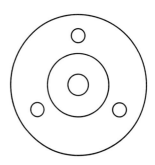

Figure 9–22
The Front View

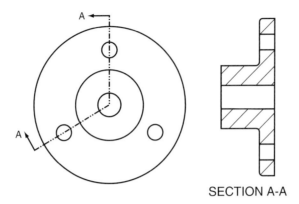

SECTION A-A

Figure 9–23   Aligned Section View

Creating an Aligned Section View

This segment of the tutorial will create the aligned full section view shown in Figure 9–23. The cutting plane for the section will match the path of the cutting plane line shown. The cross section defining this section view will be created within Drawing mode within this segment of the tutorial. Perform the following steps to create this view:

Step 1: Select INSERT >> DRAWING VIEW.

Step 2: Select PROJECTION as the view type.

Step 3: Place the view.

Step 4: Prepare to edit the view with the Drawing View dialog box (Edit >> Properties).

Step 5: On the Drawing View dialog box, select the SECTION category.

The Section category is used to select or create a cross section for your drawing.

Step 6: Select the 2D CROSS SECTION option.

Pro/ENGINEER allows you to either create a new cross section or retrieve one that was previously created. In this tutorial you will create the cross section as an offset.

Step 7: ✛ Select the Add Cross Section icon.

Step 8: Select OFFSET from the Cross Section Create menu.

Step 9: Select BOTH SIDES >> DONE.

Step 10: In Pro/ENGINEER's textbox, enter A as the name for the section view.

Step 11: Switch to the model's window.

If necessary, use the Windows taskbar to switch to the part from which the drawing is being created. When constructing an offset section, for projection views, you are required to select a sketching plane on the actual model. When selecting a sketching plane from a projection view, ensure that the model from which the drawing is being produced is in the active window. You must switch to this window at this step in the view creation process.

Step 12: As shown in Figure 9–24, select the top of the part to use as the sketching plane then select OKAY to accept the direction of viewing.

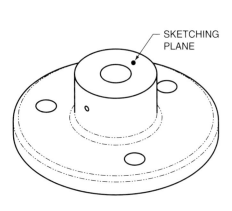

SKETCHING PLANE

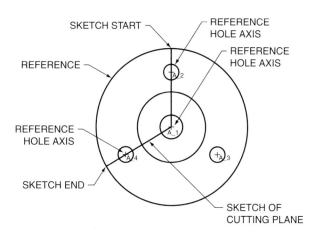

SKETCH START — REFERENCE HOLE AXIS

REFERENCE — REFERENCE HOLE AXIS

REFERENCE HOLE AXIS

REFERENCE HOLE AXIS

SKETCH END —

SKETCH OF CUTTING PLANE

Figure 9–24 Sketching Plane

Figure 9–25 Sketched Cutting Plane

INSTRUCTIONAL POINT If your part model was not opened before starting this segment of the tutorial, Pro/ENGINEER will open it automatically within its own subwindow. You will have to use the menu bar's Sketch >> References option to define the necessary references.

STEP 13: Select DEFAULT from the Sketch View menu to accept the default sketching environment orientation.

STEP 14: Specify the four references shown in Figure 9–25 (Sketch >> References).

Notice in the figure how the three axes are specified as references. These axes will be used to define the path of the cutting plane. Dynamically rotating the model will make the selection of each axis easier (middle mouse button). You can also use the pop-up menu's Pick From List option to make selection easier.

STEP 15: Use the LINE option to sketch the cutting path shown in Figure 9–25.

When sketching the path, make sure that each line is aligned with the references that were specified.

STEP 16: Select the Continue icon to end the cutting plane's definition.

STEP 17: If necessary, switch to the Window including the drawing.

STEP 18: In the Sectioned Area box on the Drawing View dialog box, select FULL(ALIGNED) as the section type.

This option will create a full section aligned about a selected axis.

STEP 19: To define the reference element for the aligned section, pick an axis of revolution (Figure 9–26).

STEP 20: In the Arrow Display option, select a view to locate the cutting plane line (Figure 9–23).

STEP 21: Select OK to exit the Drawing View dialog box.

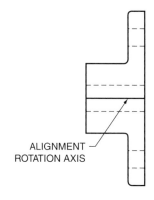

ALIGNMENT — ROTATION AXIS

Figure 9–26
Axis Selection

PARTIAL SECTIONED VIEW

This segment of the tutorial will create the Partial Sectioned view shown in Figure 9–27. Partial views are available with section and nonsection views.

STEP 1: Select INSERT >> DRAWING VIEW.

STEP 2: Select PROJECTION as the view type.

STEP 3: Project the view to the left of the front view.

STEP 4: Prepare to edit the view with the Drawing View dialog box (Edit >> Properties).

STEP 5: On the Drawing View dialog box, select the SECTIONS category.

STEP 6: Select the 2D CROSS SECTION option.

STEP 7: ✛ Select the Add Cross Section icon.

STEP 8: Select PLANAR >> SINGLE >> DONE on the Cross Section Create menu.

STEP 9: In Pro/ENGINEER's textbox, enter B as the name for the cross section.

STEP 10: In the front view of the drawing, select the datum plane that runs vertically through the drawing (see Figure 9–28).

The datum as shown in the figure will serve as the planar surface that defines the cross section. Select a corresponding datum plane.

STEP 11: Ensure that FULL is defined as the Sectioned Area.

STEP 12: Select the VISIBLE AREA category on the dialog box.

STEP 13: Select PARTIAL as the View Visibility.

The Partial View option requires you to sketch a spline to define the area that will be included in the view.

STEP 14: On the new left-side view, within the defining partial area, pick a reference point on an edge or boundary (Figure 9–29).

On the workscreen, in the left-side view, select a point on an entity that will reside in the partial section view. Pro/ENGINEER uses this selection to regenerate the section.

Note: If you have trouble selecting a point, dynamically zoom in on the view (Control key and middle mouse button).

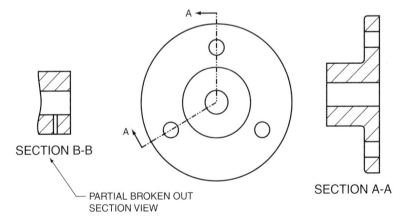

Figure 9–27   Partial Broken Out Section

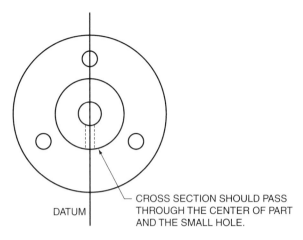

DATUM

CROSS SECTION SHOULD PASS
THROUGH THE CENTER OF PART
AND THE SMALL HOLE.

Figure 9–28 Cross Section Definition

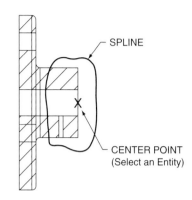

SPLINE

CENTER POINT
(Select an Entity)

Figure 9–29 Sketching the
Boundary

STEP 15: **As shown in Figure 9–29, sketch a spline that will define the partial view.**

The left mouse button is used to select points on the spline, while the middle mouse button is used to terminate and close the spline.

Note: If you have trouble selecting a point, dynamically zoom in on the view (Control key and middle mouse button).

STEP 16: **APPLY your changes, then close the dialog box.**

CENTERLINES AND DIMENSIONS

Within this segment of the tutorial, you will first show the centerlines shown in Figure 9–30. Second, you will display the parametric dimensions that define the part. Finally, you will set specific display modes for each view.

STEP 1: Select the SHOW/ERASE icon on the drawing toolbar.

STEP 2: **On the Show/Erase dialog box, select the SHOW option and the AXIS item type (Figure 9–31, on the next page).**

STEP 3: **On the Show/Erase dialog box, select the SHOW ALL option, then select YES to confirm the selection.**

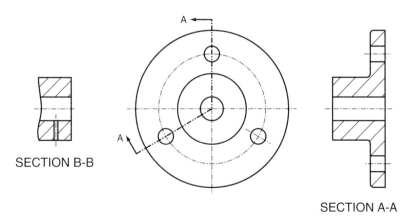

SECTION B-B

SECTION A-A

Figure 9–30 Centerlines

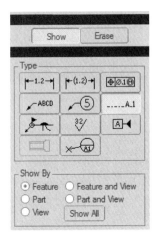

Figure 9–31
The Show/Erase Dialog Box

The Show All option will show all of a selected item type. In this example, all axes from the part will be projected as centerlines. Make sure that the Never Shown option is selected.

Note: If the With Preview option is selected, accept all of the centerlines shown on the workscreen.

STEP 4: **Use the Show/Erase dialog box to show all available dimensions.**

STEP 5: **Close the Show/Erase dialog box.**

STEP 6: **Use the mouse to drag dimensions and the mouse's pop-up menu's MOVE ITEM TO VIEW and FLIP ARROWS options to reposition the dimensions to match Figure 9–32.**

The mouse pop-up menu is available by first selecting the text of a dimension with your left mouse button, followed by reselecting the same dimension text with your right mouse button. (*Note:* A slight delay in releasing the right mouse button is necessary.) A variety of modification and manipulation tools are available through the revealed pop-up menu.

Use the following options to reposition your dimensions on the workscreen:

• **Mouse Move and Drag** Once selected, dimensions, including text and witness lines, can be moved and dragged on the screen with your mouse.

• **Move Item to View** The Move Item to View option is used to switch a dimension from one view to another. It is available under the pop-up menu. Multiple dimensions can be selected by a combination of the Control key and the left mouse button. Once dimensions are preselected, the pop-up menu is available through the right mouse button.

• **Flip Arrows** The Flip Arrows option is used to flip dimension arrowheads. It is available under the pop-up menu.

STEP 7: **Use mouse pop-up menu's ERASE option to hide dimensions not shown in Figure 9–32.**

Dimensions can be erased with the Show/Erase dialog box or the pop-up menu's Erase command.

MODELING POINT An important point to remember in Pro/ENGINEER is that any item erased is temporarily removed, while any item deleted is permanently removed. Erased dimensions can be redisplayed with the Show/Erase option.

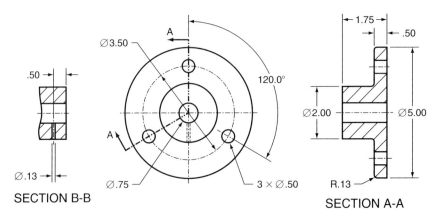

Figure 9–32   Drawing Dimensions

Step 8: 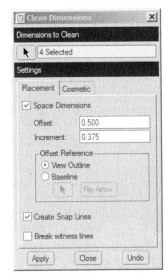 Select the Clean Dimensions icon to set dimension spacing.

The Clean Dimensions dialog box is used to consistently space dimensions.

Step 9: On the workscreen, use the Control key to select the two section views.

Step 10: Select OK on the Select menu.

Step 11: On the Clean Dimensions dialog box, set the options shown in Figure 9–33, then APPLY the settings.

Step 12: CLOSE the Clean Dimensions dialog box.

Step 13: Using the Control key, on the workscreen pick the two section views.

The next exercise in this tutorial will have you set the display mode for each view. In this example, you will set the two section views with a No Hidden display. You will also set the front view with a Hidden display and with a No Display Tangent display.

Step 14: With the two section views still picked, select the mouse pop-up menu's PROPERTIES option.

Step 15: Select the VIEW DISP option on the View Modify menu.

Step 16: Select NO HIDDEN >> DONE on the View Display menu.

Step 17: Select DONE on the View Modify menu.

Step 18: Pick the front view, then select the mouse pop-up menu's PROPERTIES option.

You will set the front view to display hidden lines and to not display tangent edges.

Step 19: Select VIEW DISP on the View Modify menu.

Step 20: Select HIDDEN LINE >> NO DISP TAN >> DONE.

Step 21: Select DONE on the View Modify menu.

Step 22: With your left mouse button, pick the 0.50 diameter hole dimension then access the pop-up menu with your right mouse button.

The next exercise in this tutorial will require you to modify the dimension text of the 0.500 diameter hole dimension to match Figure 9–34.

Step 23: Select the PROPERTIES option on the pop-up menu.

Step 24: On the Dimension Properties dialog box, select the Dimension Text tab (Figure 9–35).

Step 25: Modify the dimension parameters in the Dimension Text box to add 3 X in front of the existing text.

Figure 9–33 Clean Dimensions Dialog Box

3 × ⌀.50

Figure 9–34
Dimension Text Modification

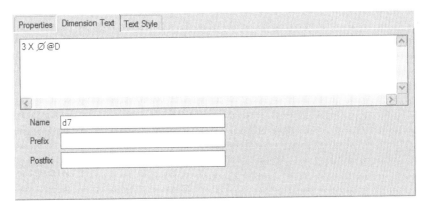

Figure 9–35 Dimension Text Tab

STEP 26:   Select OK to exit the Dimension Properties dialog box.

STEP 27:   Save your drawing file.

TITLE BLOCK NOTES

In this segment of the tutorial, create the notes for the title block as shown in Figure 9–36. Use the Insert >> Note menu option to create the required text and the pop-up menu's Text Style option to make text style and text height adjustments.

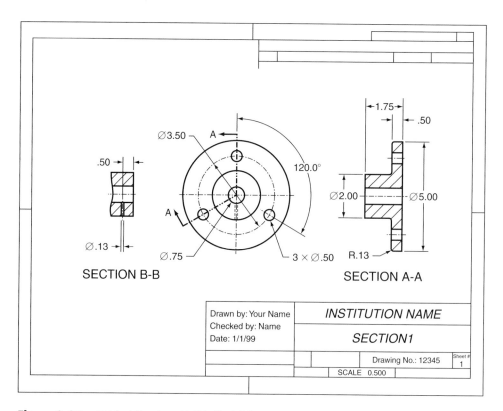

Figure 9–36   Finished Drawing with Title Block Information

ADVANCED DRAWING TUTORIAL 2

This tutorial exercise will provide instruction on how to create the drawing shown in Figure 9–37. The first step in this tutorial will require you to model the part from which the drawing will be created. Within this tutorial, the following topics will be covered:

- Starting a drawing.
- Establishing drawing setup values.
- Creating a broken view.
- Creating a partial auxiliary view.
- Annotating a drawing.
- Modifying dimension values.

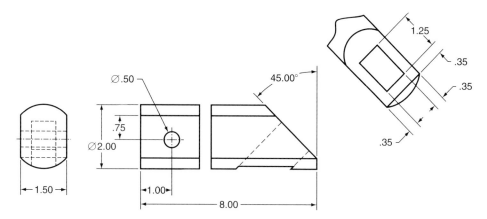

Figure 9–37   Completed Views

CREATING THE PART

Using Part mode, model the part shown in the drawing in Figure 9–37. Name your part file *auxiliary1*. Start the modeling process by creating Pro/ENGINEER's default datum planes. When defining features, the dimensioning scheme shown in the figure matches the design intent for the part. Incorporate this intent into your model.

STARTING A DRAWING

This section of the tutorial will create the new object file for the drawing to be completed. Do not start this section until the part shown in Figure 9–37 is complete.

STEP 1: Start Pro/ENGINEER.

STEP 2: Set an appropriate Working Directory.

STEP 3: Select FILE >> NEW.

STEP 4: In the New dialog box, select DRAWING mode, then enter *AUXILIARY1* as the name of the drawing file.

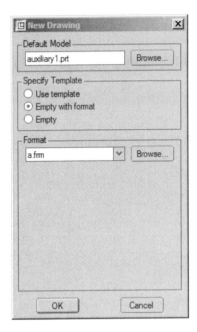

Figure 9-38   New Drawing Dialog
Box

STEP 5:   **Using the Default Template file, select OK on the New dialog box.**

After you select OK on the dialog box, Pro/ENGINEER will reveal the
New Drawing dialog box. This dialog box is used to select a model to create
the drawing from and to set a sheet size and format.

STEP 6:   **Under the Default Model option, select the BROWSE option and locate
the *AUXILIARY1* part (Figure 9–38).**

Use the Browse option to locate the *auxiliary1* part created in the first
section of this tutorial. This part will serve as the default model for the
creation of drawing views.

STEP 7:   **Under the Specify Template option, select the EMPTY WITH
FORMAT option.**

You will use the Empty with Format option to select an A size format.

STEP 8:   **Under the Format option on the New Drawing dialog box, select the
BROWSE option (Figure 9–38).**

The Browse option will allow you to browse the directory structure to
locate an existing sheet format.

STEP 9:   **Open *a.frm* as the standard sheet format (or select a format as assigned).**

STEP 10:   **Select OK on the New Drawing dialog box.**

After you select OK, Pro/ENGINEER will launch its Drawing mode.

ESTABLISHING DRAWING SETUP VALUES

This section of the tutorial will temporarily set Pro/ENGINEER's drawing settings. The
Advanced >> Draw Setup option will be used to change the default settings for this specific
drawing.

STEP 1:   **Select FILE >> PROPERTIES >> DRAWING OPTIONS.**

After you select Drawing Options, Pro/ENGINEER will open the active
drawing setup file with an Options dialog box.

STEP 2: Change the text and arrowhead values for the current drawing.

Within Options dialog box, make the changes shown below.

Drawing Setup Item	New Value
drawing_text_height	0.125
text_width_factor	0.750
dim_leader_length	0.175
dim_text_gap	0.125
draw_arrow_length	0.125
draw_arrow_style	FILLED
draw_arrow_width	0.0416
radial_pattern_axis_circle	YES

STEP 3: Apply the settings, then close the Options dialog box.

STEP 4: Select DONE/RETURN to exit the File Properties menu.

CREATING THE BROKEN FRONT VIEW

This section of the tutorial will create a broken general view as the front view of the drawing (Figure 9–39).

STEP 1: ⊬ Select the CREATE GENERAL VIEW tool on the drawing toolbar.

STEP 2: On the workscreen, select the location for the front view.

The drawing under creation in this tutorial will consist of a broken front view, a left-side view, and an auxiliary view.

STEP 3: Use the Geometry References option to orient the view to match Figure 9–22 on the next page.

On the Drawing View dialog box, use the appropriate geometric references to set the view shown in Figure 9–40. Because of the nature of the part, you will have to select a datum plane as a reference. Pay careful attention that your hole locations match those shown in the illustration.

STEP 4: Enter .750 as the scale value for the view.

STEP 5: Select the Visible Area category on the Drawing Properties dialog box.

STEP 6: Set BROKEN VIEW as the View Visibility.

STEP 7: ✛ Select the ADD BREAK option (Figure 9–41).

Figure 9–40 Oriented View

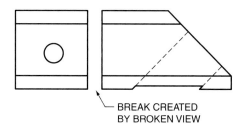

Figure 9–39
The Front View

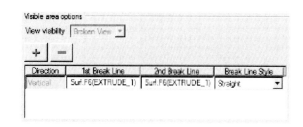

Figure 9–41
Add/Del Break menu

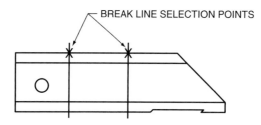

Figure 9–42   Break Line Selection

The Add Break option will add a pair of break lines to the view while the Delete option (-) will remove a pair. Break lines can be vertical or horizontal. In this tutorial, you will create a pair of vertical break lines.

STEP   8:   Pick the first edge shown in Figure 9–42, then sketch the vertical break line.

STEP   9:   Pick the second edge shown in Figure 9–42, then sketch the vertical break line.

STEP 10:   Select OK on the dialog box.

STEP 11:   Save your drawing.

PARTIAL AUXILIARY VIEW AND LEFT-SIDE VIEW

This section of the tutorial will create the partial auxiliary view and the left-side view shown in Figure 9–43.

STEP   1:   Select INSERT >> DRAWING VIEW >> AUXILIARY.

STEP   2:   Select the edge shown in Figure 9–44.

When creating an auxiliary view, Pro/ENGINEER requires the selection of an existing edge or axis. The auxiliary view will be projected from this edge.

STEP   3:   On the workscreen, pick the location for the auxiliary view (Figure 9–43).

The drawing under creation in this tutorial will consist of a broken front view, a left-side view, and a partial auxiliary view.

STEP   4:   Edit the auxiliary view's properties (Edit >> Properties).

STEP   5:   Select the Visible Area category on the Drawing View dialog box.

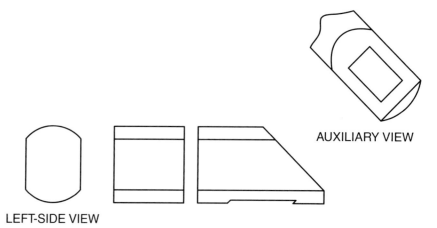

AUXILIARY VIEW

LEFT-SIDE VIEW

Figure 9–43   Partial Auxiliary View and Left-Side View

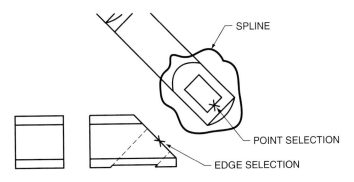

Figure 9-44 Auxiliary View Creation

STEP 6: Select PARTIAL VIEW as the View Visibility.

STEP 7: On the auxiliary view, select a point on an entity that will lie within the boundary of the partial view (Figure 9–44).

STEP 8: On the auxiliary view, sketch a spline that will serve as the boundary for the partial view (Figure 9–44).

Use the left mouse button to pick spline points and the middle mouse button to close the spline.

STEP 9: Select APPLY to create the auxiliary view.

STEP 10: CLOSE the dialog box and reposition your view as necessary.

STEP 11: Create a projected full view left-side view (Figure 9–43).

STEP 12: Reposition the views as shown in Figure 9–43.

STEP 13: Save your drawing.

ADDING DIMENSIONS AND CENTERLINES

This segment of the tutorial will add dimensions and centerlines to the drawing (Figure 9–45).

STEP 1: Select the Show/Erase icon on the drawing toolbar.

STEP 2: Select SHOW and the DIMENSION and AXIS item types (Figure 9–46).

The Dimension item type will show parametric dimensions while the axis option will create centerlines from existing axes.

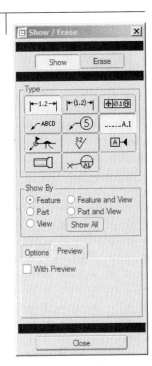

Figure 9-46 Show/Erase Item Types

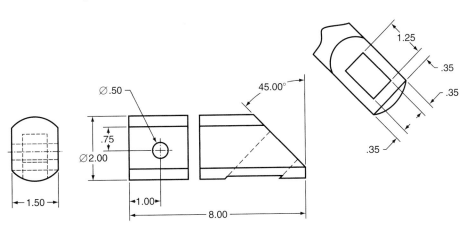

Figure 9-45 Dimensions and Centerlines

STEP 3:   Under the Preview Tab, deselect the WITH PREVIEW option (Figure 9–46).

STEP 4:   On the Show/Erase dialog box, select the SHOW ALL option, then select YES to confirm the selection.

STEP 5:   Close the Show/Erase dialog box.

STEP 6:   Use the mouse to drag dimensions and the mouse's pop-up menu's MOVE ITEM TO VIEW and FLIP ARROWS options to reposition the dimensions to match Figure 9–45.

The mouse pop-up menu is available by first selecting the text of a dimension with your left mouse button, followed by reselecting the same dimension text with your right mouse button. (Note: A slight delay in releasing the right mouse button is necessary.) A variety of modification and manipulation tools are available through the revealed pop-up menu.

Use the following options to reposition your dimensions on the workscreen:

- **Mouse Move and Drag** Once selected, dimensions, including text and witness lines, can be moved and dragged on the screen with your mouse.

- **Move Item to View** The Move Item to View option is used to switch a dimension from one view to another. It is available under the pop-up menu. Multiple dimensions can be selected by a combination of the Control key and the left mouse button. Once dimensions are preselected, the pop-up menu is available through the right mouse button.

- **Flip Arrows** The Flip Arrows option is used to flip dimension arrowheads. It is available under the pop-up menu.

STEP 7:   Select the Clean Dimensions icon on the drawing toolbar.

You will use the Clean Dimensions dialog box to consistently space the dimensions on the drawing.

STEP 8:   Using the Control key, pick the front, left-side, and auxiliary views, then select the middle mouse button.

STEP 9:   On the Clean Dimensions dialog box, take the default values of 0.500 (Offset) and 0.375 (Increment), then select the APPLY option.

The offset value will set the distance of the first dimension from the model. The Increment option will set the distance between each dimension.

STEP 10:   Close the Clean Dimensions dialog box.

STEP 11:   Use the mouse to drag dimensions and the mouse's pop-up menu's MOVE ITEM TO VIEW and FLIP ARROWS options to reposition the dimensions to match Figure 9–45.

The Clean Dimensions dialog box will not perfectly locate every dimension. Depending on the complexity of a drawing, you will probably have to reposition dimensions after using the Clean Dimensions option.

STEP 12:   On the toolbar, select the HIDDEN LINE display option to display hidden lines on each view.

STEP 13:   Save your drawing.

DIMENSION VALUE MODIFICATION

On a drawing, dimensions displayed by using the Show/Erase dialog box are parametric dimensions that are associated with other modes of Pro/ENGINEER. In this example, the dimensions displayed on the drawing are associated with the dimensions from the model in Part mode. Pro/ENGINEER is a biassociative parametric modeling application. Dimension

DIMENSION
SELECTION
1.25
AUXILIARY VIEW

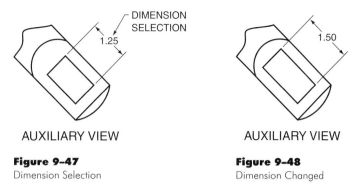

Figure 9–47
Dimension Selection

1.50
AUXILIARY VIEW

Figure 9–48
Dimension Changed

value changes made in one mode of Pro/ENGINEER will reflect in all modes. As an example, if you change the 8.00-in length of the part used in this drawing, this change will be reflected in the part and in the drawing. In this segment of the tutorial, you will use the 1.25-in dimension found in the Auxiliary view.

STEP 1: In the Auxiliary view, with your left mouse button, select the 1.25-in dimension (Figure 9–47).

STEP 2: Access the pop-up menu by selecting the dimension with your right mouse button.

STEP 3: On the pop-up menu, select the MODIFY NOMINAL VALUE option.

The Modify Nominal Value option is used to modify dimension basic or nominal values. The Properties option with the Dimension Properties dialog box can be used to modify dimension values also.

STEP 4: In Pro/ENGINEER's textbox, change the dimension value to equal 1.50.

STEP 5: **Regenerate the model.**

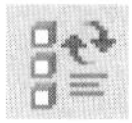

The changes in the part and in the drawing will be reflected after regenerating the model (Figure 9–48).

STEP 6: Open the object in Part mode to observe the changes to the model.

STEP 7: In Drawing mode, use the MODIFY NOMINAL VALUE option to change the value of the previously modified dimension back to a value of 1.25.

You can also modify the value in part mode.

STEP 8: **Regenerate the model.**

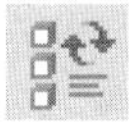

STEP 9: Save your drawing.

TITLE BLOCK INFORMATION

Use the Insert >> Note option to create the title block information shown in Figure 9–49. Use the pop-up menu's Text Style option to make changes to text style and height. Your final drawing should appear as shown in Figure 9–50.

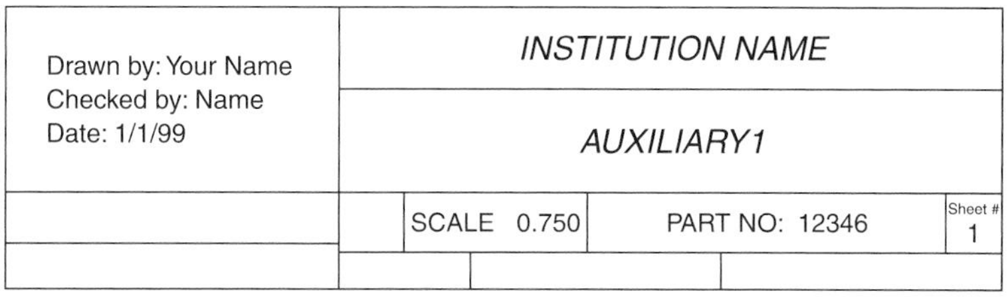

Drawn by: Your Name
Checked by: Name
Date: 1/1/99
INSTITUTION NAME
AUXILIARY1
SCALE 0.750
PART NO: 12346
Sheet #
1

Figure 9–49 Title Block Information

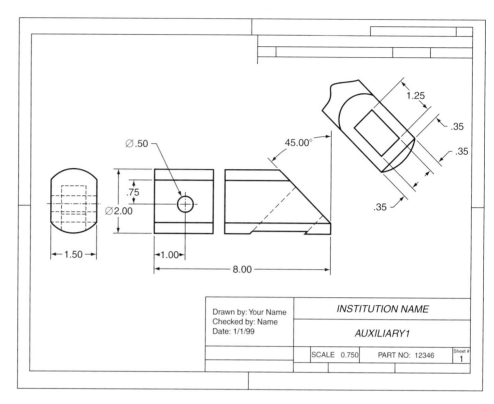

Figure 9–50   Final Drawing

PROBLEMS

1. Model the part shown in Figure 9–51, then create a detailed drawing of the part. When completing this problem, meet the following requirements:

    •   The dimensions shown in the figure should meet design intent. During part modeling, incorporate these dimensions.

    •   Create an engineering drawing with front and top views. The front view should be a full section view. (*Note:* Figure 9–51 is a half section view.)

    •   Use an A size sheet and format.

    •   Fully dimension the engineering drawing using the part's parametric dimensions.

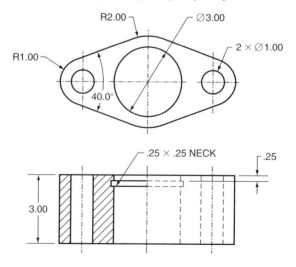

Figure 9–51   Problem 1

2. Model the part shown in Figure 9–52, then create a detailed drawing of the part. When completing the problem, meet the following requirements:

   • The dimensions shown in the figure should meet design intent. During part modeling, incorporate these dimensions.

   • Create an engineering drawing with front and top views. The front view should be an offset full section view.

   • Use an A size sheet and format.

   • Fully dimension the engineering drawing using the part's parametric dimensions.

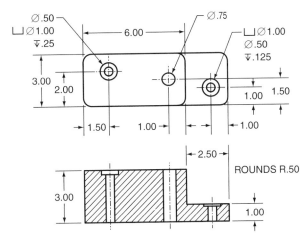

Figure 9–52 Problem 2

3. Model the part shown in Figure 9–53, then create a detailed drawing of the part. When completing this problem, meet the following requirements:

   • The dimensions shown in the figure should meet design intent. During part modeling, incorporate these dimensions.

   • When modeling the part, use the Radial Hole and Pattern commands to create the bolt-circle pattern.

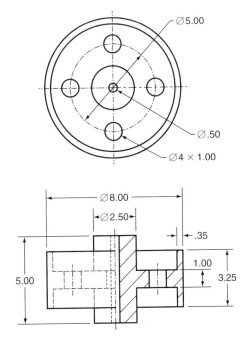

Figure 9–53 Problem 3

- Create an engineering drawing with front and top views. The front view should be a half section view.

- Use an A size sheet and format.

- Fully dimension the engineering drawing using the part's parametric dimensions.

4. Model the part shown in Figure 9–54, then create a detailed drawing of the part. When completing this problem, meet the following requirements:

- The dimensions shown in the figure should meet design intent. During part modeling, incorporate these dimensions.

- When modeling the part, use the Radial Hole and Pattern commands to create the bolt-circle pattern.

- Create an engineering drawing with front, top, right-side, and auxiliary views. Create the auxiliary view with the Of Surface view type option.

- Use an A size sheet and format.

- Fully dimension the engineering drawing using the part's parametric dimensions.

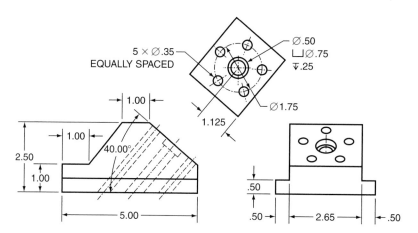

Figure 9-54   Problem 4

QUESTIONS AND DISCUSSION

1. Describe possible uses of a section view.

2. Describe different section view types available within Pro/ENGINEER.

3. What is the difference between a total cross section and an area cross section?

4. What is the purpose of a cutting plane line?

5. In Pro/ENGINEER, how is the cutting plane created for an offset section?

6. In Pro/ENGINEER, what is the difference between a broken view and a broken out view?

7. What is a revolved section when used on an engineering drawing?

8. Describe uses of an auxiliary view.

10

SWEPT AND BLENDED FEATURES

Introduction

Covered previously in this text, the Extrude tool and the Revolve tool are the two basic feature creation options available within Pro/ENGINEER's part mode. Each tool has the capability of creating protrusions, cuts, and surface features. This chapter expands on these tools by introducing the Sweep and Blend commands. In addition, datum curves, datum points, and coordinate systems will be covered. Upon finishing this chapter, you will be able to create:

- A swept protrusion feature.
- A blended protrusion feature.
- A swept cut feature.
- A swept protrusion feature.
- A Datum curve.
- A Datum point.
- A Coordinate system.

DEFINITIONS

Blended feature   A feature with two or more user-sketched sections.

Swept feature   A feature with a user-sketched section and a user-defined trajectory.

Trajectory   The extrude path of a swept feature. Sweep trajectories can be either sketched by the user or selected on the workscreen.

SWEEP AND BLEND FUNDAMENTALS

Previously covered in this textbook, the Extrude and Revolve commands can be used to create positive space, negative space, or surface features. This chapter expands on these commands by introducing the Sweep and Blend options. In comparison, while the Extrude command creates a feature by protruding a section along a straight trajectory, the Sweep command creates a section along a user-defined trajectory (Figure 10–1). This trajectory can be either user-sketched or selected on the workscreen. The Blend option can also be compared to the Extrude option. Primarily, the Blend option creates a feature by protruding along a straight trajectory between two or more user-defined sections. A partially revolved blend can also be created.

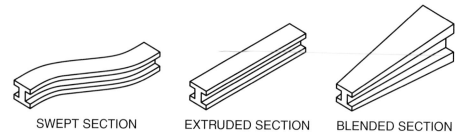

SWEPT SECTION   EXTRUDED SECTION   BLENDED SECTION

Figure 10-1   Swept, Extruded, and Blended Features

SWEPT FEATURES

Protruding a section along a user-defined trajectory gives the geographic definition of a swept feature. Figure 10–1 shows an example of a sweep. The section in this feature is protruded along a curved trajectory. Also in Figure 10–1, the same section is shown protruded along a straight trajectory with the Extrude option. Like the Extrude and Revolve options, the Sweep option is available with the Protrusion and Cut commands.

The Sweep option requires the definition of one section and one trajectory. The order of operation requires the creation of the trajectory first followed by the creation of the section. A sweep's trajectory can be either sketched or selected. For most situations, sketching the trajectory is the preferred method. For sketching the trajectory, normal sketching tools are used. The trajectory can be opened or closed. Any planar surface or datum plane can be used to sketch the trajectory. Because of the nature of the sketching environment, a sketched trajectory is two-dimensional only. A selected trajectory can be used to define a three-dimensional path. A part edge or datum curve can be used as the selected trajectory.

Within the Sweep option, when a section is swept, cross sections of the section are created normal to the trajectory's path. Along a curved path, when the feature is regenerated, these normal sections cannot overlap. An overlap can occur by having a section that is too large for its corresponding trajectory, or it can occur by having a trajectory that is too small for its corresponding section. Both situations often occur when a radius in a trajectory is too small. Figure 10–2a shows a swept feature with a small section. When the section is increased to a larger size (Figure 10–2b), an overlap in the feature's sections can occur. In Figure 10–2b, the section is on the verge of being too large. In this situation, in order to increase the section size, you will have to increase the radii of the trajectory's arcs. If the

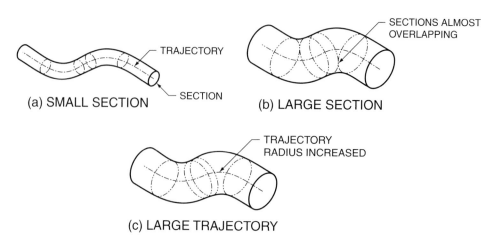

(a) SMALL SECTION   (b) LARGE SECTION

(c) LARGE TRAJECTORY

Figure 10-2   Overcoming Regeneration Failures

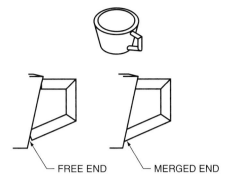

FREE END   MERGED END

Figure 10–3   Trajectory Attributes

section should overlap in the bend of the trajectory, you will get a failed regeneration. There are three ways to resolve this problem:

- Create a smaller section.
- Create a larger trajectory.
- Make the arcs on a trajectory larger.

When a swept feature's open trajectory meets with an existing feature or features, the user is provided with the option of either merging the ends of the trajectory or leaving the ends free (Figure 10–3). The Merge Ends option will smoothly merge the new swept feature with adjacent solid features. The trajectory has to be aligned with the existing features.

CREATING A SWEEP WITH A SKETCHED SECTION

A sweep can be used to create protrusions, cuts, thin protrusions, thin cuts, and surface features. Examples of a swept cut and a swept thin cut are shown in Figure 10–4. The following guide will demonstrate how to create the swept protrusion feature shown in Figure 10–5. The same steps are valid for cut and thin features. When creating any swept feature (e.g., sweep, swept blend, helical sweep, variable-section sweep), the first step is to define the trajectory. In this example, the sketching plane for the trajectory will be one of Pro/ENGINEER's default datum planes.

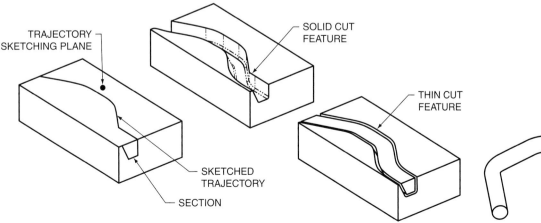

TRAJECTORY SKETCHING PLANE

SOLID CUT FEATURE

THIN CUT FEATURE

SKETCHED TRAJECTORY

SECTION

Figure 10–4   Sweep Options

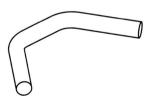

Figure 10–5   Sweep Feature

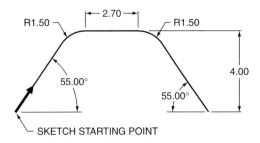

Figure 10–6   Trajectory Sketch

STEP 1:   Select INSERT >> SWEEP >> PROTRUSION on the menu bar.

STEP 2:   Select SKETCH TRAJ as the method of trajectory definition.

Sweep trajectories can be either user-sketched or selected on the workscreen. All trajectories, regardless of method of creation, must be two-dimensional.

STEP 3:   Select a sketching plane (Datum Plane Front).

The process for selecting a sketching plane is the same within the Sweep option as it is within most Pro/ENGINEER feature options. Any planar surface or datum plane can be selected.

STEP 4:   Select OKAY to accept the default view direction for your sketch.

STEP 5:   Select TOP, then select a datum plane to orient toward the top of the workscreen.

STEP 6:   Sketch the Trajectory (Figure 10–6).

Use normal Pro/ENGINEER sketching tools to create the trajectory. A trajectory's path can be open or closed. When sketching the trajectory, avoid small-radius arcs. Keep in mind that the section for the feature will be sketched at the starting point of the trajectory and normal to the trajectory at this point. By allowing you to preselect a vertex, the pop-up menu provides an option for defining the picked vertex as the starting point for the section.

INSTRUCTIONAL NOTE   This guide demonstrates the creation of a sweep as the first geometric feature of a part. The Free Ends/Merge Ends option will not be available for this example.

STEP 7:   Select the Continue icon to exit the trajectory sketching environment.

STEP 8:   If requested by Pro/ENGINEER, select either FREE ENDS or MERGE ENDS. (Not available for first geometric feature.)

The Merge Ends option will smoothly merge the swept feature with existing geometry. The trajectory has to be aligned with an existing feature. This option is not available for the first geometric feature of a part.

STEP 9:   Sketch the section for the swept feature (Figure 10–7).

Upon entering the sketching environment for the swept feature's cross section, the environment will be oriented normal to the trajectory's start point. Often, it is helpful to dynamically rotate the sketching environment to better visualize the position of this starting point. Use normal sketching tools to create the section. The section for a Sweep has to be closed.

STEP 10:   ✔ Select the Continue icon to exit the sketching environment.

STEP 11:   Preview the sweep on the Feature Definition dialog box (Figure 10–8).

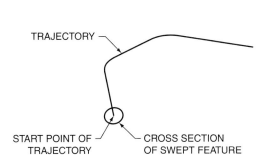

TRAJECTORY

START POINT OF
TRAJECTORY

CROSS SECTION
OF SWEPT FEATURE

Figure 10–7 Sketching Cross Section

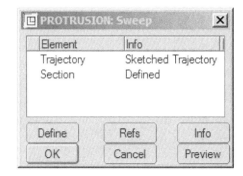

Figure 10–8 Sweep Definition Dialog Box

Previewing a swept feature is important. If a failed regeneration occurs, either enlarge your trajectory (especially radii) or decrease the size of the section.

STEP 12: **Select OK on the dialog box.**

BLENDED FEATURES

A blend is a feature created from two or more planar sections. Blend sections are joined together at their edges to form one feature. Three types of blends are available: parallel, rotational, and general. An example of a parallel blend is shown in Figure 10–9. With this type of blend, each section of the feature is sketched in the same sketching environment. The sections defining the feature are shown in the figure. Additionally, Pro/ENGINEER provides the attribute option for creating either a smooth blend or a straight blend. Both examples are shown in Figure 10–9.

The second blend type is a rotational blend. Rotational blends are created by sketching two or more sections with a rotational angle defined between each section. Within each sec-

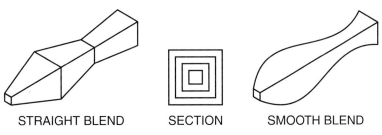

STRAIGHT BLEND   SECTION   SMOOTH BLEND

Figure 10–9 Parallel Blend Examples

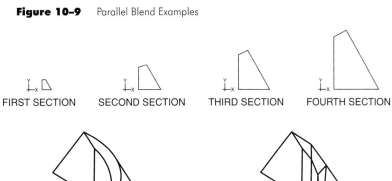

FIRST SECTION   SECOND SECTION   THIRD SECTION   FOURTH SECTION

SMOOTH BLEND   STRAIGHT BLEND

Figure 10–10 A Rotational Blend

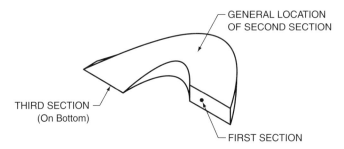

Figure 10–11   A General Blend

tion, the user must create a coordinate system. This coordinate system defines the pivot point for each section with sections revolved around the Y-axis of the coordinate system. Figure 10–10 shows an example of a rotational blend with four sections. Each section has an angular spacing of 45 degrees. The maximum possible angle between two sections is 120 degrees. Smooth and straight blended features for these sections are shown.

The Parallel Blend option will create a feature by joining two or more parallel sections and the Rotation Blend will create a feature by joining two or more sections by revolving the sections about a coordinate system. The third blend option, General, will create a feature by rotating sections about all three axes of each section's coordinate system. The construction process for a general blend is similar to the rotational blend. Unlike a rotational blend, where sections are rotated about the Y-axis, the General option will allow simultaneous rotations about all three axes. Figure 10–11 shows an example of a general blend with three sections. The second section is rotated about the Y-axis with an angle of 90 degrees. The third section is rotated about the X-axis and Y-axis an angle of 90 degrees for both axes.

CREATING A PARALLEL BLEND

This tutorial guide will demonstrate the modeling of the blended feature shown in Figure 10–12. This feature will be sketched on one of Pro/ENGINEER's default datum planes. There are several points to remember when you are creating a parallel blend:

- A blended feature must have two or more sections. A user-defined distance will separate each section.

- In most situations, each section of a blend must have the same number of entities. The only exception to this rule is if one section has a single point entity. Notice in Figure 10–12 that the first two sections consist of line and arc entities. The outside section has eight line entities, while the middle section consists of four lines and four arcs. The third section is composed of a single point entity.

- The starting point for each section should be in the same general location and normally should be pointing in the same direction. Figure 10–13 shows the starting point for each section (the point section does not have a visible starting point). Notice the arrows that graphically display the starting point and direction for each section. Figure 10–13 shows what would happen to a feature if the starting points were not placed correctly.

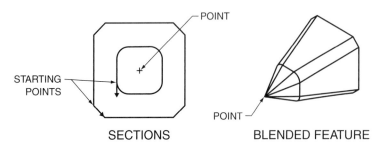

SECTIONS   BLENDED FEATURE

Figure 10–12   Blended Feature

Figure 10–13   Twisted Blend

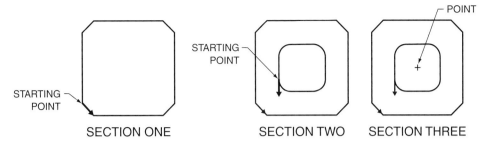

Figure 10–14 Blended Feature Sections

Perform the following steps to create the blended feature shown in Figure 10–12. (*Note:* Dimensions have been omitted from each illustration. You can select the size for the sections and the depth values.)

STEP 1: Select INSERT >> BLEND >> PROTRUSION on the menu bar.

STEP 2: Select PARALLEL >> REGULAR SEC >> DONE on the Blend Options menu.

In a parallel blend, all defined sections are parallel, which in turn creates a straight trajectory.

STEP 3: Select STRAIGHT >> DONE on the Attributes menu.

STEP 4: Select a sketching plane (Datum Plane Front).

The process for selecting a sketching plane is the same within the Blend option as it is within most Pro/ENGINEER feature options. Any planar surface or datum plane can be selected.

STEP 5: Select OKAY to accept the default view direction for your sketch.

STEP 6: Select TOP, then select a datum plane to orient toward the top of the workscreen.

STEP 7: Use appropriate sketching tools to sketch Section One (Figure 10–14).

STEP 8: Use the pop-up menu's START POINT option to place the section's starting point as shown. (You must preselect the new start point's vertex location before accessing the pop-up menu.)

On the menu bar, the SKETCH >> FEATURE TOOLS >> START POINT option is also available for changing a sections start point. As with the pop-up menu's Start Point option, the vertex for the new starting point has to be preselected.

STEP 9: On the menu bar, select SKETCH >> FEATURE TOOLS >> TOGGLE SECTION (or select Toggle Section on the pop-up menu).

Within the Parallel option, sections are sketched within the same sketching environment. The Toggle Section option is used to move from one section to another. To end the creation of sections for a parallel blend, select Toggle Section without sketching any entities.

STEP 10: Use appropriate sketching tools to sketch Section Two (Figure 10–14).

STEP 11: Use the pop-up menu's START POINT option to place the section's starting point as shown in the figure.

STEP 12: On the menu bar, select SKETCH >> FEATURE TOOLS >> TOGGLE SECTION.

Note: The Toggle Section command is also on the pop-up menu.

STEP 13: Use the POINT icon to create the point entity shown in Section Three.

The Point option can be found behind the Coordinate System icon on the sketcher toolbar.

STEP 14:   ✔ Select the Continue icon to exit the sketching environment.

STEP 15:   Enter a depth value to define the distance between the first section and the second section.

STEP 16:   Enter a depth between the second section and the third section.

STEP 17:   Preview the feature on the Blend Feature Definition dialog box.

STEP 18:   Select OK on the dialog box.

DATUM CURVES

Datum curves are used extensively for the creation of swept features. Additionally, they are used for the creation of surface features. Datum curves are considered features within Pro/ENGINEER and are labeled with a *Curve_id* on the model tree.

Datum curves can be sketched with normal sketching tools. For surfacing operations, it is often necessary to construct single curves made of many segments. Additionally, datum curves may be opened or closed. The following curve options are available:

SKETCH TOOL

The Sketching environment can be used to create a trajectory or curve.

THROUGH POINTS

The Through Points option places a datum curve through existing datum points. The curve created can be splined or can have user-defined radii. Individual datum points can be picked or a datum point array can be selected using the Whole option.

FROM FILE

This option imports a datum curve from an IGES, VDA, SET, or Pro/ENGINEER *.ibl file.

USE XSEC

The Use Cross Section option is used to create a datum curve at the intersection of a planar cross section and the outer boundary of a part.

FROM EQUATION

The From Equation option is used to create a datum curve through the use of a parametric equation. The parametric equation is defined in terms of T, with T varying from 0 to 1.

THROUGH POINTS DATUM CURVE

Perform the following steps to create a datum curve through datum points. This guide assumes the existence of two or more datum points. The Through Points option will create a datum curve through multiple single points/vertices or through a datum point array.

STEP 1:   Select the Datum Curve icon on the datum toolbar.

Step 2:   Select THRU POINTS >> DONE on the Curve Options menu.

Step 3:   Select a connection type.

The following connection types are available:

- **SPLINE** The defined curve is created as a spline through selected points.

- **SINGLE RAD** The datum curve is defined as a straight line between selected points with a bend at the intersection of line segments. The radii for all bends are equal.

- **MULTIPLE RAD** The datum curve is defined as a straight line between selected points with a bend at the intersection of line segments. The radii for bends can have unique values.

Step 4:   Select points.

Single points and points within a datum point array can be selected.

Step 5:   For single-radius or multiple-radius curves, enter a radius value.

Step 6:   For multiple-radius curves, pick points for the redefinition of the radius values.

Step 7:   Select DONE on the Connection Type menu.

Step 8:   Select OK on the Curve Feature Definition Dialog box.

PROJECTED AND WRAPPED CURVES

Projected and wrapped curves are datum curves that are constructed by sketching a section or selecting an existing datum curve and projecting it onto one or more surfaces. Both options are available under the Edit menu.

PROJECTED DATUM CURVE

A projected datum curve forms a true projection of the curve being projected. Figure 10–15 shows how a projected datum curve appears on the receiving surface.

WRAPPED DATUM CURVE

A wrapped datum curve does not form a true projection when projected onto the receiving surface. As shown in Figure 10–15, the shape of the original curve is pasted onto the receiving surface as opposed to an actual projection.

The following are terms associated with projected and wrapped curves:

NORMAL TO SURFACE

The Normal to Surface option projects a section or curve perpendicular to the plane or surface receiving the projection. This option is used with projected datum planes.

ALONG DIRECTION

This option projects a curve along a specified direction. Planes, curves, edges, axes, and coordinate systems can be used as the defining direction.

DATUM POINTS

Datum points are useful for numerous modeling situations. Like all datums, points are considered part features and can be redefined. The following reference options are available for creating datum points:

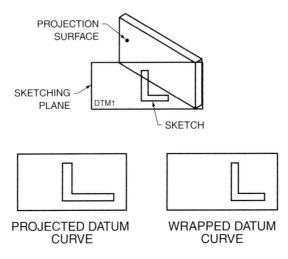

PROJECTION
SURFACE

SKETCHING
PLANE

DTM1

SKETCH

PROJECTED DATUM
CURVE

WRAPPED DATUM
CURVE

Figure 10–15   Projected and Wrapped Datum Curves

ON SURFACE

The On Surface reference option places a datum point on an existing surface or plane. The user must pick the placement surface as the reference then the drag the point's location markers to two reference edges. The distance to each location reference can be redefined.

CURVE INTERSECTION SURFACE

The Curve at the Intersection of a Surface reference places a datum point at the intersection of a curve and surface. The following elements are available for selection as a curve or surface:

• **Curve** Part edge, part curve, part axis, or datum curve.

• **Surface** Part surface or datum plane.

ON VERTEX

This option places a datum point at the vertex of a datum curve or part edge.

THREE SURFACES

The Three Surfaces option places a datum point at the intersection of three surfaces and/or planes.

ON CURVE

This reference option places a datum point on an existing curve. Options are available for offseting the point as a ration of the curve length or providing a real distance value.

ON CURVE/EDGE

The On Curve/Edge reference option places a datum point on a curve (Figure 10–16). The curve may be a part edge, part curve, or datum curve. Three options are available:

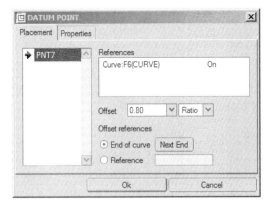

Figure 10-16 On Curve/Edge Reference Option

- **Reference** When utilizing the Reference edit box (See Figure 10–16), after picking the curve on which to place the datum point, you are required to select a planar surface from which to reference the point location.

- **Ratio** When you are utilizing the Ratio option, a ratio value is entered to represent the location on the curve of the datum point from one of the curve's vertices. The ratio value must be between 0.0 and 1.0.

- **Real** The Real option locates the datum point on the selected curve at a user-provided distance from one of the curve's vertex. The distance cannot be greater than the curve length.

CURVE INTERSECTING CURVE

The Curve Intersecting Curve option places a datum point on a selected curve at a location that represents the minimum distance from a second selected curve. The created datum point represents the location where the two selected curves are at their closest point to each other. Curves under this option do not have to intersect.

OFFSET POINT

This reference option offsets a datum point from an existing point or vertex. The direction of offset is defined by a straight datum curve or part edge.

OFFSET COORDINATE SYSTEM

The Offset Coordinate System reference option creates an array of datum points offset from an absolute Cartesian, spherical, or cylindrical coordinate system. By default, modifiable dimensions are provided for each datum point. The Convert To Non Parametric Array suboption will remove dimension values. Additionally, points may be read from an ASCII file. (See Figure 10–17.)

MODELING POINT The Offset Coordinate System option is used to create a datum point array. Datum point arrays are multiple datum points created within the same step.

SKETCHED DATUM POINT

The Sketched reference option creates one or more datum points in a sketching environment. Unlike a standard sketching environment, geometry options such as line and circle are not available.

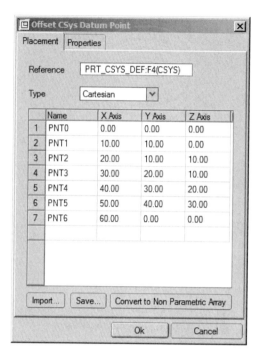

Figure 10–17   Datum Point Array

COORDINATE SYSTEMS

Coordinate systems are useful for a variety of applications. Because of their feature-based modeling and sketching capabilities, most parametric design applications, such as Pro/ENGINEER, only passively use coordinate systems. Even though Pro/ENGINEER does not utilize a coordinate system as a primary means of creating geometry, many modeling tasks do require the presence of one. A computer numeric control machine tool is another example of a manufacturing application that utilizes this type of coordinate system. The following are some of the Pro/ENGINEER tasks that require a coordinate system:

- Many modeling tools, such as the Move/Rotate option under the Copy command, can utilize a coordinate system. Many such applications also provide options that do not require a coordinate system.

- Many finite-element analysis applications require a coordinate system.

- Pro/NC (Pro/ENGINEER's manufacturing module) requires a coordinate system to help in the definition of tool paths.

- Datum point arrays require the use of a coordinate system.

- Model mass property calculations can utilize a coordinate system.

TYPES OF COORDINATE SYSTEMS

There are three types of coordinate systems utilized within Pro/ENGINEER. Of the three, the Cartesian coordinate system is the primary one used. The following is a description of each coordinate system.

CARTESIAN COORDINATE SYSTEM

A Cartesian coordinate system is defined by three orthogonal axes. A typical Cartesian coordinate system's axes are labeled X, Y, and Z. Each axis of the coordinate system is divided into system units. Each axis intersects at a common point, and units along each axis may be positive or negative, depending on the direction of the axis. Elements, such as datum points, are defined by a distance along each axis.

CYLINDRICAL COORDINATE SYSTEM

The cylindrical coordinate system is built around a type of Cartesian coordinate system. Like the Cartesian coordinate system, this coordinate system also uses three values to define a location. Instead of three axes meeting at one common absolute point, this system is defined by a radius value, a theta value, and a Z value. The radius value is the distance from the absolute point on the X-Y plane and the theta value is the angle from a selected reference. The Z value is identical to the Cartesian coordinate system's Z value.

SPHERICAL COORDINATE SYSTEM

Like the cylindrical coordinate system, the spherical coordinate system is built around a type of Cartesian coordinate system. Also like the cylindrical coordinate system, the spherical system uses a radius value and a theta value to define the first two values of the element point. The third value is defined by phi. Phi equals an angle from the Z-axis.

COORDINATE SYSTEM REFERENCES

The Coordinate System dialog box is used to define coordinate system features (Figure 10–18). The definition of a coordinate system requires the proper selection of references. The following reference options are available:

THREE PLANES

The Three Planes reference option requires the selection of three planes. The planes selected can be datum planes or part surfaces. Planes selected do not have to be perpendicular. The Origin tab, shown in Figure 10–18, displays the creation of a coordinate system with this option.

MODELING POINT The Three Planes option is commonly used to create a coordinate system at the intersection of Pro/ENGINEER's default datum planes. It is also used to place a coordinate system at the corner of a part.

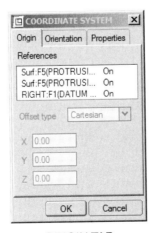
ORIGIN TAB

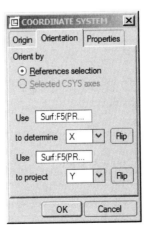
ORIENTATION TAB

Figure 10–18 Coordinate System Dialog Box

POINT AND TWO AXES

The Point Plus Two Axes option will place the absolute point of a coordinate system at an existing datum point. The datum point is the only reference with this option. Two of the three coordinate system axes are defined on the Orientation tab (Figure 10–18) by selecting existing datum axes with the available Use edit box. These datum axes do not have to pass through the datum point.

TWO AXES

With the Two Axes option, the coordinate system is defined by two selected axes. The origin of the coordinate system will lie at the intersection of the two axes. The user is required to define the orientation of a plane through the origin and first axis.

OFFSET COORDINATE SYSTEM

The Offset Coordinate System reference option creates a coordinate system referenced from a second coordinate system. Offset values are entered along each axis of the referenced coordinate system. A suboption is available for orienting the Z-axis of the coordinate system normal to the workscreen.

PLANE AND TWO AXES

The Plane Plus Two Axes option places the origin of the coordinate system at the intersection of a selected plane and axis. The user is also required to select a third axis or an edge to orient the coordinate system.

CREATING A CARTESIAN COORDINATE SYSTEM

Perform the following steps to create a coordinate system.

STEP 1: Select the Coordinate System icon on the datum toolbar.

STEP 2: Use the Origins tab of the Coordinate System dialog box to select references for the origin of the coordinate system.

Refer to the previous section for a discussion of coordinate system references. Remember to use the Control key to pick multiple references.

STEP 3: Use the Orientation tab of the Coordinate System dialog box to orient the coordinate system's axes.

When using the References Selection option on the Orientation tab, a reference direction can be set as the X-, Y-, or Z-axis. A Flip option is available for changing the direction of an axis.

STEP 4: Select OK when the coordinate system is defined.

SUMMARY

The Extrude and Revolve tools are useful for creating basic geometric features. Many parts, though, are composed of shapes that do not follow normal extrude or revolve directions. The Sweep and Blend options are available for more advanced shapes. Both options can be compared to the Extrude option. While Extrude protrudes a section in a straight direction, the Sweep option protrudes a section along a user-defined trajectory. The Blend option protrudes two or more sections to form a feature. Later in this text, the Sweep-Blend option will be introduced. This option combines the functions of the Sweep option and the Blend option.

BLEND TUTORIAL

This tutorial exercise will provide instruction on how to create the part shown in Figure 10–19. This part consists of an extruded protrusion, two blends, and an extruded cut. The first step in this tutorial will require you to construct the base extruded protrusion. Within this tutorial, the following topics will be covered:

- Creating a base protrusion.
- Creating a blend.
- Creating a section with the Use Edge option.

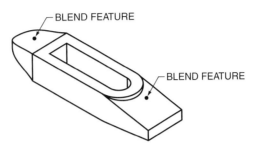

Figure 10-19   Completed Part

CREATING THE BASE FEATURE

Use Pro/ENGINEER's default datum planes to create this part. Name the part *Blend1*. Create the feature shown in Figure 10–20 as the base geometric feature of the part. Sketch the feature's section on datum plane FRONT. Extrude the feature one direction a distance of 2.00 in.

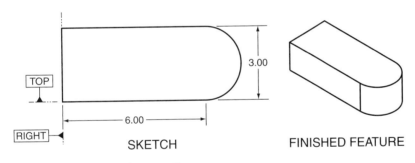

Figure 10-20   Base Geometric Feature

CREATING A BLEND

This segment of the tutorial will create the blended feature shown in Figure 10–21. The sketching plane for this feature will be an on-the-fly datum plane constructed through the two tangent edges formed on the base feature.

STEP 1:   **Select INSERT >> BLEND on the menu bar.**

STEP 2:   **Select PROTRUSION.**

A blend can be created as a protrusion, cut, or surface feature.

STEP 3:   **Select PARALLEL >> REGULAR SEC >> DONE on the Blend Options menu.**

A parallel blend consists of two or more sections extruded along a straight trajectory.

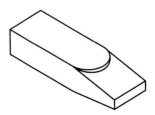

Figure 10-21
Blended Feature

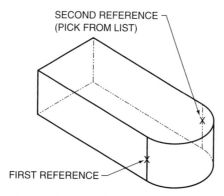
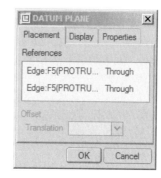

Figure 10–22   Edge Selection

STEP 4:   **Select STRAIGHT >> DONE on the Attributes menu.**

A blended feature joins the vertices of two or more sections. The connecting lines between vertices can be straight or they can be smooth.

STEP 5:   **Select the Datum Plane icon on the datum toolbar.**

You will create an on-the-fly datum plane through the tangent edges shown in Figure 10–22.

STEP 6:   **Using the Control key, pick the first reference location shown in Figure 10–22.**

The Through constraint option will place a datum plane through an axis, edge, curve, point, plane, or cylinder. In this tutorial, you will place the datum plane through an edge. The edge selection requires a paired constraint option.

STEP 7:   **Right-mouse select over the second reference location and select the PICK FROM LIST option on the pop-up menu.**

Elements and entities are often hard to select for complex or hidden features. In this example, you could dynamically rotate the part or use the Pick From List option to select the tangent edge. Pick From List will reveal available entities, features, and elements at a selection location.

STEP 8:   **On the Pick From List dialog box, select the Edge:F5(PROTRUSION) element, then ACCEPT the selection.**

The Pick From List option will reveal valid elements at a selected location. In this example, three possible elements are available: an edge and two surfaces.

STEP 9:   **When your Datum Plane dialog box looks similar to Figure 10–22, select OK to create the datum plane.**

Notice in Figure 10–22 how two edges have been selected as references. Two edge paired references will define the datum plane through each edge.

STEP 10:   **If necessary, use the FLIP option to select a direction of feature creation that will extrude the feature away from the first protrusion feature.**

The arrow displayed on the workscreen defines the direction of feature creation. Your arrow should point toward the circular end of the base protrusion feature.

STEP 11:   **When your direction of feature creation is correct, select OKAY on the Direction menu.**

STEP 12:   **Orient the sketching environment to match Figure 10–23.**

STEP 13:   **In the sketcher environment, close the References dialog box and confirm the missing references warning.**

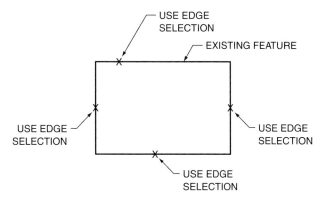

Figure 10-23   First Section Creation

STEP 14:   On the Sketcher toolbar, select the USE EDGE icon.

The Use Edge option is used to project feature elements onto the sketching plane for use in the current sketching environment. In this tutorial, you will pick the edges of the existing part feature and turn these edges into sketcher entities.

STEP 15:   Pick the four edge locations shown in Figure 10-23.

On the workscreen, pick each edge once and notice the change to the sketching environment.

STEP 16:   CLOSE the Type dialog box and select the Pick icon on the toolbar.

STEP 17:   If necessary, preselect the start point vertex shown in Figure 10-24, then utilize the START POINT option on the pop-up menu to set this vertex as the section's start point.

Blended features consist of two or more sections. The start points for each section should be in approximately the some location for each section. Within this segment of the tutorial, the start point will be in the lower left-hand corner of each section. By preselecting a vertex, the pop-up menu has an option for changing a section's start point.

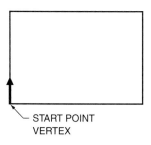

Figure 10-24
Start Point Selection

STEP 18:   On the menu bar, select SKETCH >> FEATURE TOOLS >> TOGGLE SECTION. (Select only once.)

Within the construction of a blended feature, the Toggle Section option will be used to toggle between feature sections. The pop-up menu also provides an option for toggling between sections.

Parallel blend sections are created in one sketching environment. The Toggle Section option will allow you to switch to the sketching environment for the next section of the blend. When you select Toggle Section, notice the change of the existing sketcher entities.

NOTE: Select Toggle Section only once.

STEP 19:   Use the RECTANGLE icon to sketch the entities shown in Figure 10-25.

The design intent for this feature requires the left and right edges of this section to be aligned with the left and right edges of the base feature. Use the Start Point option on the pop-up menu to set this section's start point to the vertex shown in the figure. Remember to preselect the vertex before attempting to access the menu.

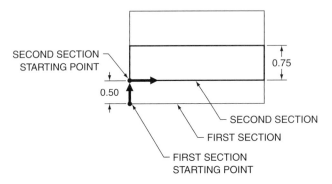

Figure 10–25   Second Section

STEP 20:   **Modify the second section's dimensioning scheme and values to match Figure 10–25.**

STEP 21:   ✔ **Select the Continue icon to exit the sketcher environment.**

STEP 22:   **Select BLIND >> DONE.**

STEP 23:   **Enter 4.00 as the depth of the second section.**

At this point in the Parallel Blend option, Pro/ENGINEER will require you to enter the distance between each section. Since there are only two sections, you will enter one depth distance.

STEP 24:   **Preview the feature on the Feature Definition dialog box.**

NOTE: If your direction of extrusion is the wrong direction, you can use the DIRECTION option of the feature definition dialog box to change the direction.

STEP 25:   **Select OK.**

CREATING A SECOND BLEND

This segment of the tutorial will create the second blended feature shown in Figure 10–26. This blend will consist of four sections. The first section will be created from existing feature edges while the second and third sections will consist of sketched rectangular entities. The fourth section will consist of a single point entity.

STEP 1:   **Select INSERT >> BLEND >> PROTRUSION on the menu bar.**

STEP 2:   **Select PARALLEL >> REGULAR SEC >> DONE.**

STEP 3:   **Select SMOOTH >> DONE.**

While the Straight option will blend sections with straight surfaces, the Smooth option will create a smooth blend between each section.

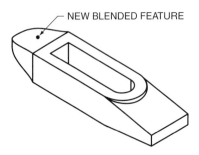

Figure 10–26   Blend Feature

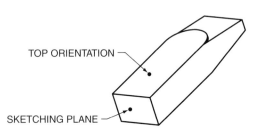

Figure 10–27   Sketching Plane and Orientation

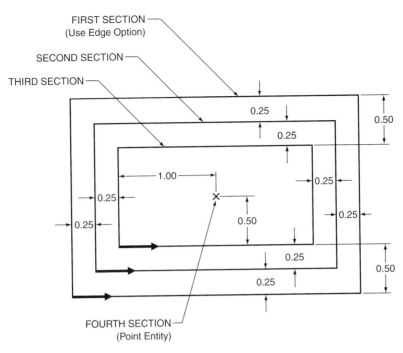

Figure 10-28   Blend Sections

STEP 4:   **Pick the sketching plane shown in Figure 10–27.**

Dynamically rotate the part to select the sketching plane or use the Pick From List option.

STEP 5:   **If necessary, use the FLIP option to define a direction of feature creation that will extrude the blend away from the part.**

STEP 6:   **Select OKAY to define the direction of feature creation.**

STEP 7:   **Orient the sketching environment as shown in Figure 10–27.**

STEP 8:   **Close the References dialog box.**

STEP 9:   **Create the four sections shown in Figure 10–28.**

Sketch the four sections that compose the blended feature. Utilize the Use Edge option to create the first section and the Rectangle option to create the second and third sections. The Toggle-Section option is required to switch between sections. Ensure that each section's start point is in the location shown in the figure. Create the fourth section with the Point option.

STEP 10:   ✔ **After the creation of the four required sections, select the continue option to exit the sketching environment.**

STEP 11:   **Enter a blind depth distance of 1.00 in between each section.**

STEP 12:   **Preview the Feature on the Feature Definition dialog box, then select OK.**

CREATING A CUT FEATURE

This section of the tutorial will create the cut extrusion shown in Figure 10–29. You will utilize the Offset Edge option to create the sketch.

STEP 1:   **Setup a sketching environment for a one directional cut with the sketching plane and orientation shown in Figure 10–29.**

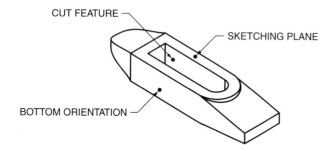

Figure 10–29   Cut Feature

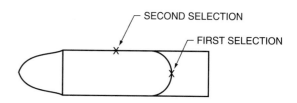

Figure 10–30   Offset Edge Option Selection

STEP   2: Within the sketching environment, close the References dialog box.

STEP   3: Select the OFFSET EDGE icon on the Sketcher toolbar.

NOTE: The Offset Edge icon is located behind the Use Edge icon.

STEP   4: On the Type dialog box, select the CHAIN option, then pick the two locations shown in Figure 10–30.

STEP   5: On the Choose menu, use the NEXT and PREVIOUS options to switch between possible chain options, then ACCEPT the chain that will create the correct Cut.

Your chain should loop around the perimeter of the base protrusion feature.

STEP   6: Enter an offset distance of 0.75 in (enter –0.75 if necessary).

When entering the offset distance, observe the offset side on the workscreen. You can enter a negative value to change the side of offset.

STEP   7: Select the Continue option to exit the sketching environment.

STEP   8: Select the Extrude option on the toolbar.

STEP   9: On the extrude dashboard, select the Cut option.

STEP 10: Change the direction of extrusion to cut through the part.

STEP 11: On the dashboard, select Through All as the extrusion depth.

STEP 12: Build the feature.

STEP 13: Save your part.

SWEEP TUTORIAL 1

This tutorial exercise will provide instruction on how to create the part shown in Figure 10–31. The first segment of this tutorial will consist of creating the pipe feature through the use of a swept protrusion. The flange features will be created as extruded protrusions with a patterned radial hole. Within this tutorial, the following topics will be covered:

- Creating a swept protrusion.
- Creating an extruded protrusion.
- Creating a patterned radial hole.

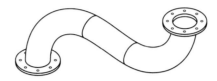

Figure 10–31   Completed Part

CREATING THE BASE FEATURE

This segment of the tutorial will create the swept protrusion feature shown in Figure 10–32. Start a part model utilizing the default template and named *Sweep1*. You will create this feature through the use of a sketched trajectory and a sketched section.

Figure 10–32
Swept Feature

STEP 1: **Select INSERT >> SWEEP >> PROTRUSION on the menu bar.**

A Sweep feature can be defined as a protrusion, cut, or surface.

STEP 2: **Select the SKETCH TRAJ (Sketch Trajectory) option.**

Within the Sweep option, the trajectory is created first, followed by the Sweep's section. A trajectory can be sketched in a sketcher environment or it can be selected on the workscreen. Datum curves make valuable features that can be selected as trajectories.

STEP 3: **Select datum plane FRONT as the sketching plane, accept the default direction of viewing, then orient datum plane TOP toward the top of the sketcher environment.**

STEP 4: **In the sketching environment, use the ARC and LINE options to create the section shown in Figure 10–33.**

When starting the sketch for a trajectory, notice the arrow that is created. This arrow denotes the trajectory's start point and the direction of sweep. A starting point can be changed with the Sketch >> Feature Tools >> Start Point option or with the pop-up menu's Start Point option.

The dimensioning scheme shown in Figure 10–33 matches the design intent for this feature. Use the Dimension option to adjust your dimensions to match this scheme. Use the Modify option to change your dimension values.

MODELING POINT   In this tutorial, this swept feature is the first geometric feature of the part. If other features were to exist at this point, you would have the option of creating the sweep with merged or free ends. The Merge End option will merge the sweep with any attached feature.

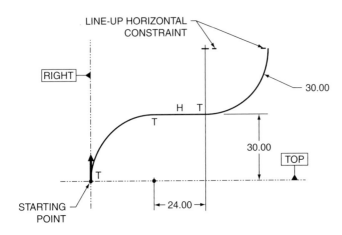

Figure 10–33   Trajectory Sketch

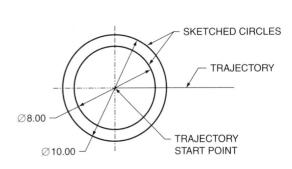

Figure 10–34   Cross Section of Sweep

STEP   5:   When the section is complete, select the Continue option to exit the sketcher environment.

STEP   6:   Sketch the swept feature's cross section as shown in Figure 10–34.

Upon finishing the trajectory's section, Pro/ENGINEER will open a sketching environment for the creation of the feature's cross section. Within this sketching environment, Pro/ENGINEER will orient the workscreen normal to the trajectory at the trajectory's start point. Often, it is useful to dynamically rotate the workscreen (middle mouse button) to better visualize the start point's location (Figure 10–35).

Use the Circle option to create the two circles shown in Figure 10–34.

STEP   7:   Select the Continue option to exit the sketching environment.

STEP   8:   Preview the feature on the Feature Definition dialog box (Figure 10–36).

Notice on the dialog box the Trajectory and Section elements. The sweep's trajectory can be redefined with the Trajectory option and the feature's cross section can be redefined with the Section option.

STEP   9:   On the dialog box, select OK to create the feature.

STEP 10:   Save your part.

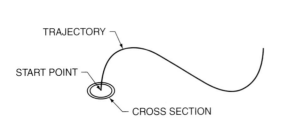

Figure 10–35   Cross Section Start Point

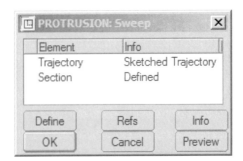

Figure 10–36   The Feature Definition Dialog Box

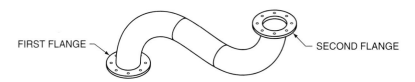

Figure 10–37 Flange Features

FLANGE FEATURES

This part of the tutorial will create the two flanges shown in Figure 10–37. The base feature of each flange will be constructed as an extruded protrusion. The bolt-circle hole pattern will be created from a patterned radial hole.

STEP 1: Set up the sketching plane shown in Figure 10–38.

Be sure that you pick the end of the pipe that incorporates the default datum planes. Orient the sketching environment to match Figure 10–39.

STEP 2: Sketch the section shown in Figure 10–39.

Specify the inside diameter of the swept pipe feature and datum planes RIGHT and FRONT as references. Sketch the two circles shown in Figure 10–39. Notice that the inside hole of the flange is the same diameter as the inside diameter of the swept pipe feature. The design intent calls for these two diameters to remain the same. Incorporate this into your sketch.

STEP 3: Select the Continue icon to exit the sketching environment.

STEP 4: Extrude the feature a blind distance of 1.00 in.

STEP 5: Build the feature.

STEP 6: Select the Hole option on the toolbar.

STEP 7: Create the radial hole feature as shown in Figure 10–40.

Create the hole shown in the figure. This hole will be patterned to create the remaining holes on the flange. Use the following options for the hole:

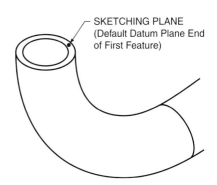

Figure 10–38 Sketching Plane

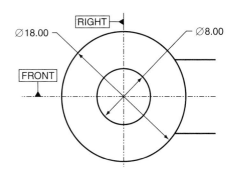

Figure 10–39 Sketched Section

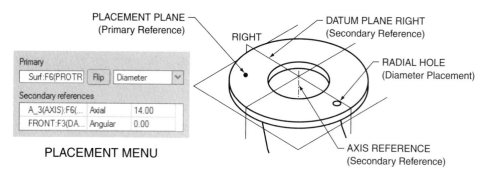

PLACEMENT MENU

Figure 10–40 Radial Hole

- Straight, simple hole.
- The hole placement plane is the primary reference (Placement menu).
- Use diameter as the hole placement type (Placement menu).
- The axis in Figure 10–40 is a secondary reference (Placement menu).
- Datum Plane RIGHT as the secondary reference (0.0 degree angle).
- Bolt-circle diameter = 14.00 in (See Placement menu).
- Through Next depth option.
- 1.00 in diameter hole.

STEP 8:   Use the Pattern tool to create a pattern of the radial hole (Figure 10–41).

Preselect the hole before selecting the pattern option or select the pattern command on the model tree's pop-up menu. Use the angular dimension that defines the radial hole as the leader dimension for the pattern. Increment eight holes an angular distance of 45 degrees.

STEP 9:   Use the methods from the previous steps of this tutorial segment to create the second flange (Figure 10–42).

As with the first flange, use the Extrude tool to create the base feature of the flange. Your references will be slightly different from the first flange feature. Pattern a radial hole to create the flange holes.

STEP 10:   Save your part.

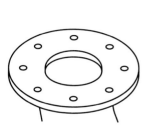

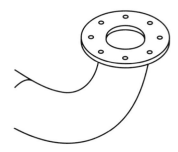

Figure 10–41 Patterned Hole

Figure 10–42
Second Flange Feature

SWEEP TUTORIAL 2

This tutorial exercise will provide instruction on how to create the wheel part shown in Figure 10–43. The first segment will consist of creating the hub and grip for the wheel. Both features will be created as revolved protrusions. The second segment will consist of creating a datum curve. This curve will be used as the trajectory for the swept spoke feature created in the final segment. Within this tutorial, the following topics will be covered:

- Creating a revolved protrusion.
- Creating a datum curve.
- Creating a swept feature.
- Creating a round.

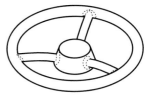

Figure 10–43 Completed Part

CREATING THE BASE FEATURE

The base geometric feature of the part, the hub of the wheel, is created as a revolved protrusion. Figure 10–44 shows the sketch for the feature. Perform the following options when creating this feature:

- Name the part file *sweep2*.
- Choose a one-directional revolved protrusion.
- Select datum plane RIGHT as the sketching plane.
- Sketch the section as shown in Figure 10–44.
- Make a 360 degree revolution.

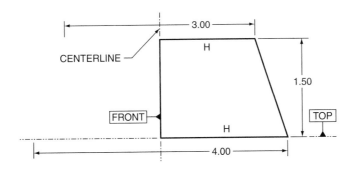

Figure 10–44 Hub Sketch

CREATING THE WHEEL HANDLE

This segment of the tutorial will create the handle of the wheel part. As with the hub feature, this feature will be created as a revolved protrusion. Figure 10–45 shows the sketch for the feature. Perform the following options when creating the feature:

- Choose a one-direction revolved protrusion.
- Select sketching plane as datum plane RIGHT.
- Sketch section as shown in Figure 10–45.
- Make a 360 degree revolution.

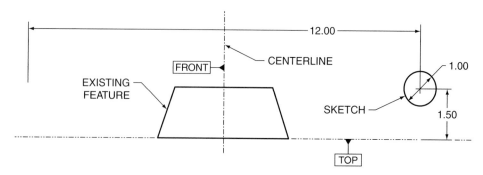

Figure 10–45   Wheel Handle Feature

CREATING A DATUM CURVE

If a datum can be defined as a theoretically exact object, a datum curve is not actually a true datum. Datum curves are used in Pro/ENGINEER primarily for the construction of features. As an example, datum curves are often used for trajectories in swept features. While Pro/ENGINEER provides multiple ways of creating datum curves (Through Points, Projected, Wrapped, etc.), the Sketch option is probably the most used. This segment of the tutorial will create a datum curve that will be used as the trajectory for the swept spoke feature of the wheel part. This trajectory could be selected within the Sweep option. An advantage of creating it as a separate feature is the flexibility it provides when you have to adjust the trajectory's path. Perform the following steps to create the datum curve feature:

STEP 1:   Select the Sketch Tool icon on the datum toolbar.

STEP 2:   **Set up the sketching environment.**

Your sketching environment should appear as shown in Figure 10–46. Establish the following options on the Sketch dialog box for creating the datum curve's sketch:

- Sketch on datum plane RIGHT.
- Orient datum plane TOP to the LEFT of the workscreen.

STEP 3:   **While in the sketcher environment, turn off the display of datum planes and set Hidden as the model's display mode.**

STEP 4:   **In the sketcher environment, using the References dialog box(Sketch >> References), specify the four references shown in Figure 10–46.**

Specify datum planes RIGHT and TOP as references. Also, specify the two feature edges shown in Figure 10–46 as references.

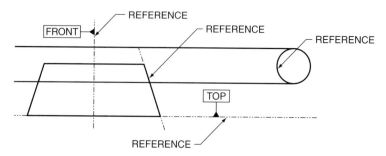

Figure 10–46   Specifying References

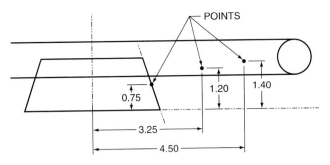

Figure 10–47   Point Creation

MODELING POINT   If you inadvertently close the References dialog box or realize that it is needed at a later point in the sketching process, it can be accessed with the Sketch >> References option on the menu bar.

STEP 5:   Use the Point option to create the three points shown in Figure 10–47.

The datum curve will be sketched with the Spline option. While a spline can be sketched without points, the establishment of points allows for better control of the definition of the spline. Create the three points with the dimensioning scheme shown in Figure 10–47. Notice that one point is aligned with the existing hub feature.

STEP 6:   Use the Spline option to create the spline entity shown in Figure 10–48.

When creating the spline, sketch from the left to the right by sketching through each control point. Align the right end of the spline with the wheel entity at the quadrant shown in the figure. After sketching the spline, the dimensions defining the control points can be used to adjust the spline geometry.

STEP 7:   Select the Continue icon to exit the sketching environment.

STEP 8: Save your part.

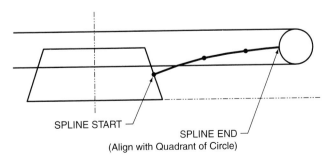

Figure 10–48   Spline Creation

SWEEP CREATION

This segment of the tutorial will create the swept feature that defines one spoke of the wheel. The datum curve created previously in this tutorial will be used as the trajectory for this feature. Later in this tutorial, this swept feature will be copy-rotated to create the remaining spoke features.

STEP 1: Select INSERT >> SWEEP >> PROTRUSION on the menu bar.

STEP 2: Pick the SELECT TRAJ (Select Trajectory) option on the Sweep Trajectory menu.

This tutorial will require you to select the datum curve created previously as the trajectory for this swept feature.

STEP 3: With the ONE-BY-ONE option selected, on the workscreen, pick the datum curve feature.

The One By One selection option requires you to select individual entities of a chain. In this example, the datum curve is only one entity.

STEP 4: On the Select menu, select the OK option.

Notice after selecting OK that Pro/ENGINEER will display an arrow to denote the starting point of the sweep. The start point defines the location where the sweep's section will be sketched. The Start Point option on the Chain menu is available for changing this location.

STEP 5: Select DONE on the Chain menu.

STEP 6: From the Attributes menu, select the MERGE ENDS option, then select DONE.

The Merge Ends option will merge the ends of the swept feature with any existing features that the trajectory is aligned with. In this tutorial, the datum curve trajectory is aligned with the hub and the wheel features.

STEP 7: In the sketching environment, create the circle entity shown in Figure 10–49.

Note: Figure 10–49 defines the section with a starting point coincident with the hub feature. An alternative would have your start point coincident with the wheel feature. Either location will work for this feature.

No additional references are needed for this section. If the orientation of the sketching environment is unclear, dynamically rotate the workscreen to see

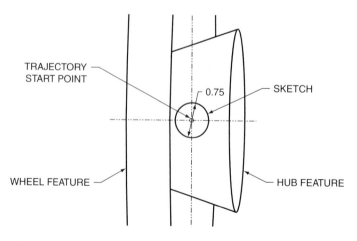

Figure 10–49   Sweep Section

the starting point of the section. Sketch your circle's center coincident with the start point. The Orient Sketch icon on the sketcher toolbar will return you to the normal two-dimensional sketching environment. Align the center of the sketched circle with the center of the trajectory.

STEP 8: ✔ **Select the Continue icon to exit the sketching environment.**

Do not attempt to exit the sketcher environment until your sketch matches Figure 10–49.

STEP 9: **Preview the Sweep, then select OK on the Feature Definition dialog box.**

STEP 10: **Save your part.**

CREATING A ROUND

In this segment of the tutorial, use the Round command to create the two fillets shown in Figure 10–50. Use the following options when creating the fillets:

- Create a Simple Round.
- Create each fillet as a single feature.
- Select the edges shown in Figure 10–50.
- Use a constant round radius of 0.25 in.

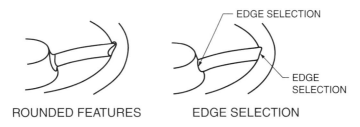

ROUNDED FEATURES   EDGE SELECTION

Figure 10–50   Fillet Creation

GROUPING FEATURES

The sweep and round will be copy-rotated to create the three instances of the spoke. In this tutorial, you will group these features and use the Copy/Rotate option to make the two copies. Within Pro/ENGINEER, grouping the features is not a requirement of the Copy command. You will group the features in this example for two reasons. First, a copy of multiple features creates a group of the copies. In other words, the copied instances of the features will be groups. Second, grouping the features helps to keep them organized on the model tree.

STEP 1: **On the model tree, use your control-key to pick the swept protrusion and round features.**

STEP 2: **Select EDIT >> GROUP on the menu bar to group these two features.**

STEP 3: **Observe the changes to your features on the model tree.**

COPYING THE SPOKE GROUP

This segment of the tutorial will create two copies of the spoke group (Figure 10–51). Perform the following steps to create each copy:

STEP 1: Select EDIT >> FEATURE OPERATIONS >> COPY.

STEP 2: Select MOVE >> DEPENDENT >> DONE on the Copy Feature menu.

> The Rotate option is located under the Move menu option. The Dependent option will make the copied feature's dimensions dependent on its parent feature. In other words, any copy of the spoke group will remain the same size as its parent group's dimensions.

STEP 3: On the model tree or on the workscreen, pick your group feature then select DONE on the menu.

STEP 4: Select ROTATE on the Move Feature menu.

STEP 5: Select the CRV/EDG/AXIS option then select the axis shown in Figure 10–52.

> As the name implies, the Curve/Edge/Axis option will rotate a feature around a curve, edge, or axis. In this example, you will rotate the features around the center axis of the model.

STEP 6: If necessary, FLIP the direction of rotation indicator arrow to match Figure 10–52.

> The Rotate option uses the right-hand rule to rotate features. Using your right hand, your thumb points in the direction of the arrow on the workscreen and your fingers point in the direction of rotation.

STEP 7: Select OKAY to accept the rotate direction shown in Figure 10–52.

STEP 8: In Pro/ENGINEER's textbox, enter 120 as the number of degrees to rotate the group.

STEP 9: Select the DONE MOVE option on the Move Feature menu to finish the move process.

STEP 10: Select the DONE option on the Group Vary Dimension menu.

> This step of the Copy command allows individual dimensions to be varied during the copy process. Within this tutorial, you will not vary any dimensions.

STEP 11: Select OK on the Feature Definition dialog box.

> The Copy command does not have a Preview option on its Feature Definition dialog box.

STEP 12: Create the second spoke copy.

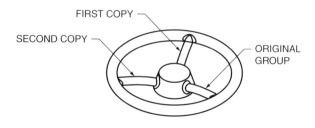

Figure 10–51   Copied Features

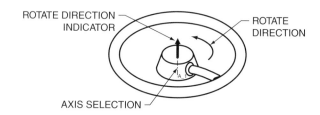

Figure 10–52   Copy Creation

Use the previous steps of this tutorial to create the second copy of the spoke. You can use either the original group or the newly copied group as the group to copy.

STEP 13: **Save your part.**

STEP 14: **Use the FILE >> DELETE >> OLD VERSIONS option to delete old versions of the part.**

PROBLEMS

1. Use Pro/ENGINEER's Part mode to model the part shown in Figure 10–53.

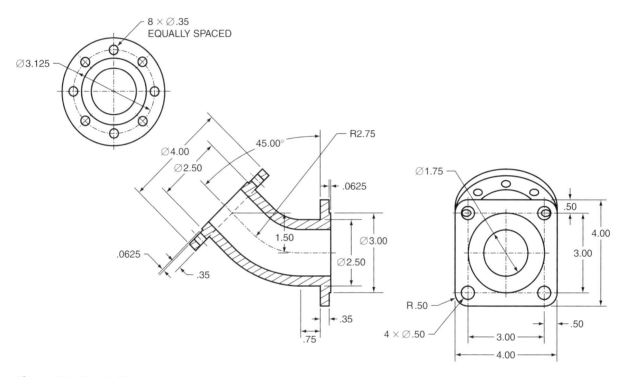

Figure 10–53   Problem 1

2. Use Pro/ENGINEER's Part mode to model the part shown in Figure 10–54. Using the Sweep command, create this part with only one geometric feature. Sketch the trajectory as an oval. With a closed trajectory, Pro/ENGINEER will give you the attribute option of adding or not adding inside faces (Add Inn Fcs and No Inn Fcs). Select the Add Inn Fcs option.

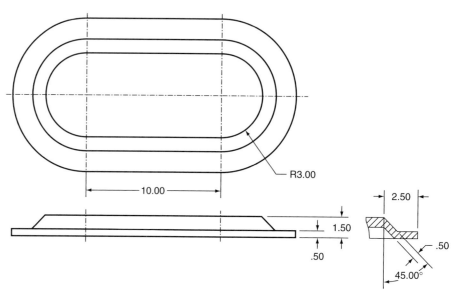

Figure 10–54   Problem 2

3. Use Pro/ENGINEER's Part mode to model the part shown in Figure 10–55.

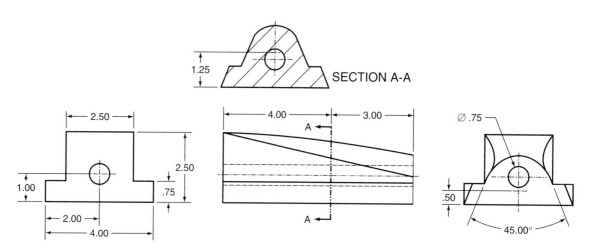

Figure 10–55   Problem 3

4. Use Pro/ENGINEER's Part mode to model the propeller part shown in Figure 10–56. The part consists of a revolved feature (the hub) and three blended features (the propellers). Sketch the hub feature and the first propeller feature on Pro/ENGINEER's default datum plane FRONT using appropriate dimensions of your choosing. For the blended propeller feature, use the two sections shown in the figure with 24 in separating each section. Use the Copy/Rotate option to create the remaining propeller features.

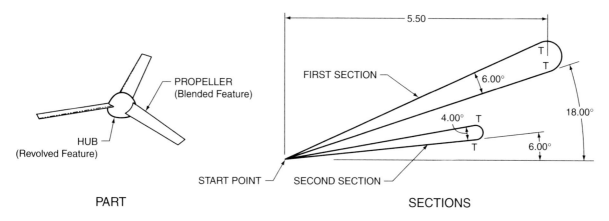

Figure 10–56 Problem 4

QUESTIONS AND DISCUSSION

1. What two methods are available for defining a sweep's trajectory? Describe appropriate situations for using each method.

2. During the creation of a sweep, if a regeneration error is encountered during the preview of the feature, what are some possible causes of this failure?

3. What is the difference between a merged ends sweep and a free ends sweep?

4. During the creation of a swept feature, which is defined first, the trajectory or the section?

5. Describe the three types of blends available under the Blend command.

6. Describe several points that should be adhered to when creating a parallel blend.

7. When sketching the sections for a parallel blend, what menu option is used to switch from one section to the next? What process is used to move between sections for a rotational blend?

8. What is a datum curve and what are some of its uses?

9. Describe the difference between a projected datum curve and a wrapped datum curve.

11

ADVANCED MODELING TECHNIQUES

Introduction

Within Part mode, solid creation options are available to perform basic modeling operations. Included are the Extrude, Revolve, Sweep, Blend, and Use Quilt options. Also included are options for creating advanced solid features. Available are advanced feature commands such as Variable-Section Sweep, Swept Blend, and Helical Sweep. Upon finishing this chapter, you will be able to:

- Create a swept blend.
- Create a variable-section sweep.
- Create a spring feature with the helical sweep option.
- Create a bolt with the helical sweep option.

DEFINITIONS

Normal to Trajectory	The Normal to Trajectory option keeps the feature's cross sections normal to a selected trajectory. This option is available under Swept Blend and Variable Section Sweep.
Normal to Origin Trajectory	The Normal to Origin Trajectory option keeps the feature's cross. sections normal to the defined origin trajectory. This option is available under the Swept Blend and Variable-Section Sweep options.
Pitch	On a thread, the pitch is the inverse of the number of threads per inch and is defined as the distance between a point on one thread to the corresponding point on the next thread. The Helical Sweep option requires the definition of either a constant pitch or a variable pitch.
Pivot Direction	The Pivot Direction option keeps the feature's cross sections normal to a selected planar pivot plane, edge, curve, or axis. This option is available under Swept Blend and Variable Section Sweep.

SWEPT BLEND COMMAND

A swept blend is a combination of a sweep and a blend. A swept feature is a section protruded along a defined trajectory. This trajectory can be either sketched or selected. A parallel blended feature is a feature protruded along a straight trajectory between two or more user-defined sections. A swept blend feature is two or more sections protruded along a user-defined trajectory (Figure 11–1). As with a swept feature, a swept blend's trajectory can be either sketched or selected. In addition, sections can be either sketched or selected.

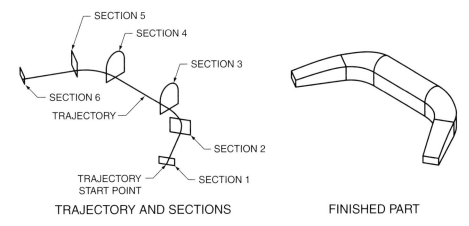

Figure 11–1   Swept Blend Creation

Pro/ENGINEER provides three Swept Blend options. The following is a description of each.

NORMAL TO TRAJECTORY

The Normal to Trajectory section control option keeps each of the feature's cross sections normal to the trajectory of the feature (Figure 11–2). This type of swept blend requires the definition of a trajectory (sketched or selected) and the definition of one or more sections. Each section is created normal to a vertex of the trajectory or normal to a datum point on the trajectory.

CONSTRAINT NORMAL DIRECTION

The Constraint Normal Direction option keeps the feature's cross sections parallel to a selected plane or normal to an edge, curve, or axis (Figure 11–2). Like the Normal to Trajectory option, this type of Swept Blend requires the definition of a trajectory and the definition of one or more sections. As shown in the figure, each section of the feature is created normal to the picked normal trajectory.

NORMAL TO PROJECTION

The Normal to Projection option keeps the feature's cross sections normal to the 2D projection of the origin trajectory along a specific trajectory (Figure 11–3). The section's trajectory can be sketched or selected. This option requires the definition of a direction reference.

CREATING A SWEPT BLEND

The Swept Blend tool is available for creating protrusions and cuts. This guide will demonstrate the creation of the swept blend feature shown in Figure 11–4. Pro/ENGINEER's default datum planes will be used as the base for the feature. Perform the following steps to create this feature:

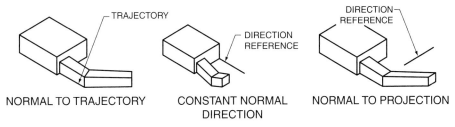

Figure 11–2   Swept Blend Types

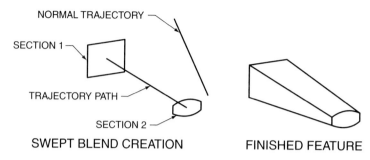

NORMAL TRAJECTORY

SECTION 1

TRAJECTORY PATH

SECTION 2

SWEPT BLEND CREATION   FINISHED FEATURE

Figure 11–3   Normal to Trajectory Creation

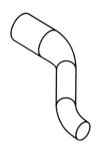

Figure 11–4
Swept Blend Feature

STEP 1: Use the sketch tool to create the trajectory shown in Figure 11–5.

Using appropriate sketching tools, sketch the trajectory's section as shown in Figure 11–5. The dimensions shown in the figure match the design intent for the feature.

STEP 2: Select INSERT >> SWEPT BLEND on the menu bar.

STEP 3: On the dashboard, select the SOLID icon to create a solid swept blend feature.

STEP 4: On the workscreen, pick your previously created trajectory.

After picking the trajectory, notice the startpoint arrow and the label of your trajectory as origin. Currently, your trajectory is serving as the origin trajectory for your swept blend's cross sections. The startpoint arrow defines the location of your first section.

STEP 5: On the dashboard, open the Reference menu option to observe its default values.

STEP 6: Under the References menu option, ensure that NORMAL TO TRAJECTORY is selected as the section plane control.

STEP 7: Open the Sections menu on the dashboard.

STEP 8: Ensure that SKETCHED SECTIONS is selected on the Sections menu.

The sections for a swept blend can be either sketched or selected. The default Sketched Sections option will allow you to sketch each section within Pro/ENGINEER's sketching environment.

STEP 9: Change your select filter to VERTEX.

Each of your sections will be created at a vertex on the feature's trajectory. The Vertex filter setting will ensure that you can pick each vertex.

STEP 10: On your workscreen, pick the vertex defined by the trajectory's startpoint arrow.

In order starting from the startpoint end of the trajectory, you will create a section at each available vertex.

STEP 11: Select SKETCH on the Sections menu then sketch the first section of the swept blend.

It is often helpful to dynamically rotate the sketching environment to get a better understanding of the sketch plane in relation to the trajectory. Also notice on the trajectory in the figure the location of each cross section. In

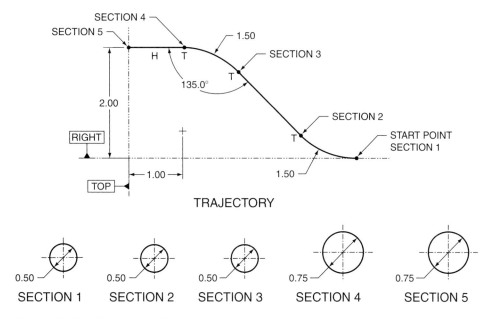

Figure 11–5 Trajectory and Sections

this example, there will be five sections. As shown in the figure, the first section is a 0.500 diameter circle.

STEP 12: ✓ When your section is complete, select the Continue icon to exit the sketching environment for the first section.

STEP 13: On the sections menu, enter the Z-axis rotation value for the section (usually 0.00).

STEP 14: Select INSERT on the sections menu then pick the vertex for the location of the second section (see Figure 11–5 and Figure 11–6).

With your selection fillet set to vertex, this selection should be fairly easy. You will know that you've had a successful selection when the section menu's Section Location collector displays an End Curve element.

STEP 15: Sketch the second section.

Sketch the second section according to the intent of the design. If necessary, dynamically rotate the sketching environment.

STEP 16: ✓ Select the Continue icon to exit the sketching environment for the second section.

STEP 17: Use the Sections menu to create any remaining sections.

Use the Insert then sketch menu options to create each remaining section.

STEP 18: Select the Build Feature option then select OK.

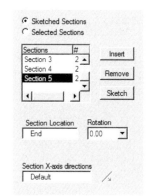

Figure 11–6
Swept Blend Definition

VARIABLE-SECTION SWEEP

The Variable-Section Sweep option is an advanced feature modeling tool used for the creation of complex geometric shapes. This option sweeps a section along one or more trajectories. It is available for protrusion, cut, and surface features.

Pro/ENGINEER provides several degrees of control with the Variable-Section Sweep option. First, the orientation of a section can be controlled with the Normal to Trajectory, Normal to Projection, and Constant Normal Direction reference options. Second, sections can be aligned to one or more trajectories to vary the shape of the section (see Figure 11–7). Finally, the varying of a section can be controlled with the Constant and Variable options. The Constant option is used to maintain a section's size.

Figure 11–7 shows the construction of a variable-section sweep feature. This feature was constructed with the Normal to Trajectory option. Notice in the figure the origin trajectory. All variable-section sweeps require the selection of an origin trajectory that defines the direction of sweep. This trajectory is similar to the trajectory of a swept feature. When you pick references, the first selected reference is defined as the origin trajectory. Like a swept feature, this trajectory requires the definition of a starting point. The feature in Figure 11–7 also has a defined X-trajectory. An X-trajectory defines the horizontal vector of a section. If the feature's section is aligned with an X-trajectory, as it is in this example, the section will vary along the path of the trajectory. Also shown in Figure 11–7 is the section for the feature. Within the construction of a variable-section sweep feature, the origin trajectory, X-trajectory, additional trajectories, and the section can be either sketched or selected.

The normality of a variable-section sweep's section can be controlled with the Normal to Trajectory, Normal to Projection, and Constant Normal Direction reference options. The following is a description of each option.

NORMAL TO TRAJECTORY

The Normal to Trajectory option keeps the feature's section normal to a defined trajectory. As seen in the Figure 11–8, this option requires the sketching of an origin

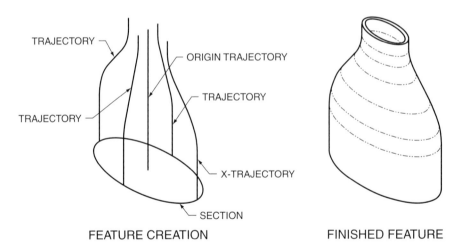

Figure 11–7   Variable-Section Sweep Construction

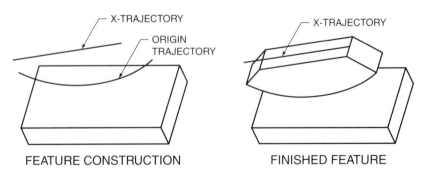

Figure 11–8   Normal to Trajectory

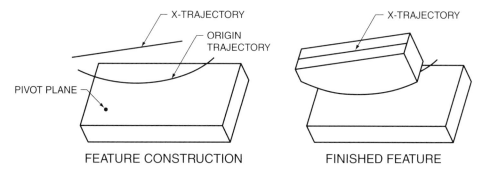

Figure 11–9   Normal to Projection

trajectory and an X-trajectory. Notice how the cross section of the finished part in this case is normal to the origin trajectory. X-trajectories or other selected trajectories can also serve as the normal trajectory.

Normal to Projection

The Normal to Projection option keeps the feature's section normal to a selected planar pivot plane, edge, curve, or axis (Figure 11–9). Like the Normal to Trajectory option, this type of variable-section sweep requires the definition of an origin trajectory and the definition of one section. Unlike the Normal to Trajectory option, the defining of an X-trajectory is optional. Without an X-trajectory, this option performs similarly to the Swept Blend/Pivot Direction option. Figure 11–9 shows the construct of a variable-section sweep with the Normal to Projection option. Notice, in the finished feature, how the feature's cross section remains normal to the selected pivot plane. Notice, in the finished feature, how the feature's cross section remains normal to the selected pivot plane.

Constant Normal Direction

The Constant Normal Direction option keeps the section's normal vector parallel to a selected plane, edge, curve, or axis (Figure 11–10). This option requires the definition of an origin trajectory, a direction reference, and one section. Notice in Figure 11–10 how the feature's section remains normal to the selected direction reference. This occurs because the section's normal vector remains parallel to the direction reference.

Creating a Variable-Section Sweep

The Variable-Section Sweep option can be found under the Advanced menu option of the Protrusion and Cut commands. This option is used to sweep a section along multiple trajec-

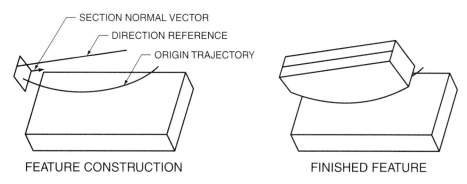

Figure 11–10   Constant Normal Direction

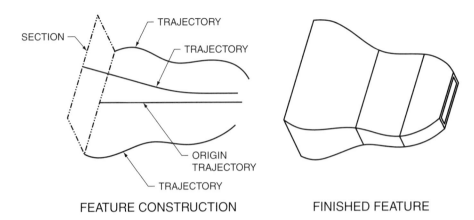

FEATURE CONSTRUCTION   FINISHED FEATURE

Figure 11–11   *Variable-Section Sweep Feature*

tories. Like all protrusion and cut features, the variable-section sweep feature can be created solid or thin. The following guide demonstrates the creation of the feature shown in Figure 11–11. This feature is constructed by sweeping a rectangular section along the four datum curves shown in the figure. A model file (*var_sec_swp.prt*) containing the part's predefined datum curves can be found on the book's web page and will serve as the base part for this guide. Perform the following steps to create the variable-section sweep feature shown.

Step 1:   Select the Variable-Section Sweep icon on the toolbar.

Step 2:   Select the Solid option on the variable-section sweep dashboard.

A variable-section sweep can be created as a solid, surface, or thin feature.

Step 3:   Using the Control key, pick the origin trajectory shown in Figure 11–12.

The first trajectory selected is the origin trajectory. This trajectory is used to define the start point for the feature's section and the extent of the sweep. When you use the Normal to Trajectory option, as this guide demonstrates, the feature's cross section will be perpendicular to this selected entity.

Step 4:   Using the Control key, pick the X-trajectory shown in Figure 11–12.

The X-trajectory for a normal to trajectory variable-section sweep defines the X vector for the section.

Step 5:   Using the Control key, pick the remaining two trajectories.

Step 6:   On the variable-section sweep dashboard, open the references menu.

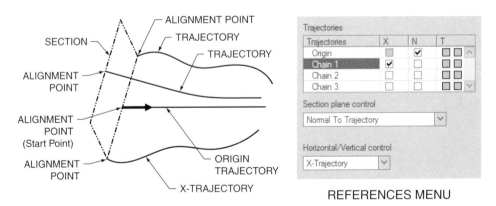

REFERENCES MENU

Figure 11–12   *Feature Construction*

STEP 7: Observing the Trajectories option under the references menu, ensure that the origin trajectory's N value and the Chain 1 trajectory's X-value are checked (Figure 11–12).

The checked N value creates the sweep normal to the origin trajectory, while the checked X value with Chain 1 defines the X trajectory for the sweep. Note that the first selected trajectory (Step 3 in this guide) is always the origin trajectory.

STEP 8: On the workscreen, pick the origin trajectory and ensure that its start point arrow is defined at the location shown in Figure 11–12.

The start point of the trajectory is denoted on the origin trajectory by an arrow. The arrow's location should match Figure 11–12. If necessary, use the pop-up menu's Flip Chain Direction option to change the start point to the opposite end of the origin direction.

STEP 9: Select the Sketch icon on the dashboard.

In this environment, unlike most sketching environments, you do not have to explicitly define the location and orientation of the section's sketching plane. For a normal to trajectory variable-section sweep, the sketching plane passes through the start point and is normal to the origin trajectory, as defined in Step 7.

STEP 10: Sketch the Section for the feature (Figures 11–12 and 11–13).

Pro/ENGINEER will provide a normal sketching environment for creating the section. When you are sketching a section for a variable-section sweep, if design intent requires it, it is important to align the sketch with the trajectories of the feature. Refer to the alignment points on Figure 11–12. Notice how they correspond to the alignment points on the sketch in Figure 11–13. By aligning with one or more trajectories, the section is allowed to vary.

MODELING POINT   When sketching the section for any sweep feature, it is often easier to sketch in a 3D orientation. Either dynamically rotate your model or use the Control key and D key combination for the default orientation.

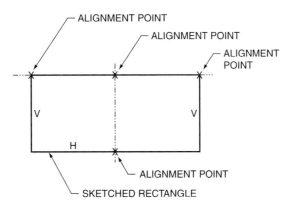

Figure 11–13   Sketched Section

STEP 11: 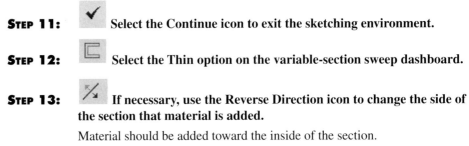 Select the Continue icon to exit the sketching environment.

STEP 12: Select the Thin option on the variable-section sweep dashboard.

STEP 13: If necessary, use the Reverse Direction icon to change the side of the section that material is added.

Material should be added toward the inside of the section.

STEP 14: On the dashboard, enter 0.125 as the thickness of the thin feature.

Your dashboard should resemble Figure 11–14.

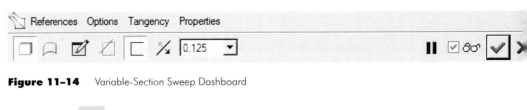

Figure 11–14   Variable-Section Sweep Dashboard

STEP 15: Build the feature.

HELICAL SWEEPS

As its name implies, the Helical Sweep option is useful for creating parts that consist of helical features. As shown in Figures 11–15 and 11–16, two features often created with the Helical Sweep options are the spring and thread. A spring feature is created with the Protrusion option and a thread feature is created with the Cut option.

Various definitions and attributes are available within the Helical Sweep option. Threads and springs are defined with a pitch definition. Gears can also be defined with the Helical Sweep option. With a thread, the pitch is the inverse of the number of threads per inch and is the distance between a point on one thread to a corresponding point on the next thread. For a spring, the pitch is the distance between a point on the wire to a corresponding point on an adjacent loop of the wire. Pro/ENGINEER provides an option for either creat-

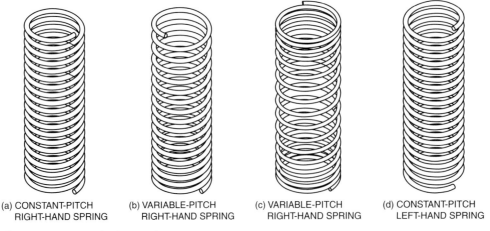

(a) CONSTANT-PITCH RIGHT-HAND SPRING

(b) VARIABLE-PITCH RIGHT-HAND SPRING

(c) VARIABLE-PITCH RIGHT-HAND SPRING

(d) CONSTANT-PITCH LEFT-HAND SPRING

Figure 11–15   Helical Sweep Options

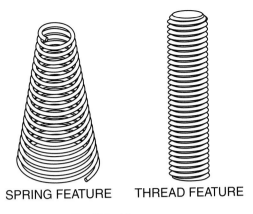

SPRING FEATURE   THREAD FEATURE

Figure 11-16   Helical Sweeps

ing a helical feature with a constant pitch or a feature with a varying pitch. Figure 11–15 shows two examples of constant pitch spring features and two examples of variable pitch spring features. In the figure, notice in the third spring feature (c) that the spring's pitch varies in the middle of the spring. When the sweep trajectory for this feature is created, points can be applied along the trajectory and the pitch at each point adjusted through the use of a pitch graph window.

Another definition available within the Helical Sweep option is the ability to create either a right-hand helical feature or a left-hand helical feature. Standard screw threads are right-handed. Some exceptions to this rule exist, such as threads assigned to an oxygen system. Figure 11–15a shows an example of a right-hand constant-pitch spring and Figure 11–15d shows an example of a left-hand constant-pitch spring. The thread shown in Figure 11–16 is a right-hand thread.

CREATING A CONSTANT-PITCH HELICAL SWEEP FEATURE

The Helical Sweep option is often used to create spring and thread features. This guide will provide general information on the steps for creating a constant pitch helical feature. Options are available for creating protrusions, cuts, and surface helical sweep features.

STEP 1:   **Select INSERT >> HELICAL SWEEP >> PROTRUSION.**

Since a spring is a positive space feature, it is created with the Protrusion command. Detail threads are created with the Cut command.

STEP 2:   **Select appropriate attributes on the Attributes menu then select DONE.**

The following attributes are available:

Constant Creates a helical feature with a constant pitch (Figure 11–15a).

Variable Creates a helical feature with a variable pitch (Figure 11–15b).

Thru Axis Creates a helical feature around an axis. The axis is sketched as the first trajectory of the feature.

Norm to Traj Creates a helical feature perpendicular to a sketched trajectory.

Right Handed Creates a helical feature swept to the right (Figure 11–15a).

Left Handed Creates a helical feature swept to the left (Figure 11–15d).

STEP 3:   **Pick, then orient, a sketching plane for the creation of the feature's profile.**

STEP 4:   **Sketch the profile of the helical sweep feature to include a required centerline (Figure 11–17).**

The first geometric entity sketched defines the profile of the sweep feature. Figure 11–17 shows two examples of profiles and their respective helical features. A helical sweep feature requires the sketching of a centerline that defines the axis of revolution. The first centerline sketched serves this purpose.

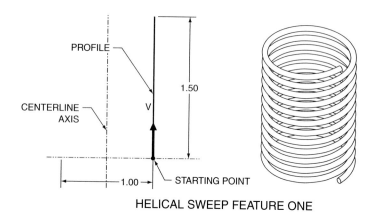

HELICAL SWEEP FEATURE ONE

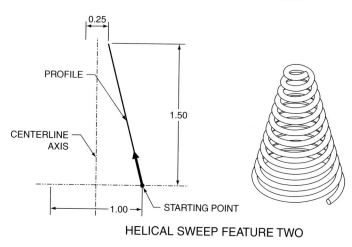

HELICAL SWEEP FEATURE TWO

Figure 11-17   Helical Sweep Feature Profile Construction

Figure 11-18   Isometric View of Cross-Section Construction

STEP 5:   ✔   Select the Continue icon to exit the sketching environment.

STEP 6:   Enter the Pitch value for the feature.

STEP 7:   Sketch the cross section of the helical feature (Figure 11–18).

Use appropriate sketching tools to create the section. Cross sections of helical sweep features are typically sketched around or near the start point of the profile of the feature. When you are sketching the cross section, the sketch plane is behind the sketching environment used to create the profile. It is often helpful to dynamically rotate the object to better visualize the sketching environment.

STEP 8:   ✔   Select the Continue icon to exit the sketcher environment.

STEP 9:   Preview the feature on the dialog box, then select OK.

SUMMARY

Most protrusion and cut features in Pro/ENGINEER can be created with relatively intuitive options such as Extrude, Revolve, Sweep, and Blend. Other common feature construction commands include Hole, Round, Chamfer, and Rib. While these commands can create most standard geometric shapes, complex shapes are more difficult to model. Pro/ENGINEER provides the Variable-Section Sweep, Swept Blend, and Helical Sweep commands for the construction of unique and/or more complex features.

SWEPT BLEND TUTORIAL

This tutorial exercise will provide instruction on how to create the part shown in Figure 11–19. This part consists of three features: Two extruded protrusions and one swept blend. Within this tutorial, the following topics will be covered:

• Creating an extruded protrusion.
• Creating a swept blend.

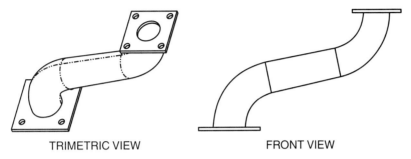

TRIMETRIC VIEW   FRONT VIEW

Figure 11–19   Completed Part

THE FIRST FEATURE

Name the part file *swept-blend1*. The first feature created in this part will be the base extruded protrusion. Create this feature on Pro/ENGINEER's default datum planes extruded in one direction with the dimensions shown in Figure 11–20. Sketch the section on datum plane TOP. Notice in the figure how the feature consists of four small holes and one large hole. Create all five holes and the base rectangular feature within the same section.

THE SECOND FEATURE

The second feature is similar to the first feature and is constructed in a similar manner (Figure 11–21). This feature's section is sketched on an on-the-fly datum plane offset

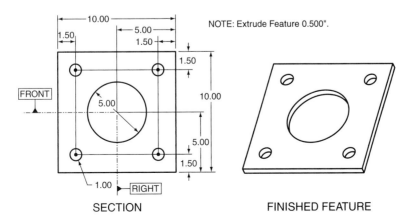

SECTION   FINISHED FEATURE

Figure 11–20   First Feature

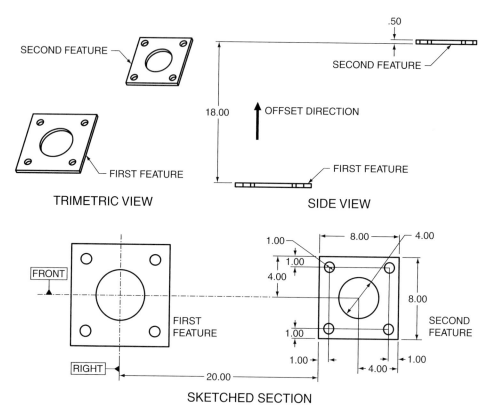

Figure 11-21   Offset Second Feature

18 -inches from the top of the first feature. As shown in the figure, the second feature's section is smaller than the first. Use the information shown in Figure 11–21 to create the second extruded protrusion feature. The dimensions shown match design intent.

SWEPT BLEND FEATURE

This segment of the tutorial will use the Swept Blend tool to create the connection between the first and second features. The Swept Blend tool is used to sweep two or more sections along a user-defined trajectory. The trajectory can be either sketched or selected. Along the trajectory's path, vertexes and datum points are used to define the location of each section. With the exception of the first and last vertex, the user has the option of accepting or rejecting a datum point or vertex for use in the swept blend. In this exercise, only the first and last vertexes will be used.

Perform the following steps to create the swept blend feature shown in Figure 11–22.

STEP 1:   Use the sketch tool to define the trajectory shown in Figure 11–23.

The trajectory of a swept blend feature can be either sketched or selected. Existing feature edges and datum curves are useful features for selecting. This tutorial will first require the sketching of the trajectory.

STEP 2:   Select datum plane FRONT as the sketching plane, then orient the sketching environment to match Figure 11–23.

STEP 3:   Select SKETCH >> REFERENCES on the menu bar to access the References dialog box.

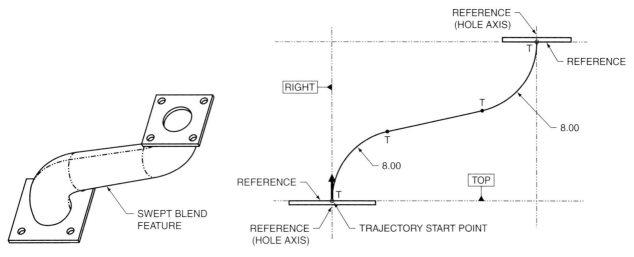

Figure 11-22   Swept Blend Feature   **Figure 11-23**   Trajectory Sketch

Step 4:   **Use the References dialog box to specify the references shown in Figure 11–23.**

Four specific references are important for this feature: the top of the base protrusion, the bottom of the second protrusion feature, the center axis of the base protrusion's hole entity, and the center axis of the second protrusion's hole entity. If necessary, selecting Sketch >> References on the menu bar will allow you to access the References dialog box.

Step 5:   **Use appropriate sketching tools to create the sketch of the trajectory (Figure 11–23).**

The dimensioning scheme and constraints shown in the figure match the design intent for the feature. When sketching the trajectory, align the start and end of the path with the axes from the large center holes of the first two features. In addition, your arc entities should be constrained tangent to the center axes of their respective holes.

Step 6:   **Select the Continue icon to complete the trajectory's section.**

Step 7:   **Select INSERT >> SWEPT BLEND on your menu.**

The object under construction in this segment is a pipe feature (not to be confused with the Pipe command). The pipe effect will be created with the Thin Protrusion option.

Step 8:   **On the dashboard, select the SOLID icon to create a solid swept blend feature.**

Step 9:   **On the workscreen, pick your previously created trajectory.**

After picking the trajectory, notice the startpoint arrow and the label of your trajectory as origin. Currently, your trajectory is serving as the origin trajectory for your swept blend's cross sections. The startpoint arrow defines the location of your first section.

Step 10:   **On the dashboard, open the Reference menu option to observe its default values.**

STEP 11: On the References menu option of the dashboard, ensure that NORMAL TO TRAJECTORY is selected as the section plane control.

As the name implies, the Normal to Origin Trajectory option will create the feature with each cross section normal to the feature's trajectory.

STEP 12: Open the Sections menu on the dashboard and ensure that SKETCHED SECTIONS is selected.

The sections for a swept blend can be either sketched or selected. The default Sketched Sections option will allow you to sketch each section within Pro/ENGINEER's sketching environment.

STEP 13: Change your select filter to VERTEX.

Each of your sections will be created at a vertex on the feature's trajectory. The Vertex filter setting will ensure that you can pick each vertex.

STEP 14: On your workscreen, pick the vertex defined by the trajectory's startpoint arrow.

In order starting from the startpoint end of the trajectory, you will create a section at each available vertex.

STEP 15: Select SKETCH on the Sections menu to enter the sketching environment for creating your first sketch.

It is often helpful to dynamically rotate the sketching environment to get a better understanding of the sketch plane in relation to the trajectory. Also notice on the trajectory in the figure the location of each cross section. In this example, there will be five sections. As shown in the figure, the first section is a 0.500 diameter circle.

STEP 16: Select the USE EDGE icon.

The first section will be created from the edge of the first feature's largest hole. The design intent for this feature requires the inside diameter of the pipe to equal the selected hole's diameter. The Use Edge option will help you to meet this intent.

STEP 17: Pick the edges of the large diameter hole in the two locations shown in Figure 11–24 (pick each location only once).

You will have to select the edge of the hole in two locations. After selecting the locations, notice the start point arrow. This arrow serves the same purpose as the start point for a typical blended feature. Like the Blend option, the Swept Blend option requires the following rules:

- Start Point arrows should be in the same general location for each section.

- In most situations, start point arrows have to point in the same direction.

- Each section of the swept blend has to have the same number of entities.

STEP 18: ✓ Select the Continue icon when the section is complete.

STEP 19: Select INSERT on the sections menu then pick the vertex on the end of the trajectory opposite from the start point.

If necessary, set your selection filter to vertex. You will know that you've had a successful selection when the section menu's Section Location collector displays an End Curve element.

STEP 20: Select SKETCH to define the second section.

STEP 21: In the sketching environment, select the USE EDGE option to create the section shown in Figure 11–25.

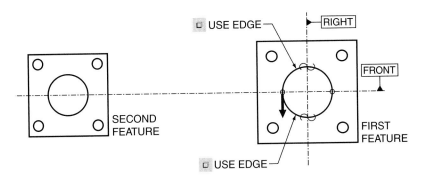

Figure 11–24 First Section Creation

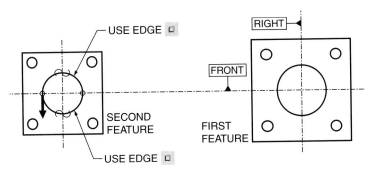

Figure 11–25 Second Section Creation

As with the first section, you will utilize the Use Edge option to create the section. On the second feature (Figure 11–25), select the edge of the largest-diameter hole. Make sure that your start point matches the start point from the first cross section.

STEP 22: Select the Continue icon to exit the sketching environment for the second section.

STEP 23: On the dashboard, select the Thin feature option.

Use the Insert then sketch menu options to create each remaining section.

STEP 24: Enter 0.125 as the thickness of the thin feature.

STEP 25: Select the FLIP option to change the material creation side.

The Thin attribute allows you to define the side of the sketch that the thin layer of material will be created. Using the Flip option will flip the side of material creation. This will allow the section to serve as the pipe feature's inside diameter.

INSTRUCTIONAL NOTE If your preview reveals a twisting effect to your swept blend feature, then the start points for each section were not in the same general location. On your Feature Definition dialog box, Redefine one of the sections (the Sections element) to set a start point corresponding to the other section's start point.

STEP 26: Select the Build Feature option then select OK.

SPRING TUTORIAL

The spring feature shown in Figure 11–26 was created with the Helical Sweep option. Helical sweeps are created by first constructing (sketching or selecting) a profile (or trajectory), then sketching the helical feature's cross section. Helical sweeps can be created with a constant pitch or a variable pitch. Within a spring feature, the pitch is the distance between corresponding points on the wire. Notice in the figure how this feature will have a variable pitch. Perform the following steps to create this feature.

STEP 1: **Start a new part file named** *spring.*

STEP 2: **Select INSERT >> HELICAL SWEEP >> PROTRUSION on the menu bar.**

STEP 3: **Select VARIABLE >> THRU AXIS >> RIGHT HANDED >> DONE on the Attributes menu.**

The Variable option creates a helical feature with a variable pitch while the Thru Axis option creates a helical sweep feature around a sketched axis. Helical sweep features can be created either right-handed or left-handed. A standard thread is an example of a right-handed helical feature.

STEP 4: **Select datum plane FRONT as the sketching plane for the springs profile.**

STEP 5: **Select OKAY to accept the default direction for viewing the sketching plane.**

STEP 6: **Select the DEFAULT orientation.**

STEP 7: **Sketch the sweep profile shown in Figure 11–27.**

The Helical Sweep option is a type of revolved feature. Because of this, it requires the sketching of a centerline that will serve as the feature's axis of revolution. Sketch the centerline entity aligned with the edge of datum plane RIGHT. In addition, notice the location of the sketched trajectory and the dimensioning scheme.

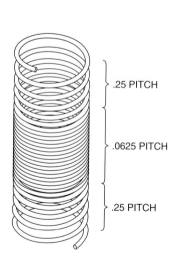

Figure 11-26 Spring Feature

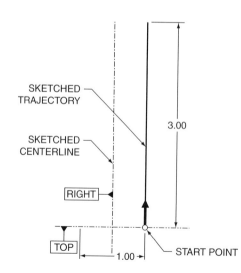

Figure 11-27 Sweep Profile Construction

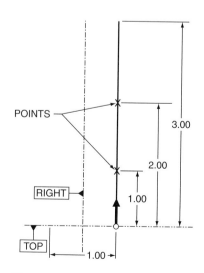

Figure 11-28 Point Construction

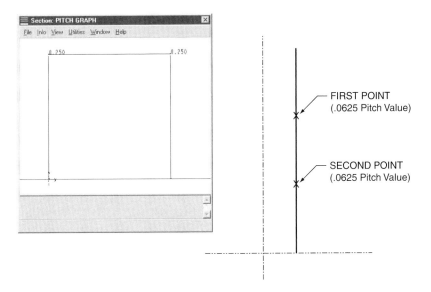

Figure 11–29   Graph Window

STEP 8:   ⌘ Use the POINT icon to create the two points shown in Figure 11–28.

Points can be used to control variable pitches on a helical feature. As shown previously in Figure 11–26 and as shown in Figure 11–28, these points will serve as the division point for the changes in pitch along the feature's trajectory.

STEP 9:   If necessary, use the pop-up menu's Start Point option to change the profile's start point to the location shown in Figure 11–27.

STEP 10:   ✔ Select the Continue icon to exit the sketching environment.

STEP 11:   In Pro/ENGINEER's Textbox, enter 0.25 as the pitch at the start of the trajectory.

This value will be the pitch of the spring at the start point of the feature.

STEP 12:   Enter 0.25 as the pitch at the end of the trajectory.

STEP 13:   On the Graph menu, select the DEFINE option, then on the workscreen select the first point (Figure 11–29).

Pro/ENGINEER provides a graph window for the definition of pitch values along the helical feature's path. The graph window serves as a tool to view the changes in the pitch values along the feature. Within the graph window, notice that the current values (0.250) of each point along the trajectory are displayed. On the workscreen (not the graph window), select the first point for the redefinition of its pitch value.

STEP 14:   In Pro/ENGINEER's textbox, enter 0.0625 as the new pitch value for the first point.

After entering the new pitch value, notice the change in the graph window.

STEP 15:   Select the second point and enter 0.0625 as its pitch value.

STEP 16:   On the Graph menu, select the DONE option.

STEP 17:   ◯ In the sketcher environment, use the CIRCLE icon to create the circle entity shown in Figure 11–30.

Sections for swept features are sketched from the opposite side of the trajectory's sketching plane. Notice on your workscreen and in

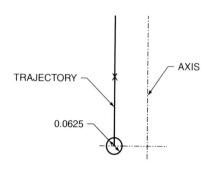

Figure 11–30   Sketched Section

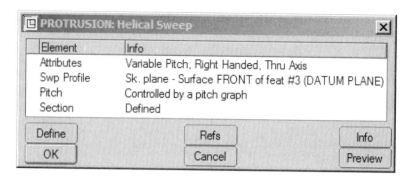

Figure 11–31   Feature Definition Dialog Box

Figure 11–30 how the trajectory's path is now to the left of the axis of revolution. Dynamically rotate the workscreen to better visualize the sketching environment. Sketch the circle's center coincident with the start point of the trajectory.

STEP 18:   **Select the Continue icon to exit the sketcher environment.**

STEP 19:   **Preview the feature on the Feature Definition dialog box then select OK to create the feature (Figure 11–31).**

Notice on the dialog box the various elements that have been defined. Each can be modified at this point with the Define option.

STEP 20:   **Save your part.**

BOLT TUTORIAL

This tutorial will create the bolt shown in Figure 11–32. Several steps are involved in the creation of this part. The first three steps involve the construction of the bolt shaft and the threads. The threads will be cut with the Helical Sweep command. The final two steps involve the modeling of the bolt's head. The following will be covered in this tutorial:

- Creating a helical sweep cut.
- Creating an extruded protrusion.
- Creating a revolved cut.

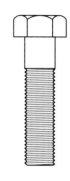

Figure 11–32
Bolt Part

CREATING THE BOLT'S SHAFT

The first feature included in the bolt consists of a revolved protrusion. Use Pro/ENGINEER's default datum planes and sketch the revolved feature on datum plane FRONT. The finished feature and the sketch are shown in Figure 11–33. The following are some pointers for creating this feature:

- Create the feature as a revolved protrusion.
- Use Pro/ENGINEER's default datum planes and align the sketch with the datum planes as shown in Figure 11–33.
- Include the chamfer within the sketch.
- Revolve the protrusion 360 degrees.

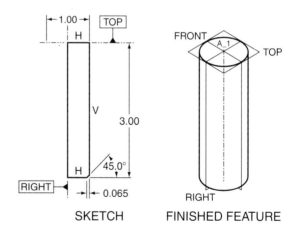

SKETCH   FINISHED FEATURE

Figure 11–33   Revolved Protrusion Feature

BOLT THREADS

This segment of the tutorial will use the Helical Sweep >> Cut option to create the threads for the bolt (Figure 11–34). The bolt will have eight threads per inch. Perform the following steps to create the helical cut:

STEP 1: Select INSERT >> HELICAL SWEEP >> CUT.

STEP 2: Select CONSTANT >> THRU AXIS >> RIGHT HANDED >> DONE as attributes for the thread.

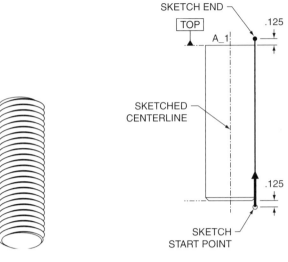

Figure 11–34
Thread Feature

Figure 11–35   Sweep Path Section

STEP 3:   Select datum plane FRONT as the sketching plane, then select OKAY to accept the default direction for viewing the sketching environment.

STEP 4:   Accept the default orientation.

STEP 5:   Use the References dialog box (Sketch >> References) to specify the top, bottom, and right edge of the shaft as references.

STEP 6:   Define the Sweep Path by creating the section shown in Figure 11–35 (including the centerline).

Sketch the single line entity shown in the figure. This line will serve as the trajectory path for the helical feature. When sketching the line, extend each end past the existing part. In addition to the line entity, sketch the centerline entity to serve as the axis of revolution.

STEP 7:   Define the dimensioning scheme shown in Figure 11–35.

STEP 8:   ✔ Select the Continue icon to exit the sketcher environment.

STEP 9:   Enter 0.125 as the pitch value of the thread.

The bolt in this tutorial will have 8 threads per inch. By definition, the pitch is the inverse of the number of threads per inch. The inverse of 8 is 0.125.

STEP 10:   Sketch the section defining the cut.

Notice in Figure 11–36 how the sketch commences at the start point of the trajectory of the feature. The profile of the sketch is also shown in the figure. The profile is a typical 60 degree United National Standard thread form. In the figure, notice the small 0.005-in gap. The pitch of the helical sweep is 0.125 in. This gap forms a flat thread ridge as would be typically found on a bolt.

STEP 11:   ✔ Select the Continue option to exit the sketcher environment.

STEP 12:   Select OKAY to accept the material removal side (or flip if necessary).

STEP 13:   Preview the feature then select OK on the Feature Definition dialog box (Figure 11–37).

STEP 14:   Save your part.

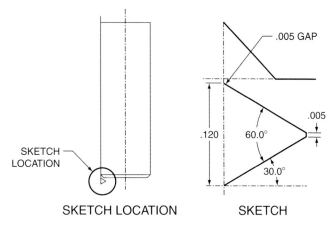

Figure 11-36 Helical Sweep Section

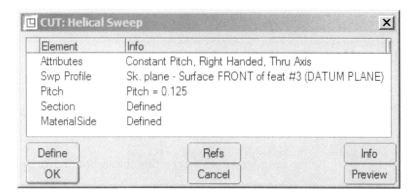

Figure 11-37 Feature Definition Dialog Box

EXTRUSION EXTENSION

The first extruded protrusion created within this part serves as the surface for forming the thread feature. The bolt in this tutorial is 4.00 in long. The original protrusion could have been created 4.00 inch in depth, but the termination of the cut helical sweep at the 3.00 point (the thread length) would result in a poor thread representation. Adding the additional protrusion partially solves this problem.

Figure 11–38 shows the finished extruded protrusion, the sketching plane, and the feature section. Create this feature, using the following options:

- Create the feature as an extruded protrusion.
- Extrude the feature 1.00 in away from the base protrusion.
- Use datum plane TOP as the sketching plane.
- Within the sketching environment, utilize the Use-Edge option to create the section.

DESIGN INTENT Utilizing the Use Edge option to create the section for this feature helps to capture the design intent of the part. Within this bolt part, the diameter of the shaft should be the same throughout the length of the bolt. The Use Edge option will tie the diameter of this feature with the first protrusion feature.

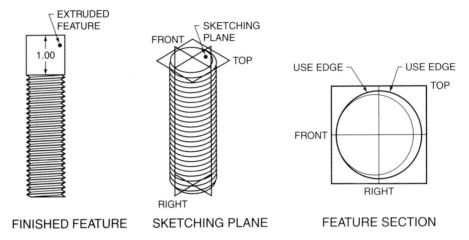

Figure 11-38 Feature Construction

BOLT HEAD CREATION

The head of the bolt requires the creation of two features. The first feature is an extruded protrusion and is shown in Figure 11–39. Use the nonthreaded end of the existing part as the sketching plane. Extrude the feature one direction away from the sketching plane a distance of 0.667 in.

Sketch the feature's section as shown in Figures 11–39 and 11–40. The sketching of the section can be challenging. The technique shown in the figures requires the sketching of one-quarter of the section, then mirroring the remainder of the section using the Mirror option. Remember, the Mirror option requires the existence of a centerline to mirror about. Entities to be mirrored must be preselected before the Mirror icon is accessible.

Notice the construction circle in both figures. A sketched entity, such as a circle, can be converted to a construct entity with the Edit >> Toggle Construction option. A construction circle will not extrude with sketched geometric entities. The construction circle in this example defines the distance across the flats of the hex-head bolt.

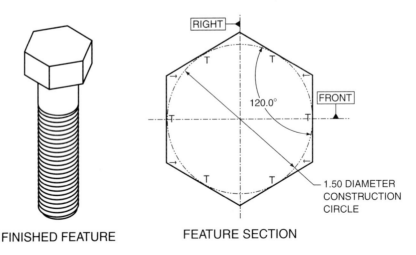

FINISHED FEATURE   FEATURE SECTION

Figure 11-39 Bolt Head Protrusion

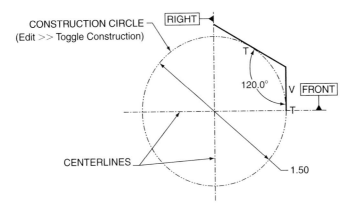

Figure 11–40 Section Construction

Bolt Head Cut

The second feature required within the construction of the bolt's head is a revolved cut (Figure 11–41). Perform the following steps to create this feature:

Step 1: Select the Sketch Tool.

Step 2: For the sketching plane, select the datum plane that extends between the corners of the bolt head's extruded protrusion (datum plane RIGHT in Figure 11–41).

Step 3: Orient the sketcher environment to match Figure 11–42.

Step 4: Sketch the section shown in Figure 11–42.

For references, specify the axis from the based revolved protrusion, the top edge of the bolt head, and right edge of the bolt head. The section requires one geometric line entity and one centerline entity. The centerline will serve as the axis of revolution.

Step 5: Select the Continue icon to exit the sketching environment.

Step 6: Select the Revolve option on the toolbar.

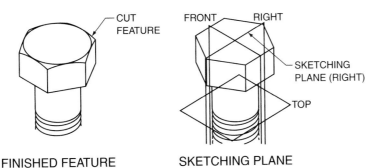

FINISHED FEATURE   SKETCHING PLANE

Figure 11–41 Bolt Head Cut

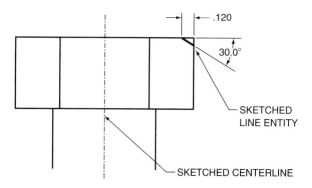

Figure 11–42   Cut Section

STEP 7:   Select the Solid option on the revolve dashboard.

STEP 8:   Select the Cut option on the revolve dashboard.

STEP 9:   Select an appropriate material removal side.

STEP 10:   Select 360 as the degrees of revolution.

STEP 11:   Build the feature.

STEP 12:   Save your part.

VARIABLE-SECTION SWEEP TUTORIAL

The Variable-Section Sweep option is used to sweep a section along multiple trajectories. It is a useful tool for creating advanced geometric shapes. This tutorial will cover the process for creating a simple variable-section sweep.

The trajectories for a variable-section sweep can be either selected on the workscreen or sketched. Datum curves are often created for use as trajectories within the construction process. This tutorial will use datum curves for this purpose (Figure 11–43). The following will be covered in this tutorial:

- Creating datum curves.
- Creating a variable-section sweep feature.
- Creating a swept protrusion.
- Creating an extruded protrusion.

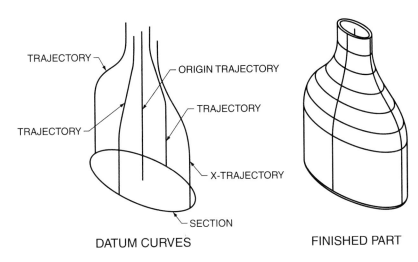

Figure 11–43   Variable Section Sweep

CREATING THE ORIGIN TRAJECTORY

In this segment of the tutorial, you will create the origin trajectory for the feature. The origin trajectory is used to define the start point of the trajectory, the direction of sweep, and the depth of the sweep. It can be either sketched or selected. In this tutorial, it will be created as a datum curve, then selected within the creation of the variable-section sweep.

Sketch the origin trajectory on datum plane RIGHT. Sketch the datum curve geometry at the intersection of datum plane RIGHT and datum Plane FRONT (Figure 11–44).

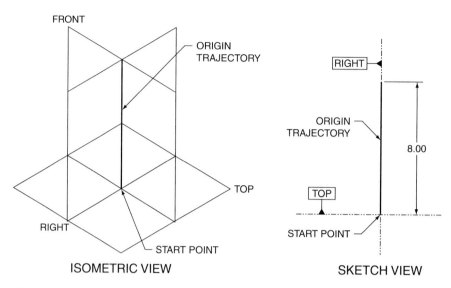

Figure 11–44   Origin Trajectory Construction

X-Trajectory Creation

The X-trajectory is used to define the variation in the horizontal vector of the variable-section sweep's section. The X-trajectory for this tutorial will be defined by a sketched datum curve that in turn will be selected during the construction of the variable-section sweep feature. Sketch the datum curve on datum plane FRONT as shown in Figure 11–45. Align the bottom of the sketch with datum plane TOP. Sketch a centerline aligned with datum plane RIGHT and create the diameter dimensions as portrayed in the figure.

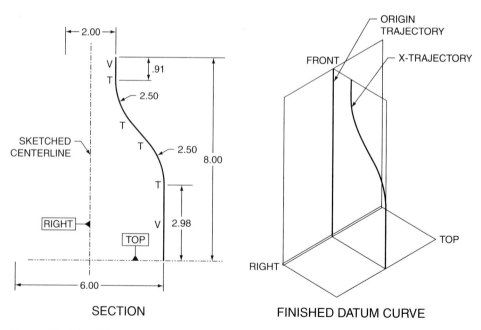

Figure 11–45   X-Trajectory Creation

TRAJECTORY CREATION

Variable-section sweep trajectories serve basically the same purpose as the feature's X-trajectory. Multiple trajectories can be selected or sketched to help define the variation of the section along the sweep's path. In this tutorial, you will create the next trajectory by copy-mirroring the X-trajectory over datum plane RIGHT (Figure 11–46). Use the following information when creating this new trajectory:

- Use the Edit >> Feature Operations >> Copy **>>** Mirror option.
- Use the Dependent suboption.
- Copy the X-trajectory over datum plane RIGHT.

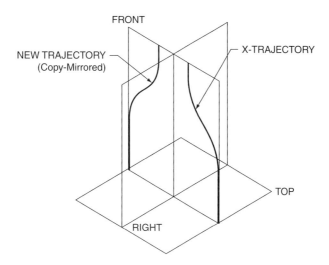

Figure 11–46 Trajectory Creation

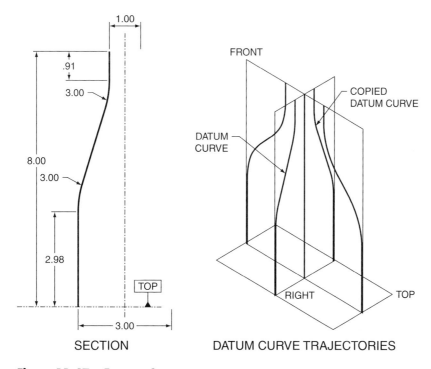

Figure 11–47 Trajectory Creation

Two additional trajectories are required and are shown in Figure 11–47. Sketch the first curve as shown in the figure, then copy-mirror this curve to create the second. The bottom of the sketched datum curve should be aligned with datum plane TOP.

Variable-Section Sweep Feature

This step of the tutorial will create the thin variable-section sweep feature shown previously in Figure 11–43. A variable-section sweep feature (with the Normal to Trajectory option) requires the definition of one origin trajectory, one X-trajectory, and one section. Multiple trajectories in addition to the origin trajectory and the X-trajectory can be defined. In this tutorial, a total of five trajectories will be utilized. Perform the following steps to create this feature.

Step 1:   Select the Variable Section Sweep option on the toolbar.

Step 2:   Using the Control key, pick the origin trajectory shown in Figure 11–48.

> The origin trajectory is always selected first with a variable-section sweep. It defines the direction and extent of your sweep.

Step 3:   Using the Control key, select the X-trajectory shown in Figure 11–48.

Step 4:   Using the Control key, select the remaining three trajectories.

Step 5:   Open the References menu on the dashboard.

Step 6:   Observing the Trajectories option under the references menu, ensure that the origin trajectory's N value and the Chain 1 trajectory's X value are checked (Figure 11–48).

> The checked N value creates the sweep normal to the origin trajectory, while the checked X value with Chain 1 defines the X-trajectory for the sweep.

Step 7:   On the workscreen, observe the origin trajectory and ensure that its start point arrow is defined at the location shown in Figure 11–12.

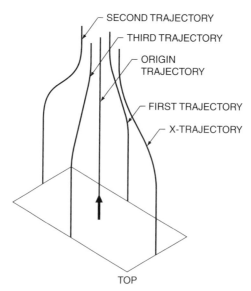

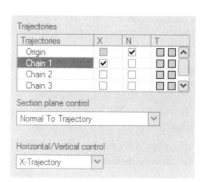

DATUM CURVE TRAJECTORIES

REFERENCES MENU

Figure 11–48   Trajectory Selection

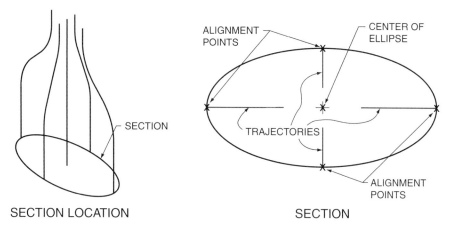

SECTION LOCATION   SECTION

Figure 11–49   Section Sketch

The start point of the trajectory is denoted on the origin trajectory by an arrow. The arrow's location should match Figure 11–48. If necessary, use the pop-up menu's Flip Chain Direction option to change the start point to the opposite end of the origin direction.

STEP 8:   Select the Solid option on the dashboard.

STEP 9:   Select the Sketch icon on the dashboard.

Unlike most sketching environments, you do not have to explicitly define the location and orientation of the section's sketching plane. For a normal to trajectory variable-section sweep, the sketching plane passes through the start point and is normal to the origin trajectory.

STEP 10:   Sketch the section using the Ellipse option (Figure 11–49).

Pro/ENGINEER will orient the sketching environment to lie normal to the origin trajectory. You can dynamically rotate the workscreen to better visualize the sketching plane. Notice in the figure the location of the alignment points.

An ellipse is created by first locating its center point followed by dragging its major and minor diameters. When locating your major and minor diameters, make sure your ellipse connects to each required alignment point.

STEP 11:   Select the Continue option when the section is complete.

STEP 12:   Select the Thin option on the dashboard.

STEP 13:   Enter 0.125 as the width of the thin feature.

STEP 14:   Use the Material Side option to ensure that material is added to the outside of the thin feature.

STEP 15:   Build the feature.

STEP 16:   Save your part.

SWEPT FEATURE

Within this segment of the tutorial, you will create the base of the part as a swept protrusion (see Figure 11–50 on the next page). As shown in Figure 11–51, the outside of the base of the existing part will be selected as the trajectory and the section will be sketched on datum plane FRONT.

STEP 1:   Select INSERT >> SWEEP >> PROTRUSION on the menu bar.

STEP 2:   Pick the SELECT TRAJ option as the Trajectory Definition option.

Within this tutorial, you will select the edge shown in Figure 11–51 as the trajectory for this swept feature.

STEP 3:   Select TANGENT CHAIN on the Chain menu then pick the trajectory edge shown in Figure 11–51.

The Tangent Chain option will select a picked entity and any other entity tangent to the selected entity. In this step of the process, you will select the outside edge of the variable-section sweep feature as shown in the figure.

Note: After selecting the edge, your model should display the start point location shown in Figure 11–52.

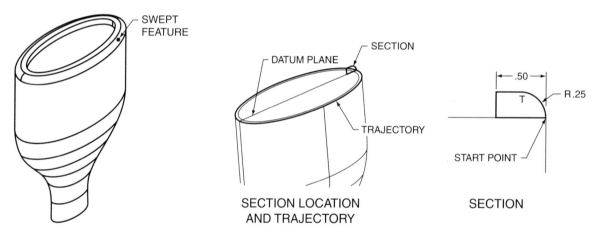

SECTION LOCATION
AND TRAJECTORY

SECTION

Figure 11-50   Swept Feature   **Figure 11-51**   Swept Feature Creation

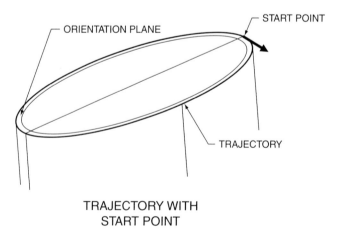

TRAJECTORY WITH
START POINT

Figure 11-52   Trajectory and Start Point Locations

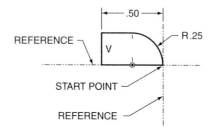

Figure 11-53　Sweep Section

INSTRUCTIONAL NOTE　The Tangent Chain option should select the entire perimeter edge of the part. If it doesn't, you can use the One By One option to continue picking necessary edges. A successful selection will reveal a start-point arrow at the location of the trajectory's start point. Observe your screen to confirm the presence of this start point.

STEP 4: Select DONE to exit the Chain menu.

INSTRUCTIONAL NOTE　The next step in this tutorial will require you to identify a planar surface to orient toward the top of the sketching environment. This is a similar process to orienting the sketching environment with the TOP option during the creation of an extruded protrusion.

STEP 5: Using the Choose menu, identify and ACCEPT the orientation plane shown in Figure 11–52.

The accepted surface will be oriented toward the top of your sketching environment.

STEP 6: Select OKAY to accept the default orientation of the cross section.

STEP 7: Turn off the display of your datum planes.

STEP 8: Ensure the presence of the two references shown in Figure 11–53.

STEP 9: Sketch the section shown in Figure 11–53.

Refer to Figures 11–51, 11–52, and 11–53 when sketching the section. Incorporate any necessary dimensions and constraints.

STEP 10: ✔ When your section matches Figure 11–53, exit the sketcher environment.

STEP 11: Preview your feature on the Feature Definition dialog box.

STEP 12: Select OK to create feature.

EXTRUDED PROTRUSION

As shown in Figure 11–54, your part has a very obvious hole that needs to be plugged. This segment of the tutorial will solve this problem by filling this hole with an extruded protrusion. Within the sketcher environment of this feature, you will utilize the Use Edge option to create all necessary sketch entities.

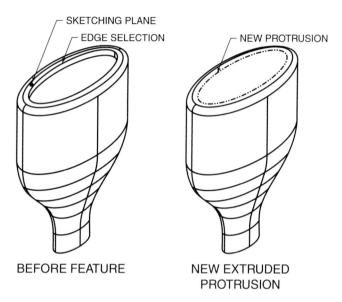

BEFORE FEATURE   NEW EXTRUDED
                          PROTRUSION

Figure 11–54   Extruded Protrusion

STEP 1:   **Set up an extruded protrusion sketching environment to create the feature shown in Figure 11–54.**

Use the following options when establishing your protrusion:

- Choose a one-sided extrusion.

- Select the sketching plane shown in Figure 11–54 (the bottom of the part).

- Orient datum plane FRONT toward the bottom of the sketcher environment as shown in Figure 11–55.

- The feature creation direction should extrude toward the inside of the part. You might have to flip the default direction.

STEP 2:   **On the Environment dialog box (Tools >> Environment), set NO DISPLAY as the Tangent Edges display style (see Figure 11–56).**

As previously mentioned, you will utilize the Use Edge option to create sketcher entities. Within your current model, tangent edges can clutter the workscreen and make edge selection difficult. By turning off the display

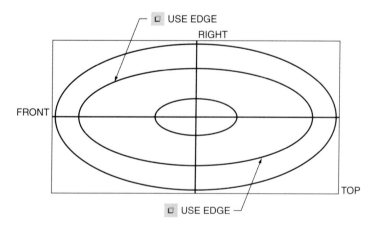

Display Style	Hidden Line
Default Orient	Isometric
Tangent Edges	No Display

Figure 11–55   Sketcher Environment and Use Edge Option   **Figure 11–56**   Tangent Edges Display

of tangent edges, the selection of the edges in this segment of the tutorial should be easier.

Step 3: On the Display toolbar, set **HIDDEN** as the model's display style.

Step 4: (Optional step.) Specify datum planes **FRONT** and **RIGHT** as references.

This step is not a requirement for this section. Since the Use Edge option will be used to create all necessary sketcher entities, you actually do not need to specify references.

Step 5: Select the **USE EDGE** icon on the sketcher toolbar.

Step 6: Pick the two edges shown in Figure 11–55 (select each edge only once).

Step 7: Exit the sketcher environment.

If you receive an "Incomplete Section" error message, you probably picked an edge more than once in the previous step. You will need to delete any extra edges. Another possible cause of the error message is an incomplete closure of your section. If this is the case, you should observe your section and use the Use Edge option to pick edges that were excluded from the previous pick process.

Step 8: Extrude your feature a blind distance of 0.25.

Step 9: Ensure that your direction of extrusion is into the existing part.

Step 10: Turn the display of tangent edges on.

Step 11: Build your feature.

Step 12: Save your part file.

Problems

1. Model the part shown in Figure 11–57.

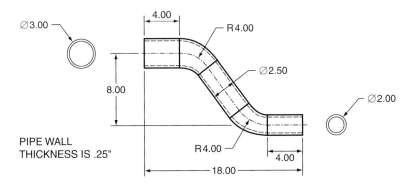

Figure 11–57   Problem 1

2. Use Pro/ENGINEER to model a 1.25-in nominal diameter hex head bolt. The hex head has a 1.875-in distance across its flats and a height of 0.844 in. The bolt length is 3.00 in with a thread length of 2.25 in. Create the thread as a cut helical sweep.

3. Use Pro/ENGINEER to model a 1.50-in nominal diameter hex head bolt. The hex head has a 2.25-in distance across its flats and a height of 1.00 in. The bolt length is 4.00 in with a thread length of 3.25 in.

4.   Model the spring feature shown in Figure 11–58.

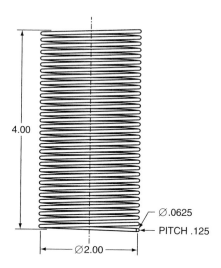

Figure 11–58   Problem 4

5.   Model the bottle part shown in Figure 11–59. Create the bottle as a thin variable-section sweep feature. The trajectories for the feature are shown in the figure. The dimensions for the X-trajectory are also shown. Create the X-trajectory as a datum curve then use the Copy >> Move >> Rotate option to create three more curves rotated 90 degrees apart. Create the remaining four trajectories by first sketching a datum curve on an on-the-fly datum plane at a 45 degree angle to one of Pro/ENGINEER's default datum planes. Create the section by projecting one of the existing datum curves with the Use Edge option. Use the Copy >> Move >> Rotate option to create the final three trajectories.

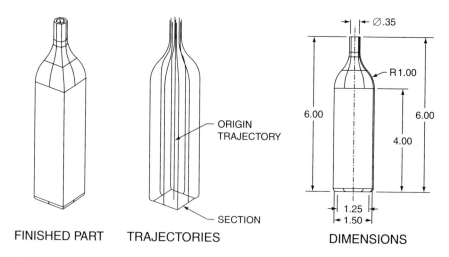

FINISHED PART   TRAJECTORIES   DIMENSIONS

Figure 11–59   Problem 5

6.   Use the hex head bolt created in Problem 2 as the generic part of a family table of hex head bolts. The following rules apply to the creation of hex head bolts:

  •   The distance across the flats of the bolt head is 1.5 times the bolt diameter.

  •   The height of the bolt head is 0.667 times the bolt diameter.

  •   The thread length is 2 times the diameter plus 0.25 in.

Refer to Chapter 7's section on family tables. Create the following instances:

- 1-in nominal size bolt with a 3.00-in bolt length.

- 2-in nominal size bolt with a 5.00-in bolt length.

- 0.5-in nominal size bolt with a 2-in bolt length.

QUESTIONS AND DISCUSSION

1. Compare and contrast the Normal to Trajectory, Normal to Projection, and Constant Normal Direction Variable-Section Sweep options.

2. Describe the basic principles behind a swept blend feature.

3. Name two methods for creating the trajectory for a swept blend feature.

4. When creating a swept blend, what is Pro/ENGINEER requesting when it is asking for the Z-axis rotation angle?

5. When creating a variable-section sweep feature, what is the first trajectory that must be defined? What is the purpose of this trajectory? What is the second trajectory that Pro/ENGINEER requires?

6. How many trajectories does Pro/ENGINEER require within the Variable-Section Sweep option? How many can be defined?

7. Define the meaning behind the pitch of a thread.

8. What two geometric features are often created with the helical sweep feature?

12

ASSEMBLY MODELING

Introduction

Pro/ENGINEER is considered a design and engineering tool. The various modules of Pro/ENGINEER provide designers, engineers, and manufacturers with the tools necessary to take a design from the conceptual stage through the final manufacturing process. One of the most powerful tools of Pro/ENGINEER is its Assembly module. Within this application, existing components can be grouped as part of an assembly or as part of a subassembly. In addition, Pro/ENGINEER is capable of completing a design from the top down by allowing parts to be modeled with Assembly mode. Upon finishing this chapter, you will be able to:

- Create an assembly through the placement of existing parts.
- Create parts within Assembly mode.
- Modify parts within an assembly.
- Apply dimensional relationships between parts in Assembly mode.
- Create assembly features.
- Create a simplified representation of an assembly.
- Create and animate an assembly mechanism.

DEFINITIONS

Assembly	A collection of components that forms a complete design or a major end item.
Bottom-up design	The placing of existing components within an assembly.
Constraint	The explicit relationship defined between components of an assembly.
Component	A part or subassembly.
Package	A component that has not been fully constrained within an assembly.
Parametric assembly	An assembly with parts constrained to other parts.
Subassembly	A collection of parts and/or smaller subassemblies that forms a subcomponent of a complete design or major end item.
Top-down design	The designing of components within an assembly.

INTRODUCTION TO ASSEMBLY MODE

Pro/ENGINEER's Assembly mode is used to group components to meet the requirements of a design. Components can consist of existing parts and subassemblies or components can be created directly within Assembly mode. Placing existing components to form an assembly is referred to as bottom-up assembly design, while creating parts within Assembly mode is a tool used within top-down assembly design.

Parts in Assembly mode maintain their associativity with their separate part files. Within Part mode, if a dimension value is modified, the part instance in Assembly mode is modified. Correspondingly, if an instance of a part is modified in Assembly mode, the component in Part mode is modified. In addition, when a part is created within Assembly mode by using top-down assembly design, a new part file is created that can be modified separately within Part mode. When a component is placed into an assembly, the component's separate part or assembly file is placed into memory and remains there until the parent assembly is erased from memory.

PLACING COMPONENTS

Pro/ENGINEER has two options for defining components within an assembly. First, existing parts and subassemblies can be placed directly into an assembly model, a process loosely referred to as bottom-up assembly design. The Insert >> Component >> Assemble and the Add Component icon are used to locate and open components. When a component is opened, its associated object file is opened into Pro/ENGINEER's memory (but not into a separate window). When an assembly is saved, objects within the assembly are saved to their separate object files. Individual components cannot be erased from memory as long as an associated assembly object is open. Pro/ENGINEER also provides top-down assembly tools for defining components directly within the assembly environment.

A component can be placed into an assembly at any point during the assembly creation process, including as the first element of an assembly. When placed as the first component and before the creation of any assembly features, the object is placed without any defined constraints. When the object is placed after a component or after an assembly feature, Pro/ENGINEER will launch the Component Placement dashboard (Figure 12–1). This dashboard has two primary component placement types: constraints and connections. The constraints option is used to establish traditional assembly constraints such as mates and aligns. The connections option is used to create mechanism joints. Constraints and connections define the relationship between components of an assembly.

Figure 12-1   Component Placement Dashboard

ASSEMBLY CONSTRAINTS

When a component is placed into an assembly by using the Assemble option, it should be fully constrained to existing components and features. Pro/ENGINEER provides a variety of constraint types for the placement of components (Figure 12–2). The following is a description of each:

AUTOMATIC

The Automatic constraint type is not actually a constraint but a quick tool incorporated within the available constraint options to expedite the constraint defining process. It is the default constraint type when you access the Component Placement dashboard. With the Automatic option, references are selected for both the component and the assembly. Pro/ENGINEER will determine the constraint to apply, but will provide you the option of selecting an alternative constraint. As an example, when mating two surfaces, with the automatic option, you must pick each surface.

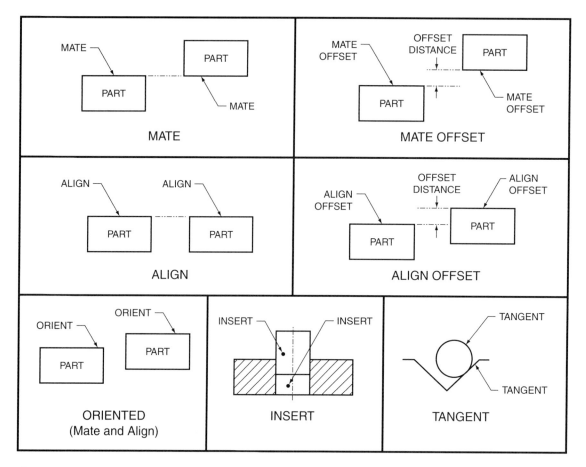

Figure 12–2   Constraint Types

Pro/ENGINEER might give you an Align constraint with the option of changing your constraint to a Mate.

MATE

The Mate constraint type is used to place two surfaces coplanar. Any datum plane, part plane, or planar surface may be used. As shown in Figures 12–2 and 12–3, selected surface faces are placed along a common plane, but do not actually have to touch.

MATE OFFSET

Offset is a suboption under the Mate constraint type. Mating surfaces are placed coincident by default with the Mate constraint. The Offset option places a user-

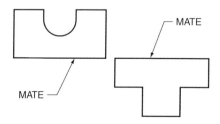

Figure 12–3   Mate Constraint

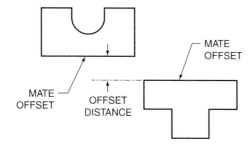

Figure 12–4   Mate Offset

specified offset distance between the selected surfaces (Figure 12–4). The distance can be modified at a later time.

ALIGN

The Align constraint type is used to place two surfaces coplanar and facing in the same direction (Figure 12–5). Like the Mate constraint, the surfaces do not have to touch. In addition, the Align constraint type is used to align axes, edges, and curves.

ALIGN OFFSET

Like the Mate Offset option, Align has a suboption for offsetting two aligned surfaces. As with the Mate constraint, aligned surfaces are placed coincident by default. These surfaces can be offset a user-specified distance (Figure 12–6).

ORIENTED

Similar to the Offset suboption, the Align and Mate constraints have a suboption for orienting two surfaces in the same direction (Figure 12–7). Unlike coincident and offset aligns and mates, two constrained surfaces are not coplanar and have no specified offset distance.

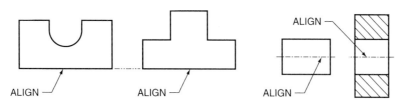

Figure 12–5   Align Constraint

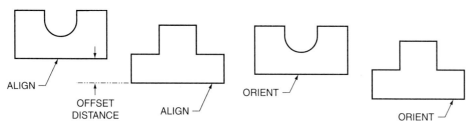

Figure 12–6   Align Offset

Figure 12–7   Orient Constraint

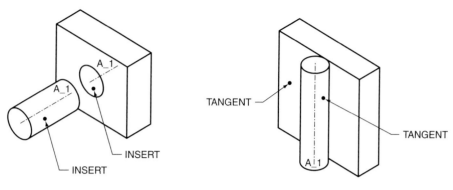

Figure 12–8 Insert Constraint   **Figure 12–9** Tangent Constraint

INSERT

The Insert constraint type makes the axes of two revolved features coincident (Figure 12–8). The user is required to select the surface of each feature. It is often used with shafts and holes for the alignment of each feature's centerline.

TANGENT

The Tangent constraint type makes a cylindrical surface tangent to another surface (Figure 12–9). The user is required to select the surfaces of each feature.

COORDINATE SYSTEM

The Coordinate System constraint type aligns the coordinate systems of two parts. Within this constraint type, the axis of one coordinate system is aligned with the corresponding axis of a second coordinate system (e.g., X-axis with X-axis).

POINT ON LINE

The Point On Line constraint type aligns a datum point with an existing edge, datum curve, or axis.

POINT ON SURFACE

The Point On Surface constraint type aligns a datum point with a surface. A surface can be any part surface or datum plane.

EDGE ON SURFACE

The Edge On Surface constraint type aligns the edge of a component with a surface. A surface can be any part surface or datum plane.

FIX

The Fix constraint type will ground a component at its current location.

MOVING COMPONENTS

When a component is placed into a scene, its default location might overlap existing components (making it hard to visualize) or it might be out of position for the type of constraints to apply. The Move menu on the Component Placement dashboard (Figure 12–10) allows a partially constrained component to be moved on the workscreen. The component can be moved only within the degrees of freedom allowed by existing constraints.

There are three motion types available: Translate, Rotate, and Adjust. The Translate option will move the component within the motion reference, while the Rotate option rotates the component around the selected motion reference. The Adjust option works similarly to available constraint options. This option allows the moving component to be mated and aligned with existing components.

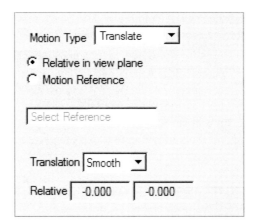

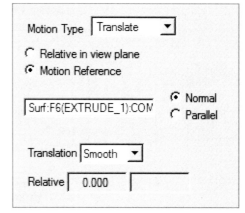

Figure 12-10 Move Tab　　　　**Figure 12-11** Move Dialog Box

When a motion type is selected, the relative motion is based on the motion reference selected. The following references are available:

- Relative in **View Plane** The motion will be relative to the current screen orientation.
- **Motion Reference** The motion will be relative to a selected reference entity such as a surface, plane, edge, or curve (Figure 12–11). Suboptions of this type include:
  - **Plane Normal or Plane Parallel** The motion will be perpendicular or parallel to a selected plane
  - **2 Points** Two selected vertices on the workscreen are used to created the relative motion.
  - **Csys** The motion will be relative to the X-axis of a selected coordinate system.

PACKAGED COMPONENTS

When a part or subassembly is placed with constraints using the Assemble option, it is considered a parametric assembly. Components of a parametric assembly have to be fully constrained. If a component is only partially constrained, it is considered a packaged component. Pro/ENGINEER provides the option of placing a component directly into a model as a package with the Package option. A packaged component is considered nonparametric.

When you use the

Insert >> Component >> Package >> Add >> Open option,

a component is placed and repositioned with the Move dialog box. Options available on this dialog box are basically the same as those under the Move menu on the Component Placement dashboard.

PLACING A PARAMETRIC COMPONENT

Perform the following steps to place a parametric component:

STEP 1: Select the Add Component icon on the toolbar.

STEP 2: Use the Open dialog box to open a part or assembly.

STEP 3:   **Use constraint types available under the Component Placement dialog box to fully constrain the component.**

A parametric assembly should be fully constrained. The Placement Status box on the dialog box tells the current constraint status of the component. The Move Tab is available for the translation and rotation of the component along existing degrees of freedom. A partially constrained component can be placed as a packaged assembly.

STEP 4:   **Select OK on the dialog box to finish the placement process.**

MECHANISM DESIGN

Placing components into an assembly by using traditional assembly constraints is a powerful but not very intuitive process. One of the disadvantages of this form of assembly modeling is the necessity to fully constrain (or fix) parent components. Fixed components have zero degrees of freedom to move. When a component is not fully constrained it is considered packaged.

Traditional component placement controls assembly by limiting the degrees of freedom of a component. When placing a component as a connection within mechanism, the component's placement is defined by the definition of a joint type, with the joint type limiting specific degrees of freedom. In many cases, the component is packaged by design. The following joint types are available:

PIN

A pin joint is the most basic connection (Figure 12–12). It is primarily defined through the alignment of two axes. An additional requirement is the mating of two planes and/or two points to define the joint's translation limitation. This joint type provides one rotational degree of freedom.

CYLINDER

The cylinder joint is similar to the pin joint (Figure 12–13). Like the Pin joint, this joint is defined through the alignment of two axes. Unlike the Pin joint, no translation limitation is provided. Two degrees of freedom are provided: one rotational and one linear.

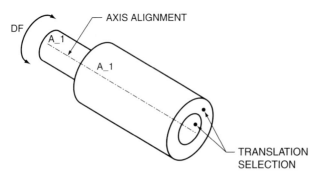

Figure 12-12   Pin Joint

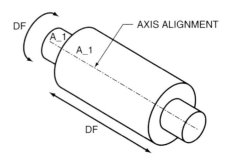

Figure 12-13   Cylinder Joint

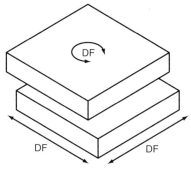

Figure 12-14 Planar Joint

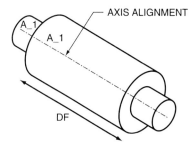

Figure 12-15 Slider Joint

PLANAR

The planar joint connects two planar surfaces (Figure 12–14). It is defined through the selection of a planar surface on each component. It provides three degrees of freedom: two linear and one rotational.

SLIDER

The slider joint is similar to the cylinder joint, except the slider joint does not provide any rotational degrees of freedom (Figure 12–15). Like the Cylinder joint, the Slider joint requires the aligning of two axes. To control the rotational degrees of freedom, two planar surfaces must be mated and/or aligned. Any defining planes must lie parallel to the selected axes. One linear degree of freedom is provided.

BALL

The ball joint provides three rotational degrees of freedom, but no linear (Figure 12–16). This joint requires the selection of two datum points.

BEARING

The bearing joint is a combination of the ball joint and the slider joint (Figure 12–17). Within this joint, the ball definition is allowed to slide along a selected axis. This joint requires the alignment of two axes and two points. A bearing joint has four degrees of freedom: one linear and three rotational.

RIGID

The rigid connection is available to group components by normal assembly constraints such as mate, align, and insert. It is used to define components that have zero degrees of freedom.

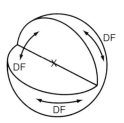

Figure 12-16 Ball Joint

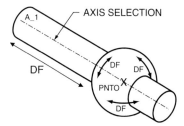

Figure 12-17 Bearing Joint

EDITING ASSEMBLIES AND PARTS

Pro/ENGINEER's Assembly mode provides a variety of options for the manipulation and editing of assembly components. Within an assembly, modifiable components can include parts, skeleton models, subassemblies, assemblies, dimensions, and exploded models. Additionally, options are available for modifying dimensions, features, and components.

The editing of a feature within a specific part is a three-step process. The first step requires the activation of the part within the context of the assembly. The Activate command can be accessed by using the model tree's pop-up menu. Once activated, a feature within an active part can be edited by first selecting it with the pop-up menu's Pick From List option. Once picked in this manner, the final step requires the selection of a feature editing command (Edit, Edit Definition, etc.).

MODIFYING DIMENSIONS

Pro/ENGINEER provides the ability to edit part dimensions directly within assembly mode. Perform the following steps to modify a dimension value:

STEP 1: Use the model tree's pop-up menu's ACTIVATE command to activate a specific component.

STEP 2: Use the mouse pop-up menu's PICK FROM LIST option to select a specific part feature.

Only features on the active part can be selected with the Pick From List command. Note that features can be edited by double-picking them on the workscreen.

STEP 3: Select the pop-up menu's EDIT command.

STEP 4: On the workscreen, select a dimension to modify, then enter an appropriate new value.

STEP 5: Regenerate the model.

STEP 6: To reactivate the assembly, select WINDOW >> ACTIVATE on the menu bar.

CREATING NEW PART FEATURES

Within Assembly mode, features can be added to parts and to skeleton models. Before a part feature can be created, the part must be activated with the pop-up menu's Activate command.

When a feature is created within a part or skeleton model, it is considered a component feature and will be created in the individual part or skeleton model object file. Most feature creation tools from part mode are available (e.g., Extrude, Revolve, Round, Variable-Section Sweep, etc.). When creating a part feature in assembly mode, other components within the assembly can be used as references. This is referred to as an **external reference**. External references should be used only when design intent dictates.

REDEFINING A COMPONENT FEATURE

The Edit Definition command is used in Part mode to modify attributes and elements associated with part features. This command is also available for the modification of part and skeleton models within Assembly mode. As with the modification of component dimensions within Assembly mode, features redefined in Assembly mode will also be redefined within their respective component object files. Perform the following steps to redefine a component feature:

STEP 1: **Use the model tree's pop-up menu's ACTIVATE command to activate a specific component.**

STEP 2: **Use the mouse pop-up menu's PICK FROM LIST option to select a specific part feature.**

Only features on the active part can be selected with the Pick From List command.

STEP 3: **Select the pop-up menu's EDIT DEFINITION command.**

STEP 4: **Redefine the feature as necessary.**

After the feature is selected for redefinition, Pro/ENGINEER will launch the Dashboard or Feature Definition dialog box associated with the selected feature.

STEP 5: **To reactivate the assembly, select WINDOW >> ACTIVATE on the menu bar.**

CREATING ASSEMBLY FEATURES

Pro/ENGINEER provides feature creation and editing toolbars to construct assembly features. Within this capability, only negative space solid features such as holes and cuts can be utilized. Datum and surface options are available also. When an assembly feature is created, its visibility level can be set at the assembly level or at the part level. When it is set at the assembly level, the feature is visible only in the assembly model. In other words, each part when opened in Part mode will not be affected. When the visibility level is set at the part level, the feature is visible when the part is open in Part mode.

When you are creating an assembly feature such as a hole or a cut, the process for constructing the feature is similar to the process for constructing the same feature in Part mode. The only additional requirement is to specify the feature's intersection components. Features with dashboards (Extrude, Revolve, etc.) provide the Intersection menu option to serve this purpose.

TOP-DOWN ASSEMBLY DESIGN

Grouping existing components to form an assembly is referred to as bottom-up assembly design. This is a common assembly construction technique when a design consists of existing parts. Figure 12–18 shows a graphical representation of this method of creating an assembly. Within such a technique, existing parts govern the final assembly design.

Rarely, in true conceptual design, do individual components govern the makeup of an assembly. As an example, suppose an aircraft company is designing a new style of light single-engine aircraft. Input would be provided from a variety of sources to include potential customers, industrial designers, aerospace engineers, manufacturing engineers, and subcontractors. The conceptual design would start with the hull design of the aircraft. The aircraft would have major subassemblies such as the power plant, the constant-speed propeller assembly, control linkages, fuel system, and seating. Within top-down assembly design, each major subassembly would be built around and within the conceptual aircraft

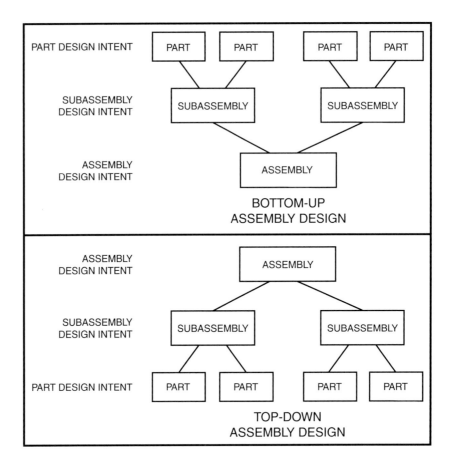

Figure 12–18   Top-down Assembly Design versus Bottom-up

hull. Each major subassembly would have smaller subassemblies. These subassemblies would be designed around and within their larger parent assemblies. At the end of the top-down design process would be parts. Parts would be designed around the lowest level subassembly in which they reside. This form of design is shown graphically in Figure 12–18.

In real-world assembly design, a combination of top-down and bottom-up assembly design is utilized. Manufacturing companies often use components common to other designs. Examples of this can be found in the automotive industry. Automotive companies often design cars that share identical major subassemblies, such as the power train. Within Pro/ENGINEER, assemblies can be designed using both top-down and bottom-up design.

Pro/ENGINEER provides multiple capabilities to enhance top-down assembly design. One of these capabilities is the ability to create components within Assembly mode. Another is the utilization of skeleton models.

CREATING PARTS IN ASSEMBLY MODE

One of the strengths of Pro/ENGINEER is its ability to create parts directly within Assembly mode. Creating parts for an assembly can be tedious at best when you are modeling within Part mode. Pro/ENGINEER provides the ability to create components directly in an assembly where they will actually function. Figure 12–19 shows an example of a part created in Assembly mode. The first illustration shows an existing part with the sketched section of the part under construction. Within the sketching environment, Pro/ENGINEER al-

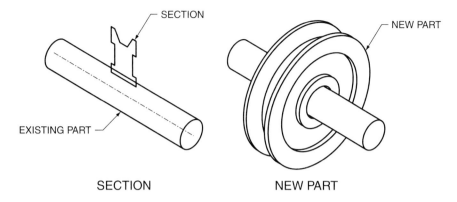

Figure 12-19 Part Creation

lows existing parts, features, datum planes, and axes to be referenced. In this example, the axis of the existing shaft will be referenced and used as the axis of rotation for the revolved pulley feature.

When creating features within Part mode, new features can reference existing features. When a feature is referenced, it becomes a parent feature for the feature under construction. This same parent-child relationship can exist between parts and features in Assembly mode. When a part within an assembly references another part, an external reference is formed. When creating a part within Assembly mode, care should be taken when creating external references. The Info >> Global ReferenceViewer option can be used to check for the existence of references between parts.

Components can be created in Assembly mode by using the Insert >> Component >> Create option. Once a component (part, subassembly, or skeleton model) has been created, it becomes a part within the assembly. When the assembly is saved, the component is saved as its own object file. Perform the following steps to create a part within Assembly mode:

STEP 1: **On the Menu bar, select TOOLS >> ASSEMBLY SETTINGS >> REFERENCE CONTROL.**

It is advisable to avoid the creation of external references within the creation of a new component. An external reference is created when one component references another. If a component does have an external reference, the component being referenced must reside in memory to pass necessary reference information.

STEP 2: **On the reference control dialog box, select NONE and deselect BACKUP FORBIDDEN REFERENCES (as shown in Figure 12–20).**

The None option will not allow any form of external reference.

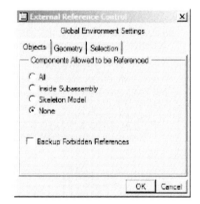

Figure 12-20 External Reference Control

STEP 3:   Select OK to exit the dialog box.

STEP 4:   **Select INSERT >> COMPONENT >> CREATE on the menu bar.**

Components consist of parts, subassemblies, and skeleton models. When you select the Create option, Pro/ENGINEER will launch the Component Create dialog box (Figure 12–21).

STEP 5:   On the Component Create dialog box, select PART (Figure 12–21).

The Component Create dialog box allows for the creation of parts, subassemblies, and skeleton models. In addition, bulk items can be created. Bulk items are features that are not suitable for modeling but are necessary for the assembly.

STEP 6:   Enter a name for the component then select OK.

STEP 7:   On the Creation Options dialog box, select a component creation method (Figure 12–22).

The following Creation Method options are available:

*   **Copy From Existing**   This option will create a new part from an existing part. If this option is selected, use the Copy From text box to enter the name of the part to copy. New parts lose their associativity with the parts from which they are copied. This option is often used with a template part file that consists of a set of default datum planes and other settings (layers, units, etc.).

*   **Create First Feature**   This option allows for the creation of the first feature of the part. Typical feature creation options such as Solid, Surface, Datum, and Protrusion are available.

*   **Locate Default Datums**   This option creates a new part with its own set of default datum planes. Suboptions are available for defining the location of the datum planes.

*   **Empty**   The Empty option creates a part definition with no geometric definition. The part will be listed on the model tree. Features can be added at a later time.

Figure 12–21   Component Create Dialog Box

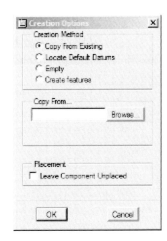

Figure 12–22   Creation Options menu

STEP 8: **If necessary, use the model tree's pop-up menu to activate the new part.**

A part has to be activated before part features can be created.

STEP 9: **Use Pro/ENGINEER's feature creation tools to construct the part.**

The feature creation tools available in Assembly mode are identical to the same tools found in Part mode. Options available include Solid, Surface, Datum, Protrusion, and User Defined. Existing planar surfaces can be used as sketching planes. Since this is the first geometry feature that defines the part, no negative space feature creation options are available. Existing components and assembly features can be used as references.

SKELETON MODELS

Skeleton models are important components for the proper utilization of the top-down design process. Skeleton models serve as a form of three-dimensional layout and are utilized in a variety of ways. The following is a discussion of the possible uses of a skeleton model:

TOP-DOWN DESIGN CONTROL

Within the realm of top-down assembly modeling, the intent of a design is passed from the upper levels of the design to the individual components. Skeleton models are used to portray and transfer the upper level design intent during the modeling of subassembly and components.

ASSEMBLY SPACE CLAIM

Top-down assembly design usually requires larger and more exterior components to be designed before the design of smaller and more specific components. As an example, an automobile's exterior would probably be designed before the front passenger seats. During the designing of the seats, a designer must work within the space allocated by the exterior design. Skeleton models can be utilized to represent claimed space for major design subassemblies.

SHARING OF INFORMATION

In large manufacturing companies, different teams design major subassemblies of a design. Skeleton models can be used to pass design information from one major subassembly to another. As an example, within the design of a new automobile, a company might elect to use an existing design for the engine, but use a new design for the transmission. The existing engine can be incorporated into the transmission design as a skeleton model. Feature information such as hole and shaft locations can be passed to the new design without the worry of creating external references.

MOTION CONTROL

The motion of an assembly can be designed and controlled through the use of skeleton models. True skeleton components constructed from datum axes, curves, and components can be created to serve as the "skeleton" of a subassembly. The relative motion of each component can be designed and modified by using the skeleton components. When the design is optimized, the actual components can be created around the skeleton.

An example of the use of a skeleton model is shown in Figure 12–23. In this example, the base skeleton model represents the external shell of a product. It was designed separately from internal components. The problem presented here requires the designing and

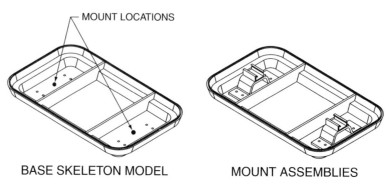

BASE SKELETON MODEL   MOUNT ASSEMBLIES

Figure 12–23   Skeleton Model

modeling of the mount assembly. Locations for the two required mount assemblies are shown in the first illustration. Within Pro/ENGINEER's Assembly mode, the base model was created in the assembly as a skeleton model, using the Copy From Existing creation option. This option places a copy of the base model as a skeleton within the assembly. This allows the mount bracket to be modeled within its allocated space and around the existing mount holes. By using the skeleton model, no external references are created between the original base model and the mount assembly.

As shown in the above example, a skeleton model can be created as a copy from an existing part. The Component >> Create option has the capability to create skeleton models from the first feature also. Most of the modeling tools available for the creation of a part in assembly mode are available for the creation of a skeleton model. Only one skeleton model can be included in an assembly.

Assembly Relations

In Part mode, the Relations option is used to create a relationship between two dimension values. In Assembly mode, the Relations option can be used to create dimensional relationships between dimensions within a part or between two parts. The rules used to define relations in Assembly mode are similar to the rules for applying relations in Part mode.

One difference between adding relations in an assembly and adding them to a part is in the dimension symbols. In Part mode, a dimension symbol consists of the letter d followed by the dimension number. In Assembly mode, the dimension symbol is followed by the Session_ID for the component (Figure 12–24).

Notice in Figure 12–24 the format for the dimension symbols in Assembly mode. The assembly consists of two parts: Part A and Part B. The Session_ID for Part A is 16 and the Session_ID for Part B is 18. When a dimensional relationship is created for a part, only the dimension symbol needs to be entered. In this example, the Part Relations (Part Rel) suboption would be selected. When a dimensional relationship is created for an assembly, the full dimension symbol, including the Session_ID, needs to be entered. In this example, the Assembly Relations (Assem Rel) suboption would be used.

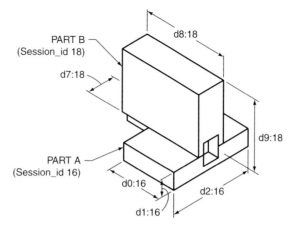

Figure 12–24 Dimension Symbols

LAYOUT MODE

Layout mode is used to create two-dimensional layouts of an engineering design. It is helpful for the capture of design intent and for developing the general layout of a design. The sketch creation options available in Layout mode work identically to the two-dimensional entity creation options found in Drawing mode. See Chapter 8 for detailed information on creating a two-dimensional drawing.

A layout can be created and used to drive the dimensional values of a part or assembly. Figure 12–25 shows an example of a layout and an associated assembly model. When a layout is constructed, dimension values are created and assigned a dimension symbol and value. Unlike Part and Assembly modes, Layout mode requires the definition of a user-defined symbol name. By declaring a layout in an assembly, selected dimensions of an assembly can be related to the dimensions within a layout. As an example, the height symbol shown in the layout is used to drive the d0:6 dimension in the assembly. When the height dimension's value is changed in the layout, the driven dimension in the assembly is changed. In Assembly mode, the File >> Declare >> Declare Lay option is used to link a layout with an assembly. Once a layout as been declared, the layout's dimension symbols can be used with Assembly mode's Relations option.

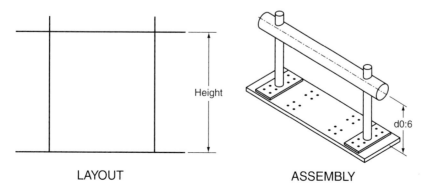

Figure 12–25 Layout with Assembly

SIMPLIFIED REPRESENTATION

Figure 12–26
View Manager Dialog Box

Assembly modeling necessitates the combining of many different parts and subassemblies. The more components added to an assembly and the more complex an assembly, the longer Pro/ENGINEER takes to regenerate and retrieve models. Assemblies are often composed of major subassemblies, minor subassemblies, and parts. During the design process, certain subassemblies and parts may not be needed. As an example, when designing an aircraft, a design team modeling a seat assembly probably would not need the power plant assembly. Pro/ENGINEER's Simplified Representation option allows selected components to be removed from the display. This allows for faster retrieval and regeneration.

Simplified representations are created in Assembly mode with the View Manager dialog box (Figure 12–26). The three most utilized representations are Master, Graphics, and Geometry. The following is a description of each:

MASTER REPRESENTATION

The Master Representation option includes selected parts or subassemblies in the simplified representation with all available features and parameters. Components with this representation are fully modifiable.

GRAPHICS REPRESENTATION

The Graphics Representation option includes selected components on the workscreen in a wireframe display format (Figure 12–27). Components with this representation cannot be modified. The wireframe display is a Pro/ENGINEER default and can be changed with the configuration file option save_model_display. This option significantly improves regeneration time for the model.

GEOMETRY REPRESENTATION

The Geometry Representation option includes selected components on the workscreen in Pro/ENGINEER's current display style. Geometry with this method can be modified, but the model does takes longer to regenerate compared to the graphics representation.

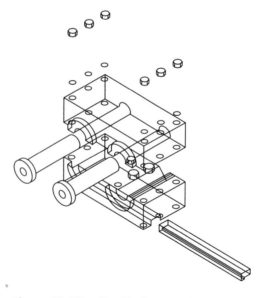

Figure 12–27   Graphics Representation

CREATING A SIMPLIFIED REPRESENTATION

Perform the following steps to create a simplified representation:

STEP 1:   Select VIEW >> VIEW MANAGER on the menu bar.

STEP 2:   Select the Simplified Representation tab on the View Manager dialog box.

STEP 3:   Select the NEW option on the dialog box and enter a name for the user-defined representation.

STEP 4:   If necessary, highlight the new representation, then select EDIT >> REDEFINE.

STEP 5:   Select the Include tab on the Edit dialog box.

STEP 6:   Select a Simplified Representation method and then pick a component in which to apply the method.

STEP 7:   Use the Exclude tab to exclude components from the display.

STEP 8:   Exit the edit dialog box.

STEP 9:   Use the SET option to set a specific representation then close the View Manager dialog box.

EXPLODED ASSEMBLIES

When components are added to an assembly, they are placed in their functional orientation and located. Often, this state of viewing an assembly can be confusing and less descriptive. Within the technical language of engineering graphics, assembly drawings are used to display the location of assembled components. To make the assembly drawing legible, the assembly can be exploded to separate components. Figure 12–28 shows an illustration of an exploded and an unexploded assembly. The View >> Explode option is used to explode a view, while the View Manager dialog box is used to create and set explode states. Multiple explode states can be created. The Set on the View Manager dialog box is used to set a specific explode state.

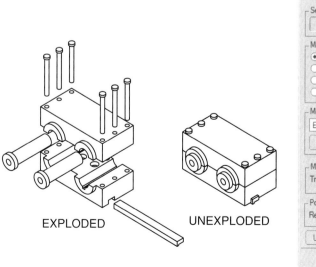

EXPLODED   UNEXPLODED

Figure 12–28   Exploded View

CREATING AN EXPLODED STATE

Perform the following steps to create an exploded state for an assembly:

STEP 1:   **Select VIEW >> VIEW MANAGER on the menu bar.**

STEP 2:   **Select the Explode tab on the View Manager dialog box.**

STEP 3:   **Select the NEW option and enter a name for the exploded state.**

STEP 4:   **Select PROPERTIES on the dialog box.**

STEP 5:   **Select the EDIT POSITION option.**

STEP 6:   **On the Explode Position dialog box, select TRANSLATE as the Motion type (Figure 12–28).**

STEP 7:   **On the Explode Position dialog box, select a Motion Reference.**

The setting of an explode state's components is similar to the movement of packaged components. Pro/ENGINEER provides the following translation options:

- **View Plane** The motion will be relative to the current screen orientation. When this type is selected, any selected components movement will be parallel to the workscreen.

- **Sel Plane** When this type is selected, the motion will be parallel to a selected plane. Planar surfaces and datum planes can be selected.

- **Entity/Edge** When this type is selected, the motion will be along the path of a selected axis, edge, or curve. This type is useful for confining the movement of a component along one axis.

- **Plane Normal** When this type is selected, the motion will be perpendicular to a selected plane.

- **2 Points** When this type is selected, two selected points on the workscreen are used to created the relative motion. This type is useful for confining the movement of a component along one axis.

- **Csys** The motion will be relative to the X-axis of a selected coordinate system.

STEP 8:   **On the workscreen, select an entity or plane relevant to the motion reference.**

Your selection on the assembly model is based on the motion reference selected in Step 7. With the exception of the View Plane reference, you will select an entity, point, plane, or axis on the workscreen relevant to the current reference. As an example, if you select the Entity/Edge motion reference, you will select either an axis, edge, or curve. The selected motion reference will remain current until it is changed.

STEP 9:   **On the workscreen, select and move a component.**

STEP 10:   **Continuing moving components on the workscreen or change motion types.**

Continue moving components on the workscreen until the explode state is complete. You can change motion types to optimize the state.

STEP 11:   **Select OK on the dialog box when the explode state is complete.**

STEP 12:   **Select the DONE/RETURN option on the Modify Explode menu.**

STEP 13:   **Close the View Manager Dialog box.**

STEP 14:   **Use VIEW >> EXPLODE to explode and unexplode the view.**

SUMMARY

Pro/ENGINEER is more than just a modeling application. It is a true engineering design package. One of the capabilities that separates Pro/ENGINEER from midrange computer-aided design and drafting applications is its ability to model an entire engineering design. With appropriate modules of Pro/ENGINEER, a product can be designed, modeled, detailed, simulated, and manufactured. While Pro/ENGINEER's Part mode is the main component for low-end design work, Assembly mode is the module for complete product design.

ASSEMBLY TUTORIAL

This tutorial will explore Pro/ENGINEER's basic bottom-up assembly modeling capabilities. In addition, the creation of exploded views and the creation of an exploded assembly drawing will be covered. The final product of this tutorial is shown in Figure 12–29. Within this tutorial, the following topics will be covered:

- Creating components for an assembly.
- Placing components into an assembly.
- Creating an exploded assembly.
- Creating an assembly drawing.

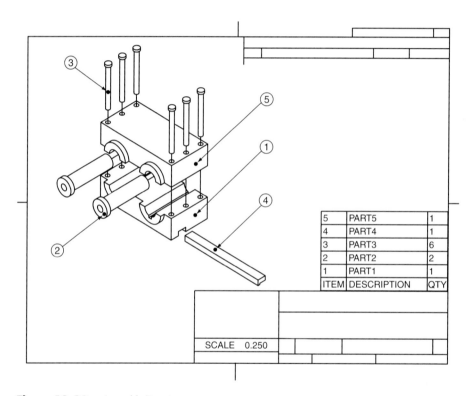

ITEM	DESCRIPTION	QTY
5	PART5	1
4	PART4	1
3	PART3	6
2	PART2	2
1	PART1	1

SCALE 0.250

Figure 12–29   Assembly Drawing

CREATING COMPONENTS FOR AN ASSEMBLY

During bottom-up assembly design, parts for an assembly are created with normal part modeling tools and techniques. As with any parametric model, the intent of the design needs to be considered. In addition, how components of an assembly will be parametrically linked is another important consideration. Often, extra datums will have to be added to a component to simulate the correct function of the design.

The first segment of this tutorial will require the modeling of the six parts that make up the assembly. Notice in Figure 12–29 that the actual assembly consists of 11 parts. Two of the parts are used multiple times. Within Pro/ENGINEER, a component can be placed multiple times into an assembly. Figure 12–30 shows the parts used in this tutorial. Use Part

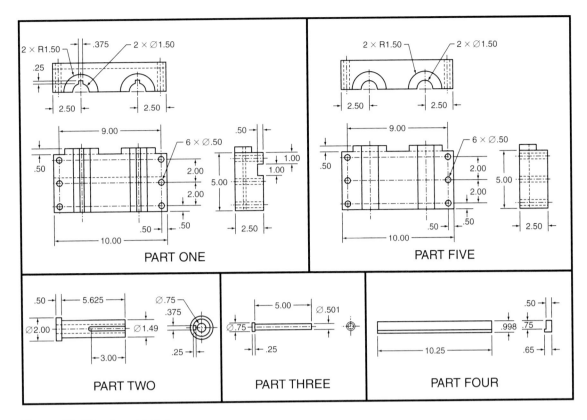

Figure 12–30 Assembly Parts

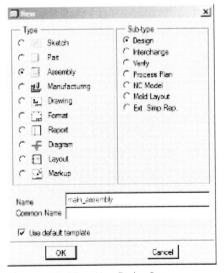

Figure 12–31 New Dialog Box

mode to model each part. The dimensions shown in each illustration represent the design intent. Incorporate this intent into each part.

MODELING POINTERS

- Notice the similarities between Part One and Part Five (Figure 12–30). The only difference between the two parts is the absence of three cut features in Part Five. If Part One is created first, you can use the New File Options dialog box as shown in Figure 12–31 to create Part Five as a copy of Part One. After creating the copy, delete the cut features. The New File Options dialog box is accessible by deselecting the Use Default Template on the New dialog box.

• Create the six 0.50-in diameter holes in Part One and Part Five as a patterned hole. Within the assembly, Part Three will be inserted into each hole. The Reference Pattern option can be used to pattern the first instance of Part Three to place the remaining five. This saves time in the component placement process and also meets the design intent for this assembly.

PLACING COMPONENTS INTO AN ASSEMBLY

This segment of the tutorial will place the parts into the assembly. Part One will be placed first followed by Part Five, Part Two, Part Three, then Part Four.

STEP 1:   **Use the New option to create a new Assembly object named *main_assembly* (Figure 12–31).**

STEP 2:   **Select the ADD COMPONENT option on the toolbar.**

Two options are available for adding components to an assembly file: Add Component and Create Component. The Add Component option is available for adding existing part and assembly models. The Create Component option allows components to be created directly within the assembly environment.

STEP 3:   **Use the Open dialog box to open the *Part_One* component.**

After opening the part, Pro/ENGINEER will launch the Component placement dashboard. This dashboard is used to constrain components within the context of the assembly. The Part_One component will be constrained to the set of default datum planes.

STEP 4:   **Select the Default assembly constraint option on the Component Placement dashboard (Figure 12–32).**

The Default option will align your component's default datum planes with the assembly model's default datum planes.

Figure 12–32   Default Location Option

STEP 5:   **Select Accept Changes option to finish the placement of the component.**

STEP 6:   **Select the ADD COMPONENT option on the toolbar.**

STEP 7:   **Use the Open dialog box to open the *Part_Five* component.**

> **INSTRUCTIONAL POINT**   When a component is opened, it is placed into the workscreen with a reference to how it was created in Part mode. The initial component placement illustrations within this tutorial may not match how your components are initially oriented.

After selecting a component for placement, Pro/ENGINEER will initially place the component within the workscreen as an unconstrained model.

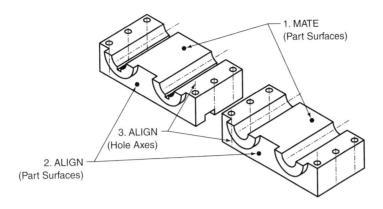

1. MATE
(Part Surfaces)

3. ALIGN
(Hole Axes)

2. ALIGN
(Part Surfaces)

Figure 12–33 Constraint Type Selection

Pro/ENGINEER will also launch the Component Placement dialog box. This dialog box is used to parametrically constrain a model to existing assembly models and/or features.

STEP 8: **Ensure that AUTOMATIC is selected as the Constraint type on the dashboard.**

Automatic is the default constraint type on the Component Placement dialog box. With this constraint, references are selected for both the component and the assembly. Pro/ENGINEER will determine the constraint to apply, but will provide you with the option of selecting an alternative constraint, such as Mate in this example.

STEP 9: **Select the two Mate surfaces shown in Figure 12–33.**

STEP 10: **Under the Constraints section of the dashboard, change the previously created constraint from an aligned constraint to a MATE constraint.**

The Automatic option will normally assign an Align straight between two selected surfaces. You should change this constraint type to a Mate.

STEP 11: ⊥ **Ensure that Coincident is selected as the constraint option.**

INSTRUCTIONAL POINT Your next assembly constraint will be the Align constraint between the two surfaces shown in Figure 12–33.

STEP 12: **Pick the Align (2. Align) surfaces shown in Figure 12–33.**

The next constraint will align the front surfaces of the parts as shown in the figure.

STEP 13: **Under the Constraints section of the dashboard, ensure that ALIGN is selected as the constraint.**

STEP 14: ⊥ **Ensure that Coincident is selected as the constraint option.**

Aligned and Mated surfaces can have an offset value. When two surfaces are offset, they will remain parallel but will be separated by the user-specified offset value.

Figure 12–34   Changing Constraint Types

Figure 12–35   Placement Status

MODELING POINT   When a constraint type is added to the component, the model and the workscreen change to reflect this constraint. Use dynamic rotation (Control key and middle mouse button) to better visualize and select the components. If the components overlap during placement, use options under the Move Tab to adjust the temporary location of the component being placed. When the component is fully constrained, the dialog box will provide the message shown in Figure 12–35.

STEP 15:   Change this constraint type to ALIGN and COINCIDENT as the offset (see Figure 12–33).

STEP 16:   Pick the Align (3. Align) axes shown in Figure 12–33.

STEP 17:   Ensure that ALIGN and COINCIDENT are entered as the constraint type and offset.

STEP 18:   Select the continue checkmark on the dashboard to complete the placement of your component.

A component that is fully constrained is a parametric model. Pro/ENGINEER does allow components to be placed that have no constraints or that are only partially constrained. This is referred to as a packaged component.

STEP 19:   Use the ADD COMPONENT option to open *Part_Two* (Figure 12–36).

STEP 20:   Using the INSERT constraint type, pick the surface of the shaft of Part Two, then the surface of the hole feature on Part One.

The Insert constraint will align the centerlines of each revolved feature.

STEP 21:   Select the dashboard's Move menu option, then select TRANSLATE and RELATIVE IN VIEW PLANE as the motion types.

This step and the following step will move your part to a location that will allow you to better pick constraint entities. The *Relative In View Plane* option will move the part parallel to the view plane with respect to existing assembly constraints.

STEP 22:   Pick PART-TWO on the workscreen, then drag to a location removed from the existing assembly.

MODELING POINT   If your components overlap, causing the selection of surfaces and entities to become difficult, you can use the Move tab to temporarily reposition components.

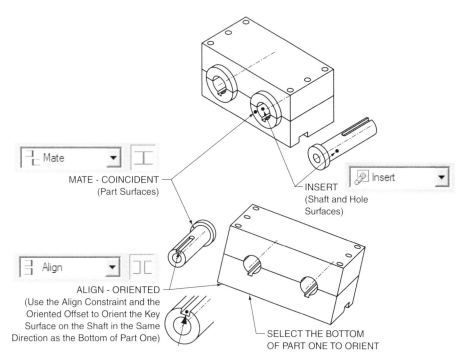

Figure 12–36   Part Two Placement

STEP 23:   Under the dashboard's Placement menu, select the NEW CONSTRAINT option.

STEP 24:   Select ALIGNED and ORIENTED as the constraint type.

STEP 25:   Pick the two Aligned-Oriented surfaces shown in Figure 12–36.

Orient the bottom surface of the key slot on Part Two with the bottom of Part One. This constraint combination will orient the surfaces in the same direction. You must manually set the Oriented option as shown in the figure.

STEP 26:   Under the dashboard's Placement menu, select the NEW CONSTRAINT option.

STEP 27:   Select MATE and COINCIDENT as the constraint type.

STEP 28:   MATE the back of the head of Part Two with the front surface of the boss feature on Part One.

STEP 29:   When your part is fully constrained, select the continue checkmark.

STEP 30:   Use the Place Existing Component option to open a new instance of Part_Two into the assembly.

STEP 31:   Select the SEPARATE WINDOW option on the dashboard.

The Separate Window option will open a window for your new component, allowing for easier selection of constraint entities.

STEP 32:   Use the same constraint types from the first instance of *Part_Two* to place the second instance of the part (Refer to Figure 12–29).

STEP 33:   When your part is fully constrained, select the continue checkmark.

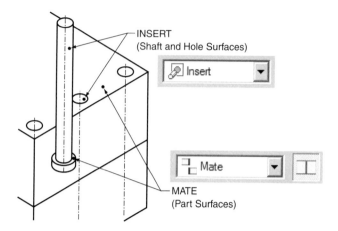

Figure 12–37   Part Three Placement

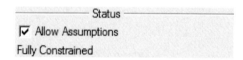

Figure 12–38   Allow Assumptions

STEP 34:   Use the ADD COMPONENT option to open *Part_Three* (Figure 12–37).

Place the constraint types in the following order:

- **Mate-Coincident** Mate the back of the head of Part Three with the top surface of Part Five.

- **Insert** Insert the surface of the shaft of Part Three into the surface of one on the six hole features on Part Five.

- Allow the assumption shown under the Placement menu in the Status area (Figure 12–38). When two revolved features are constrained together, Pro/ENGINEER will assume the constraint around the axis of revolution.

STEP 35:   Pick part *Part_Three* on the model tree and select the PATTERN command on the pop-up menu.

STEP 36:   On the Pattern dashboard, select REFERENCE as the pattern type, then build your pattern (Figure 12–39).

Six instances of *Part_Three* exist in the assembly. Instead of placing each instance of the part, you can pattern the first instance around the reference pattern used to create the holes in Part mode. This technique will work only if you created the six holes using the Pattern option. If you created the holes using a different technique, you must place each *Part_Three* instance individually.

Figure 12–39   Reference Pattern

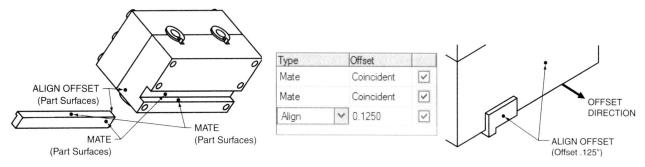

Figure 12-40 Placing Part Four

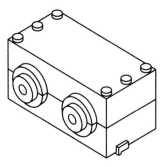

Figure 12-41 Align Offset

STEP 37: Use the ADD COMPONENT option to open *Part_Four* (Figures 12–40 and 12–41).

Place the constraint types in the following order:

- **Mate** Mate the bottom of Part Four with the bottom of the cut slot.
- **Mate** Mate the back of Part Four with the side of the cut slot.
- **Align-Offset** Use the Align constraint and the Offset suboption to offset the end of Part Four from the side of Part One.

STEP 38: **Save your assembly file.**

Your complete assembly should appear as shown in Figure 12–42. Observe the components on your Model Tree. Modify, feature creation, and feature manipulation options can be accessed by right-mouse button picking the component on the model tree.

Figure 12-42 Finished Assembly

CREATING AN EXPLODED ASSEMBLY

This segment of the tutorial will create the explode state shown in Figure 12–43. An explode state is the placement of individual components when an assembly model is exploded. Note that within an explode state that components are positioned relative to each other and not relative to the workscreen. Also, within any one assembly model, multiple explode states can be defined.

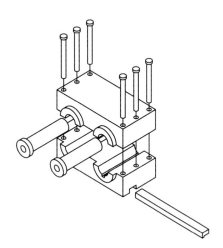

Figure 12-43 Exploded Assembly

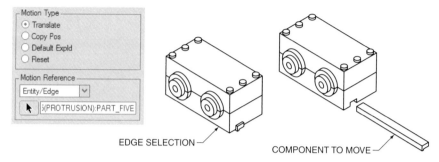

Figure 12–44   Entity/Edge Motion

STEP 1: Select VIEW >> ORIENTATION >> STANDARD ORIENTATION on the menu bar or select the CONTROL-KEY and D-KEY combination.

> **INSTRUCTIONAL POINT**   Your standard orientation may not match Figure 12–43. It is not as important to replicate the figures in this tutorial as it is to develop a well-defined exploded state and exploded view. You can dynamically rotate the model to match the figure, then use the Orient tab on the View Manager dialog box to save your new orientation.

STEP 2:   Select VIEW >> VIEW MANAGER on the menu bar.

STEP 3: Select the EXPLODE tab on the View Manager dialog box.

STEP 4: Select the NEW option, then enter *EXPLODE1* as the name for the explode state.

STEP 5: Select the PROPERTIES button on the bottom of the View Manager dialog box.

STEP 6:   Select the EDIT POSITION option.

STEP 7: On the Explode Position dialog box, select ENTITY/EDGE as the Motion Reference (Figure 12–44).

STEP 8: Select the entity edge shown in Figure 12–44 or a similar parallel edge.

STEP 9: Select TRANSLATE as the Motion Type (Figure 12–44).

STEP 10: Select and move the component as shown.

STEP 11: On the Explode Position dialog box, select the Motion Reference pick icon, then pick the entity edge shown in Figure 12–45.

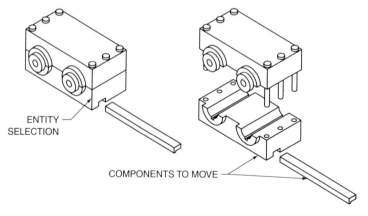

Figure 12–45   Entity/Edge Motion

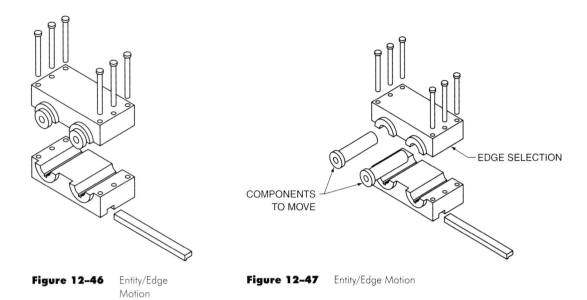

Figure 12–46   Entity/Edge
Motion

Figure 12–47   Entity/Edge Motion

This step will define a motion reference. You must first reselect the reference through the Pick icon located on the dialog box. As a note, the Entity/Edge selection works with edges and axes. Within this step, you could also pick one of the six vertical axes available within the model.

STEP 12:   Move the two components shown in Figure 12–45.

STEP 13:   Individually, move each instance of *Part Three* to the locations shown (Figure 12–46).

Once an entity motion type has been selected on the Motion Preference menu, this type remains current until changed. In this case, the previous entity edge selected remains current.

STEP 14:   Define the edge shown in Figure 12–47 as the motion reference and move the two components shown.

STEP 15:   Select OK to exit the Explode Position dialog box.

STEP 16:   Select the LIST button on the View Manager dialog box, then CLOSE the dialog box.

STEP 17:   Select VIEW >> EXPLODE >> UNEXPLODE on the menu bar.

STEP 18:   Save your assembly object file.

CREATING AN ASSEMBLY DRAWING (REPORT)

This segment of the tutorial will take the exploded assembly created in the last two segments and create an assembly drawing through the use of the Report module. Assembly views of an object can be placed into a drawing or report, using the same procedures for placing a part view. In addition to the assembly view, a bill of material and balloon notes will be created. Pro/ENGINEER, through its Pro/REPORT module, allows reports such as a bill of materials to be placed into a drawing with full associativity. In this tutorial, a bill of materials will be created with a table through the use of a repeat region. A repeat region allows a table to expand to incorporate a list of all the components of an assembly.

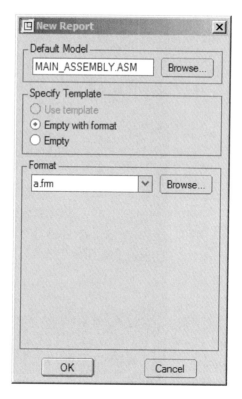

Figure 12–48   New Drawing Dialog Box

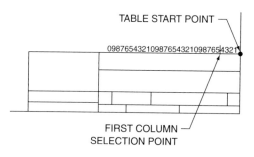

Figure 12–49   Table Creation

STEP 1:   **Start a New Report object file named *Assembly*.**

Use the File >> New option and select Report as the mode to use. Enter *Assembly* as the name of the file, then select OK.

STEP 2:   **On the New Report dialog box, select the options shown in Figure 12–48.**

Select the assembly file created in this tutorial as the default model. Select the Empty with Format option, then Browse to find an A-size format. By selecting an A-size format, you will set the drawing sheet to the size of the format.

STEP 3:   **Select OK to accept the New Report Dialog Box options.**

After you select OK, Pro/ENGINEER will launch a new report session. Notice how the options in Report mode are similar to the options in Drawing mode.

STEP 4:   **Select the Insert Table icon on the toolbar.**

The first step in creating an assembly drawing is to define the bill of materials table and repeat region.

STEP 5:   **Select the ASCENDING >> LEFTWARD >> BY NUM CHARS options on the Table Create menu.**

STEP 6:   **Select the table start point shown in Figure 12–49.**

STEP 7:   **On the workscreen, select 4 Number Characters for the first column (Figure 12–49) followed by 20 Number Characters for the second column and 5 Number Characters for the third column.**

When the By Number Characters (By Num Chars) option is selected, Pro/ENGINEER provides numerical characters on the workscreen to define the width of each column. On the workscreen, select the spacing for each column.

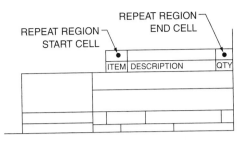

Figure 12–50   Repeat Region Creation and
Column Headers

Step 8: On the Table Creation menu, select DONE to end the creation of the table's columns.

Step 9: Using the BY NUM CHARS option, create two rows, each one character high.

Step 10: On the Table Creation menu, select DONE to end the creation of the table.

Step 11: On the workscreen, double-pick the lower left cell of the table to access the Note Properties dialog box.

Step 12: Enter *ITEM* as the text for the cell, then select OK on the dialog box.

Step 13: Create the remaining column headers shown in Figure 12–50.

For each individual cell, enter the header text shown in the figure (e.g., ITEM, DESCRIPTION, and QTY).

Step 14: Select TABLE >> REPEAT REGION on the menu bar.

Step 15: From the Table Regions menu, select the ADD option, then pick the Repeat Region Start and End cells shown in Figure 12–50.

Step 16: Select DONE on the Table Regions menu.

Step 17: On the workscreen, double-pick the cell above the ITEM header to access the Report Symbol dialog box (Figure 12–51).

Pro/ENGINEER and Pro/REPORT use report parameters to assign associative data to table cells. In this tutorial, you will assign parameters that define each components, item number (&rpt.index), description (&asm.mbr.name), and quantity (&rpt.qty). For cells defined within a repeat region, you can enter each parameter directly from a menu that is accessible by double-picking the respective cell (Figure 12–51).

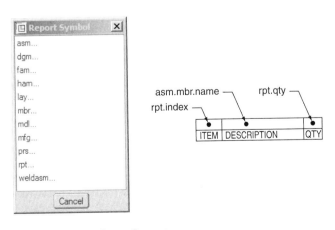

Figure 12–51   Report Parameters

STEP 18: On the Report Symbol dialog, select *RPT . . .* , then select *INDEX.*

These two selections will add the &rpt.indx parameter to the first cell of the first row of the table. Notice on the workscreen how the parameter has been added. Do not worry if the parameter crosses into the next cell.

STEP 19: On the workscreen, double-pick the cell above the DESCRIPTION header to access the Report Symbol dialog box (Figure 12–51).

STEP 20: On the Report Symbol dialog, select *ASM . . .* , then select *MBR . . .* , then select *NAME.*

These selections enter the assembly member's name parameter into the second cell.

STEP 21: Enter the *RPT.QTY* (component quantity) parameter into the cell above the QTY header (Figure 12–51).

STEP 22: ⤵ Select the CREATE A GENERAL VIEW option on the toolbar.

STEP 23: If necessary, select OK on the Select Combined Sate (or presentation state) dialog box.

STEP 24: On the workscreen, select the location for the exploded view.

STEP 25: Define the following settings for your exploded view:

- View Type Category.

  - View Name = Explode.

  - Model View Name = Standard Orientation.

  - Default Orientation = Isometric.

- Scale = 0.20.

- View States Category

  - Combined State = DEFAULT ALL

  - Explode View = EXPLODE1.

- View Display Category – Display State = No Hidden.

STEP 26: Select OK or close the dialog box.

After defining the view, notice on the workscreen how the table is expanded to include all components of the assembly. Currently, the table is displaying every component on a separate row, even if it is a duplicate. You will change this in the next step.

NOTE: You may not actually see any component names in the repeat region table. They will be revealed later.

STEP 27: Select TABLE >> REPEAT REGION on the menu bar.

STEP 28: Select the ATTRIBUTES option, then on the workscreen select the repeat region table.

STEP 29: Select NO DUPLICATES >> DONE/RETURN.

The No Duplicates option will not duplicate a component on the model tree. After selecting the Done/Return option, notice how the table shrinks. The next several steps of the tutorial will have you add the bill of materials balloon notes. Once these balloon notes are added, the bill of materials table will be updated with each component member.

STEP 30: Select the TABLE >> BOM BALLOON option on the menu bar then on the workscreen select the repeat region table.

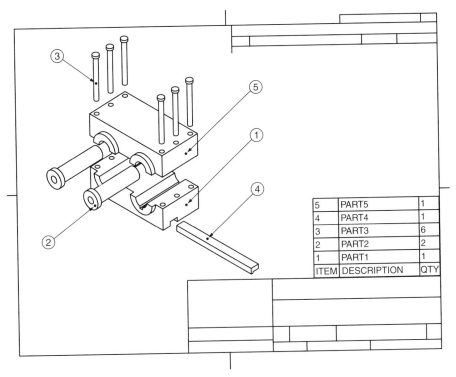

5	PART5	1
4	PART4	1
3	PART3	6
2	PART2	2
1	PART1	1
ITEM	DESCRIPTION	QTY

Figure 12–52   Finished Assembly Drawing

After selecting the region, verify in the message area that balloon attributes have been added to the repeat region.

STEP 31:   **Select CREATE BALLOON >> SHOW ALL on the BOM Balloons menu.**

STEP 32:   **Select DONE to exit the menu.**

STEP 33:   **Use pop-up menu's EDIT ATTACHMENT option to modify each balloon's attachment method.**

The pop-up menu is accessible by preselecting a balloon leader with the left mouse button, then right-mouse selecting the leader. The Attachment Type menu has options for modifying the selected leader's arrowhead type (arrowhead, dot, filled dot, etc.) and attachment point (on entity, on surface, midpoint, and intersect).

STEP 34:   **Use your mouse to reposition each balloon note to match Figure 12–52.**

STEP 35:   **Save your report object.**

The final report drawing is shown in Figure 12–52.

TOP-DOWN ASSEMBLY TUTORIAL

This tutorial will cover basic principles of Pro/ENGINEER's top-down assembly design capabilities. Assembly mode allows components (parts, subassemblies, and skeleton models) to be created within the assembly environment. The final assembly for this tutorial is shown in Figure 12–53. This tutorial will cover the following principles of top-down design:

- Creating a layout to capture design intent.
- Creating parts within assembly mode.
- Controlling external references within assembly mode.
- Declaring and using a layout within assembly model.

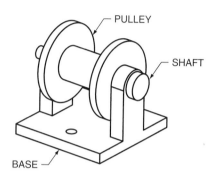

Figure 12–53   Final Assembly Model

CREATING A LAYOUT

Layouts are similar to an engineer's design notebook. They are useful for capturing design intent and for controlling an assembly from the top down. Within this tutorial, you will create a layout that will help to capture the design intent of the pulley assembly. Key dimensions within the assembly will be controlled by the layout.

This tutorial's assembly consists of three components: a base part, a shaft part, and a pulley part. As shown in Figure 12–54, you will sketch line entities within layout mode to represent these components. When sketching in layout mode, the actual sketched size of the representation is not important. What is important is creating dimensions that will be incorporated within the assembly model. Perform the following steps to create the layout shown in Figure 12–54.

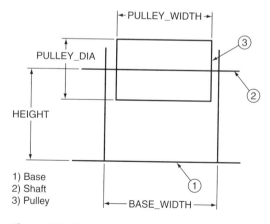

Figure 12–54   Layout of Assembly

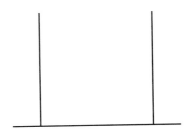

Figure 12–55 Base Sketch

Step 1: Use FILE >> NEW to create a new Layout object named *PULLEY*.

Step 2: On the New Layout dialog box, select the EMPTY WITH FORMAT option, browse to find an A-size standard sheet, and then select OK.

Step 3: Use the LINE option to create the horizontal and vertical lines shown in Figure 12–55.

The lines do not have to be a precise length. Try to manually keep the first line horizontal or use the pop-up menu's Relative Coordinates option to enter coordinate values.

Step 4: If necessary, close the Snapping References dialog box.

Step 5: Select EDIT >> GROUP >> DRAFT GROUP on the menu bar.

You will group the three previously created line entities to define them as one component.

Step 6: Select CREATE on the Draft Group menu.

Step 7: Using the Control key, pick the three line entities, then select OK.

Step 8: Enter *BASE* as the name for the group.

Step 9: Use the LINE option to create the horizontal line shown in Figure 12–56.

This line represents the shaft part.

Step 10: Select the LINE icon.

The next sketched entities will represent the pulley. In sketching of this feature, you will create a chain of line entities.

Step 11: Select the SKETCHING CHAIN icon on the toolbar.

The representation of the pulley component will consist of four line entities (see Figure 12–57). The Sketching Chain option will allow you to sketch multiple connected line entities, forming a polyline.

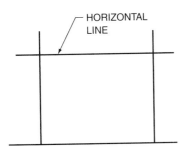

Figure 12–56 Shaft Sketch

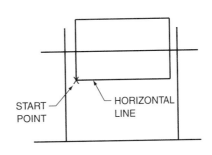

Figure 12–57 Pulley Sketch

STEP 12: Sketch the four line entities representing the pulley as shown in Figure 12–57.

After sketching the horizontal line, notice how Pro/ENGINEER switches to the Vertical Line option.

STEP 13: ▶ When the Pulley representation is complete, select the Pick icon.

STEP 14: If necessary, close the Snapping References dialog box.

STEP 15: Select INSERT >> BALLOON on the menu bar.

STEP 16: Select LEADER >> MAKE NOTE.

STEP 17: On the workscreen, pick the horizontal line entity representing the base part (see note 1, Figure 12–58).

Balloon leader note 1 will be attached to this entity.

STEP 18: Select your Middle Mouse button.

STEP 19: Enter *BASE* as the name for the first balloon note.

If the balloon note is not initially placed correctly, you can use the mouse drag capabilities to reposition it. Notice, on the workscreen, how Pro/ENGINEER creates a list of named balloon notes.

STEP 20: Create balloon notes for the *SHAFT* part and for the *PULLEY* part.

STEP 21: When your layout matches Figure 12–58, select DONE/RETURN to exit the Note Types menu.

STEP 22: ▶ Individually pick each balloon note and drag with the mouse to match Figure 12–58.

STEP 23: |↔| Use the INSERT >> DIMENSION >> NEW REFERENCES option to create the four dimensions shown in Figure 12–59.

Dimensions in Layout and Drawing modes are created in the same way dimensions are created in Part mode's sketcher environment. When creating dimensions in layout mode, the user is required to input a

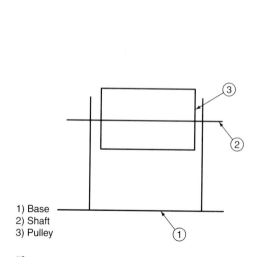

Figure 12–58 Balloon Notes

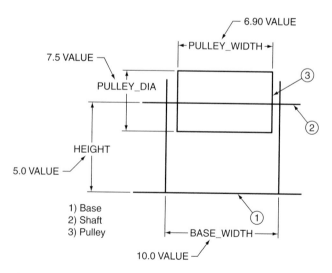

Figure 12–59 Dimension Creation

dimension symbol name first, followed by a dimension value. Referring
to Figure 12–59, use the following symbol names and values:

Symbol Name	Value
PULLEY_WIDTH	6.90
PULLEY_DIA	7.50
HEIGHT	5.00
BASE_WIDTH	10.00

STEP 24: **Use the Relations dialog box (TOOLS >> RELATIONS) to make the**
_PULLEY_WIDTH_ dimension 3.1 in less than the _BASE_WIDTH_
dimension.

This design intent will keep the width of the pulley always 3.1 in less than
the width of the base part. Enter the following equation within the Relations
dialog box:

$$PULLEY\_WIDTH = BASE\_WIDTH - 3.1$$

STEP 25: **Use the Relations dialog box to make the _PULLEY_DIA_ dimension**
1.25 times the _HEIGHT_ dimension.

Enter the following equation: $PULLEY\_DIA = HEIGHT * 1.25$.

STEP 26: **Select OK on the Relations dialog box.**

STEP 27: **Save your layout.**

CREATING A START PART

Within top-down assembly design, the first features of a component are often created by
copying a seed part. In this tutorial, the first features of the base part will be copied from an
existing start part (or template). This start part will consist of Pro/ENGINEER's default
datum planes and one datum axis (Figure 12–60). Later in this tutorial, this start part will
be used to create components.

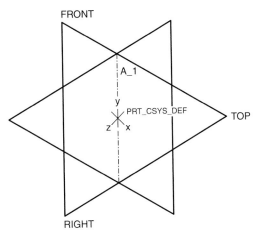

START PART   DATUM AXIS CREATION

Figure 12–60   Start Part

STEP 1:   Using Pro/ENGINEER's default template (e.g., *inlbs_part_solid*), create a new part named *START.*

The inlbs_part_solid template part file includes Pro/ENGINEER's default datum planes and default coordinate system.

STEP 2:   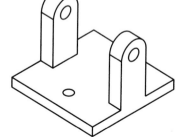 Create a datum axis at the intersection of datum planes RIGHT and FRONT.

Select the Datum Axis icon on the toolbar. You must define the first datum plane with a Through constraint option then Control key pick the second datum plane (Figure 12–60).

STEP 3:   After creating the datum axis, save the start part.

CREATING THE ASSEMBLY MODEL AND DEFININ THE FIRST COMPONENT

You will create the base part (Figure 12–61) of the assembly by copying the start part. Since it is usually important to control external references within an assembly, copying an existing component is the recommended procedure for creating a new component within assembly mode. In this tutorial, the default datum planes will be copied to start the base part.

STEP 1:   Not using a default template (Figure 12–62), create a new assembly model named *PULLEY.*

STEP 2:   On the New File Options dialog box, select EMPTY as the template, then select OK (see Figure 12–62).

STEP 3:   Select the CREATE COMPONENT option on the toolbar.

STEP 4:   On the Component Create dialog box, select the PART Type and enter *Base* as the Name for the part, then select OK (Figure 12–63).

STEP 5:   On the Creation Options dialog box, select COPY FROM EXISTING (Figure 12–64).

The Copy From Existing option will copy an existing component into the assembly model. Components selected for copying cannot have external

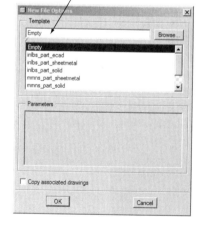

Figure 12–61   Base Part   **Figure 12–62**   New Dialog Box

Figure 12–63   Component
Create Dialog Box

Figure 12–64   Creation
Options
Dialog Box

references. Once copied into an assembly, the new component is completely independent from the component from which it was copied.

STEP 6: **Use the BROWSE option to open the *start* part, then select OK on the dialog box.**

After selecting OK, Pro/ENGINEER will drop the Base part into the assembly. Notice the Base part on the model tree.

STEP 7: **On the model tree, select the Base part with the right mouse button, then select the REFERENCE CONTROL option.**

You can set the reference control for individual items through the model tree, or you can select a global reference control by selecting Tools >> Assembly Settings >> Reference Control.

STEP 8: **On the External Reference Control dialog box, select NONE and deselect the Backup Forbidden References option (as shown in Figure 12–65), then select OK.**

When you are modeling components within assembly mode, external references can be created to other models within the assembly. Since having

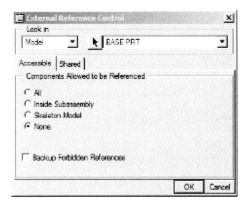

Figure 12–65   Reference Control Dialog Box

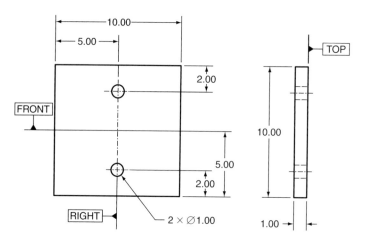

Figure 12–66   Section of First Feature

external references can limit the future usability of a part, it is usually advisable to avoid their creation.

Pro/ENGINEER provides several tools to help control references. In this example, the None option will not allow any external references. The Inside Subassembly option can be used to allow only external references between components of a subassembly, and the Skeleton Model option will allow references to skeleton models in a subassembly.

STEP 9:   **On the model tree, pick the Base part with the right mouse button, then select the ACTIVATE option.**

This step activates the base part for further feature creation. Once activated, normal feature creation tools such as Extrude and Revolve can be used to construct the geometry of the part. Part features created in this manner will be reflected in the part's individual part object file.

STEP 10:   **Use the Extrude option to create the part feature shown in Figure 12–66.**

As shown in the illustration, sketch the feature on datum plane TOP. Orient the remaining datum planes as shown. The holes can be created within the extruded protrusion or they can be created as separate features.

STEP 11:   **Create one of the two extruded features shown in Figure 12–67.**

Use the side of the base protrusion as the sketching plane.

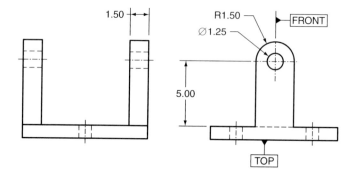

Figure 12–67   Extruded Features

Step 12: **Select EDIT >> FEATURE OPERATIONS option on the menu bar.**

The next several steps will create a dependent copy of the previously created extruded protrusion.

Step 13: **Create a dependent mirrored copy of the previously created extruded protrusion.**

Use the Copy >> Mirror >> Dependent option to create the second feature. You should be able to mirror the feature over datum plane RIGHT.

Step 14: **When the mirror is complete, select DONE/RETURN to exit the Part Feature menu.**

Step 15: **Select SETTINGS >> TREE FILTERS on the model tree.**

The next several steps of this tutorial will modify the dimension symbol names of the base part to match Figure 12–68. Modifying a dimension symbol to a more descriptive name makes identifying the dimension easier.

Step 16: **Check the FEATURES option to display part features on the model tree, then select OK.**

Step 17: **Expand the BASE part on the model tree to display its features (Figure 12–68).**

Step 18: **Access the first protrusion feature on the model tree and select the EDIT option.**

This technique should display the dimensions associated with the base protrusion feature.

Step 19: **On the workscreen, right-mouse select the *ASM_BASE_W* dimension (value of 10.0) displayed in Figure 12–68 and access the PROPERTIES option.**

This specific dimension on the workscreen should currently be displayed with a dimension value of 10.0. The Properties option will open the

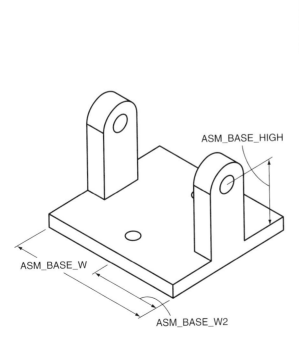
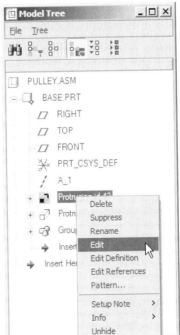

Figure 12–68 Dimension Symbols

Dimension Properties dialog box. This dialog provides an option under its Dimension Text tab for changing the name of a dimension symbol.

STEP 20: Under the Dimension Text tab of the Dimension Properties dialog box, enter *ASM_BASE_W* as the name for the dimension.

After you rename a dimension symbol in assembly mode, Pro/ENGINEER will add the Session ID to the end of the name.

STEP 21: Select OK on the Dimension Properties dialog box.

STEP 22: Rename the two remaining dimensions shown in Figure 12–68 with their respective new symbol names.

STEP 23: Use the Relations dialog box (Tools >> Relations) to make the *ASM_BASE_W2* dimension (see Figure 12–68) equal to half of the *ASM_BASE_W* dimension.

You do not have to enter the session ID (e.g., 4) when entering a part relation. Enter the equation *ASM_BASE_W2=ASM_BASE_W/2*.

STEP 24: Select OK to exit the Relations dialog box.

CREATING THE SECOND COMPONENT (SHAFT.PRT)

In this segment of the tutorial, you will create the shaft part shown in Figure 12–69. Similar to the base part, this part will be created with a set of Pro/ENGINEER's default datum planes. Instead of copying the start part, you will use the Locate Default Datums option on the Component Creation dialog box.

STEP 1: Select TOOLS >> ASSEMBLY SETTINGS >> REFERENCE CONTROL on the menu bar.

The Reference Control option located under the Tools menu will set the default reference control for all assemblies. You could have set the default reference control using this technique before the creation of the base part. This tutorial had you set the reference control individually for the base part for instructional purposes.

STEP 2: On the Reference Control dialog box, select the NONE option and deselect the Backup Forbidden References option.

STEP 3: Select OK to exit the Reference Control dialog box.

STEP 4: On the Model Tree, select the *PULLEY.ASM* file name with the right mouse button, then select the ACTIVATE option.

This option will activate the current assembly for editing purposes.

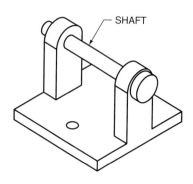

Figure 12–69   Shaft Part

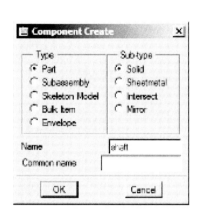

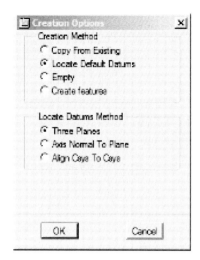

Figure 12–70   Component Create
Dialog Box

Figure 12–71   Creation Options
Dialog Box

Step 5: Select the **CREATE COMPONENT** option on the toolbar.

Step 6: On the Component Create dialog box, select the PART Type and enter *Shaft* as the Name for the part, then select OK (Figure 12–70).

Step 7: Select LOCATE DEFAULT DATUMS >> THREE PLANES >> OK on the Creation Options dialog box (Figure 12–71).

The Locate Default Datums option will place a set of Pro/ENGINEER's default datum planes. Datum planes placed using this option do not create external references. The Three Planes suboption will allow for the alignment of each plane to existing assembly model planes.

Step 8: Pick the first plane selection shown in Figure 12–72.

The planes for the default datums of the shaft part will be aligned with the planes picked in the next few steps. The first plane picked in this step will define the sketching plane for the first geometric feature of the shaft part. The second plane will be defined on the fly with the Make Datum option.

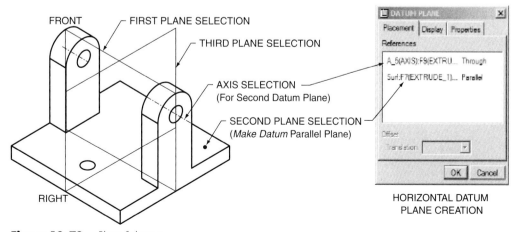

Figure 12–72   Plane Selection

STEP 9: 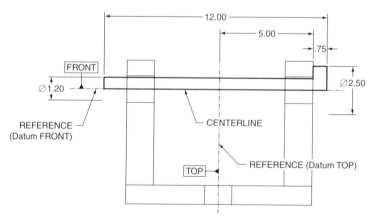 Select the Datum Plane tool to define the second plane; use the THROUGH AXIS and PARALLEL paired constraint options to define this datum plane (see Figure 12–72).

With the Through constraint option, pick the axis shown in Figure 12–72. For the Parallel constraint option, pick the part surface shown in the illustration. Make sure you use the control-key to pick each entity.

STEP 10: Select the third plane selection shown in Figure 12–72.

After defining the three planes, notice how the shaft part has been added to the model tree. Also notice on the model tree how the shaft part has been automatically activated for feature creation.

STEP 11: Ensure that the Shaft part is the active component within your model.

Your shaft part should be marked as active on the model tree. The pop-up menu can be used to activate it if necessary.

STEP 12: Start a new sketch and with datum plane DTM1 of the new part as the sketching plane for the feature and orient datum plane DTM2 toward the bottom of the sketching environment.

STEP 13: Within the sketching environment, ensure that datum planes DTM2 and DTM3 are selected as references.

The shaft part's default datums planes will be utilized as references for this feature.

STEP 14: Sketch the centerline shown in Figure 12–73.

Revolved features require a centerline within the sketcher environment. Use the Centerline icon found on the sketcher tools toolbar to create this line. Notice how this line is aligned with datum plane TOP.

STEP 15: Turn off the display of datum planes.

STEP 16: Set Hidden as the model's display mode.

STEP 17: Sketch the geometric entities shown in Figure 12–73.

Use the Line icon on the sketcher toolbar to sketch the profile shown in the illustration. The dimensioning scheme shown in the figure matches the design

Figure 12-73   Reference Dialog Box

intent for the part and for the assembly. Notice the visualization for the existing base part. You can sketch this section, or any section in top-down design, to fit around or within existing components. Since external references are prohibited for this part, you cannot reference the base part.

STEP 18: Define the dimensions shown in Figure 12–73.

STEP 19: When the sketch is complete, exit the sketcher environment.

STEP 20: Select the REVOLVE option on the toolbar.

STEP 21: Revolve the feature 360 degrees.

STEP 22: Build the feature.

STEP 23: On the model tree, activate the assembly model.

STEP 24: Save your assembly model.

CREATING THE THIRD COMPONENT (PULLEY.PRT)

In this segment of the tutorial you will create the pulley part (Figure 12–74). You will begin this part by placing the start part (*start.prt*) created earlier in this tutorial.

STEP 1: Confirm the setting of your reference control.

Use the Tools >> Assembly Settings >> Reference Control option to set your environment reference settings. You should set your component scope to None and you should deselect the Backup Forbidden References option.

STEP 2: If necessary, on the Model Tree, select the *PULLEY.ASM* file name with the right mouse button, then select the ACTIVATE option.

This option will activate the current assembly for editing purposes.

STEP 3: Select the CREATE COMPONENT option on the toolbar.

STEP 4: On the Component Create dialog box, select the part type and enter *Pulley* as the name for the part, then select OK.

STEP 5: On the Creation Options dialog box, select the COPY FROM EXISTING option, Browse to locate the *START* part, then select OK (Figure 12–75).

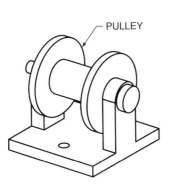

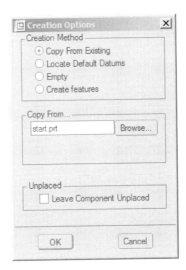

Figure 12–74   Pulley Component

Figure 12–75   Creation Options dialog box

Make sure that the display of your datum planes and axes are turned on. Your start part will be copied to your workscreen.

STEP 6: **On the Component Placement dashboard, if necessary, use the Move menu's Translate and Relative to View Plane suboptions to move the start part to a position that will allow for better entity selection.**

STEP 7: **Using the Component Placement dialog box, constrain the start part using the constraints shown in Figure 12–76.**

Use the Automatic constraint option with the following assigned constraints:

*   Align the axis feature of the start part with the axis of the shaft part. Use Query Select if necessary to pick the shaft's axis.

*   Align datum plane TOP from the start part with a vertical datum plane running through the existing assembly.

*   Allow for assumptions if necessary.

MODELING POINT While placing a part with the Component Placement dashboard, if the visibility of the part being placed is restricted by existing components, use the Move tab to adjust the temporary location of the new part.

STEP 8: **When the component is fully constrained, select the accept checkmark to exit the dashboard.**

STEP 9: **On the model tree, right-mouse select the *pulley* feature, then select the ACTIVATE option (see Figure 12–77).**

STEP 10: Setup a sketching environment with datum plane FRONT as the sketching plane and datum plane RIGHT oriented toward the bottom of the sketching environment (Figure 12–78).

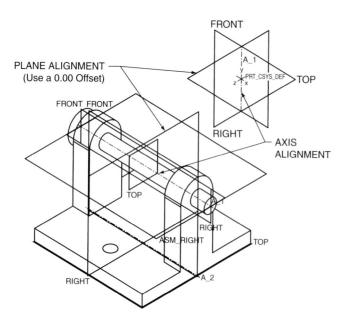

Figure 12–76   Start Part Placement

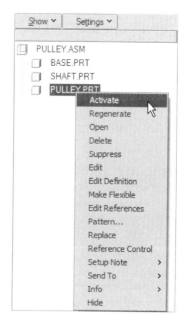

Figure 12–77   Feature Create

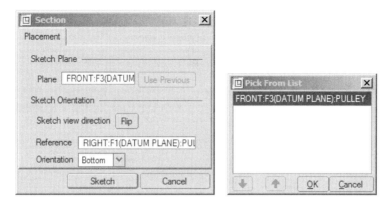

Figure 12-78   Sketching Plane Setup

If necessary, use the Pick From List pop-up menu option to pick available planes from the Pick From List dialog box. Your sketching environment should match Figure 12–79.

STEP 11:   **On the Sketch dialog box, accept the default sketching view direction, then select SKETCH to enter the sketching environment.**

STEP 12:   **Specify datum planes RIGHT and TOP as references (see Figure 12–79).**

If necessary, use the Pick From List option.

STEP 13:   **Turn off the display of datum planes.**

STEP 14:   **Select HIDDEN as the model's display style.**

STEP 15:   **Sketch the centerline shown in Figure 12–79.**

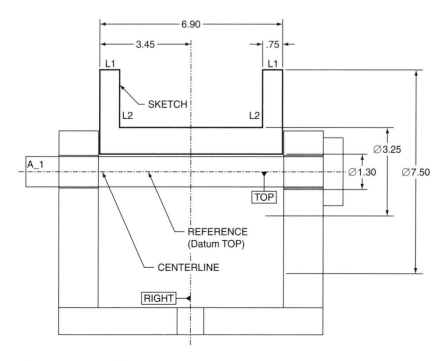

Figure 12-79   Pulley Section

STEP 16:   Sketch the geometry shown in Figure 12–79.

STEP 17:   Incorporate the dimensional design intent shown in the illustration.

STEP 18:   Add a dimensional relationship to make the 3.45 dimension in Figure 12–79 equal to half the value of the 6.90 dimension (Tools >> Relations).

STEP 19:   Exit the sketching environment.

STEP 20:   Select the Revolve option on the toolbar.

You will create a revolved protrusion sketched on the pulley part's datum plane Right.

STEP 21:   Revolve the feature 360 degrees.

STEP 22:   Build the feature.

STEP 23:   Activate the Assembly model.

Use the Activate command on the model tree's mouse pop-up menu to activate the pulley assembly.

STEP 24:   Save your assembly.

MODELING POINT   When a part is created within Assembly mode, the part is saved as a part file and can be opened individually in Part mode. If a part created in Assembly mode does not have external references, it can be copied to create a new part.

STEP 25:   Select INFO >> REFERENCE VIEWER on the menu bar.

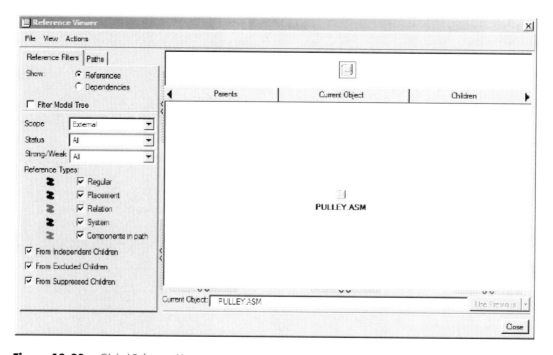

Figure 12–80   Global Reference Viewer

STEP 26: **Select the SHOW FILTERS tab, then check the settings shown in Figure 12–80.**

The Reference Viewer is used to view and beak dependencies between features, relations, and components. A component of an assembly has an external reference when it depends on another model. When you are utilizing top-down assembly design capabilities, external references should be avoided. The settings shown in Figure 12–80 will display models that have reference to the current object (References and External). The Local scope option is used to display references within an object. Notice the lack of components linked to the PULLEY.ASM name. This means that no models within the assembly have external references.

DECLARING AND USING A LAYOUT

Layouts can be used to control assembly dimension values. Within this segment of the tutorial, you will utilize the layout created previously in this tutorial to control the assembly model. To perform this function, you have to declare the layout within the assembly.

STEP 1: **Open both the assembly model and the layout model, then activate the assembly model's window.**

STEP 2: **Select FILE >> DECLARE >> DECLARE LAY within Assembly mode's menu bar.**

STEP 3: **Select the PULLEY layout on the Declare menu.**

> **MODELING POINT** In this tutorial, you will use the dimension values created within the layout to control dimension values within the assembly model. Once a layout has been declared within an assembly, the dimension values from the layout can be used in relation equations within the assembly model. Notice in Figure 12–81 the dimension symbols illustrated for the layout and for the assembly. These are the dimensions that you will relate within this tutorial.

STEP 4: **Save the assembly model.**

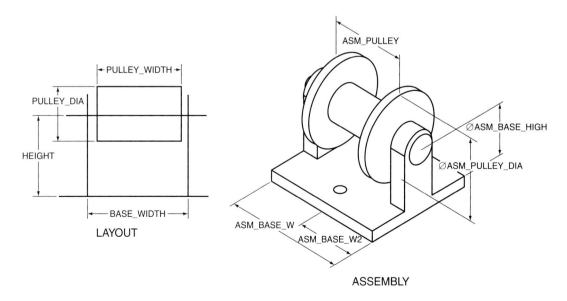

Figure 12–81 Dimension Symbols

STEP 5:   Select SETTINGS >> TREE FILTERS >> ITEM DISPLAY on the model tree.

The next several steps of this tutorial will modify the dimension symbol names of the pulley part to match Figure 12–81.

STEP 6:   Check the FEATURES option to display part features on the model tree, then select OK.

STEP 7:   Select SHOW >> EXPAND ALL on the model tree to show all available parts and features.

STEP 8:   Right-mouse select the revolved protrusion feature under the pulley part and select the EDIT option.

This technique should display the dimensions associated with the revolved protrusion feature.

STEP 9:   Edit the properties of the following pulley dimensions and rename each dimension symbol.

Match the following dimensions with their respective symbol name in Figure 12–81. Use the Dimension Properties dialog box's Dimension Text tab to modify each dimension symbol.

Dimension	Value	Symbol Name
Pulley Width	6.90	ASM_PULLEY
Pulley Diameter	7.50	ASM_PULLEY_DIA

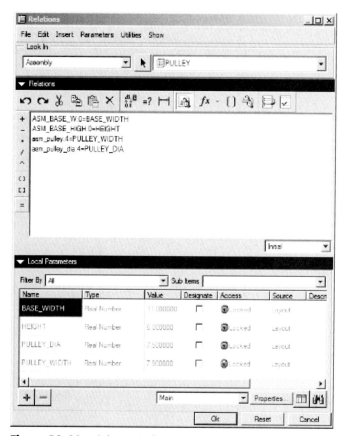

Figure 12–82   Relations Dialog Box

Step 10: **Select TOOLS >> RELATIONS and create the following dimensional relationships (refer to Figures 12–81 and 12–82).**

Add the following relations equations. Use the Local Parameters drop-down option and the INSERT >> FROM LIST option to insert parameters from your layout. Dimension symbols from the assembly model can be inserted directly from your screen.

- ASM_BASE_W:(session ID) = BASE_WIDTH
- ASM_BASE_HIGH:(session ID) = HEIGHT
- ASM_PULLEY:(session ID) = PULLEY_WIDTH
- ASM_PULLEY_DIA:(session ID) = PULLEY_DIA

NOTE: Refer to your model for each equation's session ID. An example of a session ID would be ASM_BASE_W:0. You will need to select the features, holding the above dimensions to record their dimension symbol names. Each part can have its own unique session ID.

Step 11: ☑ **Verify your relations equations, then exit the Relations dialog box.**

Step 12: **Regenerate the assembly model.**

Step 13: **Activate or open your layout model.**

Step 14: **Within the layout, use the pop-up menu's EDIT VALUE option and change the *BASE_WIDTH* dimension to a value of 11.**

Step 15: **Modify the *HEIGHT* dimension to a value of 6.**

Step 16: **Activate the assembly's window and regenerate the model.**

Your assembly model should be enlarged to match the dimensional values of the layout model (Figure 12–83).

Step 17: **Save your assembly model.**

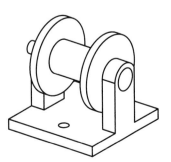

Figure 12-83 Final Model

MECHANISM DESIGN TUTORIAL

Within this tutorial, you will explore the assembly and animation capabilities of assembly mode's Mechanism option. The nutcracker model shown in Figure 12–84 will be utilized. You will model each component in part mode, then assemble each using mechanism joints.

Assembly mode provides powerful tools for modeling a complete design. Traditional bottom-up constraints necessitate fully constrained components. When a component is not fully constrained, it is considered packaged and presents assembly difficulties. The mechanism option provides tools for assembling components in a manner that replicates a real design. Joints such as pin, cylinder, slider, and ball are available. Notice in Figure 12–84 the connection linkage that exists from the handle through the connection part through to the piston part. The mechanism joints defining these parts only constrain each component within the degrees of freedom required by the design. Within this assembly, the handle can be moved, which in turn will move the piston. This will be demonstrated. The following topics will be covered in this tutorial:

- Modeling assembly parts.
- Assembling a mechanism.
- Manipulating a mechanism.
- Running a mechanism's motion.
- Animating a mechanism.

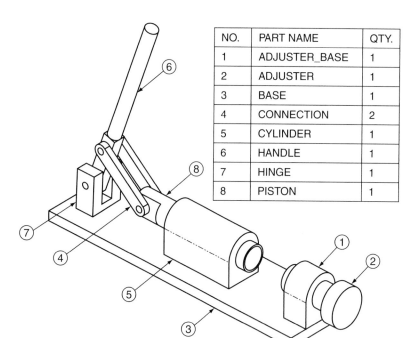

NO.	PART NAME	QTY.
1	ADJUSTER_BASE	1
2	ADJUSTER	1
3	BASE	1
4	CONNECTION	2
5	CYLINDER	1
6	HANDLE	1
7	HINGE	1
8	PISTON	1

Figure 12–84   Mechanism Assembly

MODELING ASSEMBLY PARTS

The assembly in this tutorial consists of eight different parts: base, cylinder, hinge, piston, adjuster base, adjuster, connection, and handle. There will be two instances of the connection part. Use part mode to model each of the parts as shown in Figure 12–85. When modeling each part, pay careful attention to the locations of your datum planes. For proper assembly of the mechanism, your datum planes should match the datum planes represented in each part's drawing.

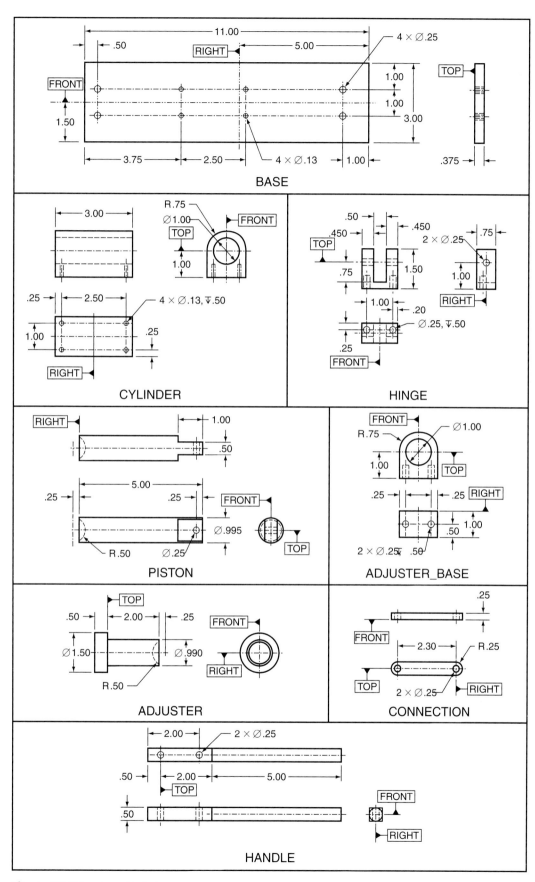

Figure 12–85 Assembly Parts

Assembling a Mechanism

Within this segment of the tutorial you will assemble the parts composing the design. Within this exercise, you will not use a template file. Do not start this segment of the tutorial until you have modeled all the parts portrayed in Figure 12–85.

Step 1: **Start Pro/ENGINEER and then select FILE >> NEW.**

Step 2: **On the New dialog box, deselect the USE DEFAULT TEMPLATE OPTION.**

Within this tutorial, do not use Pro/ENGINEER's default template.

Step 3: **Create a new Assembly object file named *NUTCRACKER.ASM*.**

Step 4: **On the New File Options dialog box, select the EMPTY template file, then select OK.**

An empty template file has no preestablished datum features, units, layers, or other settings.

Step 5: **Select the Add Component option on the toolbar.**

Step 6: **Using the Open dialog box, place the *BASE* part.**

Without any existing features or components, Pro/ENGINEER will place the first component without requiring any constraints or joints. If in this tutorial you inadvertently create Pro/ENGINEER's default datum planes, you can mate and/or align the BASE part to the default set of assembly datum planes.

Step 7: **Select the Add Component option on the toolbar and open the *CYLINDER* part.**

Step 8: **Using traditional assembly constraints, assemble the cylinder part as shown in Figure 12–86.**

To locate the cylinder part, assemble the component with one Mate constraint and two Align constraints as shown in the illustration.

There are two ways to add components to a mechanism: Fixed and By Connection. The fixed method is identical to the traditional way of assembling components in Pro/ENGINEER. Connections allow components to move according to the degrees of freedom provided by their defined joints.

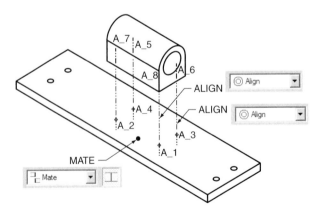

Figure 12-86   Cylinder Fixed Constraints

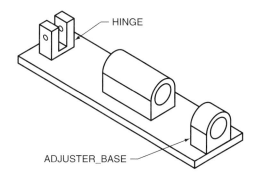

Figure 12–87   Adjuster_Base and Hinge Fixed Constraints

STEP 9:   When the cylinder part is fully constrained, select the continue checkmark to exit the Component Placement dashboard.

STEP 10:   Use the same technique for assembling the cylinder part to constrain the *HINGE* and *ADJUSTER_BASE* parts (Figure 12–87).

As with the cylinder part, use two align constraints and one mate constraint for each part. Your assembly should appear as shown in the illustration.

Next you will assemble the piston part using a Cylinder connection. Other available connections include: pin, bearing, slider, planar, and ball.

STEP 11:   Select the Add Component option on the toolbar and open the *PISTON* part.

STEP 12:   On the connections drop-down menu, select CYLINDER as the connection type (see Figure 12–88).

STEP 13:   Pick the two axes shown in Figure 12–88.

A Cylinder joint type is defined through the alignment of two axes. This joint type provides two degrees of freedom: one linear and one rotational.

STEP 14:   If necessary, select the FLIP option to point the piston's cut feature toward the hinge part (see Figure 12–88).

STEP 15:   On the dashboard, select the MOVE tab.

You will reposition the piston part to match Figure 12–89.

STEP 16:   On the Move menu, with the Translate and Motion Reference options selected, pick the axis of the piston part (see Figure 12–89).

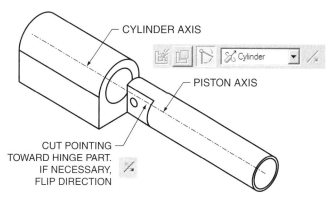

Figure 12–88   Cylinder Joint Definition

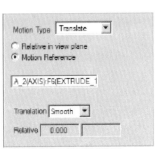

Figure 12-89 Move Option

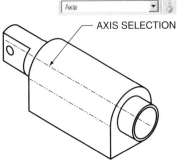

AXIS SELECTION

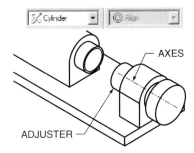

AXES

ADJUSTER

Figure 12-90 Adjuster Placement

NOTE: You may need to use Pro/ENGINEER's filter option to allow for the selection of axes.

STEP 17: **Move the Piston part's location to approximately match Figure 12–89.**

The placement status should state "Connection Definition Complete."

NOTE: You can also use the Mechanism application to move components that have mechanism joints.

STEP 18: **Select the continue checkmark to complete the placement.**

STEP 19: **Use the same technique for assembling the piston part to constrain the *ADJUSTER* part (Figure 12–90).**

Use cylinder as the connection for the component. If necessary, use the Flip option and the Move tab to position the component to match the illustration.

STEP 20: **Select the Add Component option on the toolbar and open the *HANDLE* part.**

STEP 21: **Select the PIN joint type under the Connections option (see Figure 12–91).**

Pin connections provide one rotational degree of freedom. It is defined through the alignment of two axes and the aligning or mating of two planes.

STEP 22: **Align the hole axes of the handle and hinge parts as shown in Figure 12–91.**

STEP 23: **Pick two surfaces to form a mate condition between the inside of the hinge part and the outside of the handle.**

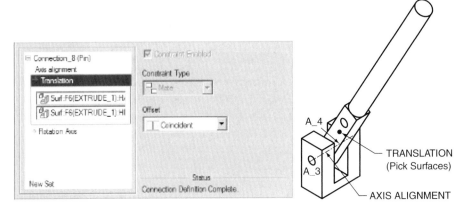

A_4

TRANSLATION
(Pick Surfaces)

A_3

AXIS ALIGNMENT

Figure 12-91 • Handle Placement

Step 24: Use the Move tab to rotate the handle to the approximate location shown in Figure 12–91.

Under the Move menu, use the Rotate and Motion Reference options. The handle should be pointing toward the center of the base part.

Step 25: Select the continue checkmark.

Step 26: Assemble the *CONNECTION* part.

Step 27: Create the PIN joint shown in Figure 12–92.

The connection part will have two joints: one pin and one cylinder. The pin joint will join the connection part to the handle part. The cylinder joint will join the connection part to the piston. If necessary, use the Flip option to create the mate translation connection.

Step 28: Use the Move menu on the dialog box to rotate the connection part to the approximate connection location shown in Figure 12–92.

Step 29: Under the Placement menu on the dashboard, select the NEW SET option.

This component will have two joints: a pin to the hinge part and a cylinder connection to the piston part.

Step 30: Create the CYLINDER joint shown in Figure 12–92.

After creating the cylinder joint, your assembly connection may not look like the illustration. This is typical of a Pro/ENGINEER looped mechanism. In a later step, you will execute the Connect option to assume a successful assembly.

Step 31: If the placement status signifies a complete connection, select OK to exit the dialog box.

Step 32: Use the same technique for assembling the first connection part to place the second instance of the connection part.

Repeat Steps 27 through 32 to place the second instance of the connection part.

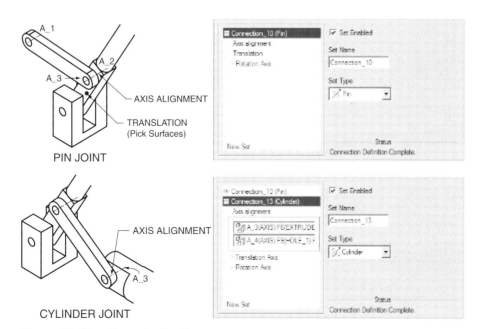

Figure 12–92 Connection Part Placement

Step 33:   Select APPLICATION >> MECHANISM on the menu bar.

Note: You must have a license for Mechanism_Design within your Pro/ENGINEER implementation. The Mechanism_Design extension provides options for defining slots, cams, gears, and component drivers.

Step 34:   Select EDIT >> CONNECT on the menu bar, then select the RUN option on the Connect Assembly dialog box.

You should get a confirmation message that your mechanism has assembled correctly with a specific tolerance value.

Step 35:   If you get a positive confirmation message, select YES to accept the successful assembly.

Step 36:   Save the assembly.

MANIPULATING A MECHANISM

This segment of the tutorial will demonstrate how components can be dragged through any defined degrees of freedom. In addition, you will create snapshots of component placements that will be used in the last segment of this tutorial to animate the mechanism. Ensure that you are in the Mechanism application before performing the following steps (Application >> Mechanism).

Step 1:   Select the Drag Package Component icon on the model display toolbar, then expand the Snapshots option on the Drag dialog box (Figure 12–94).

Step 2:   Pick the end of the handle part (see Figure 12–94).

The Drag dialog box is used to drag components on the screen. Use the Point option to select a point on the end of the handle part. After selecting the handle, you can dynamically drag the component with the mouse.

Note, once the point has been selected, you don't need to hold the left mouse button down to drag the handle. Use the left mouse button to end dragging.

Step 3:   Drag the handle to the First Position shown in Figure 12–94.

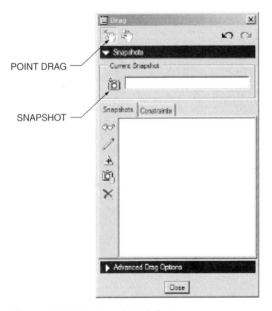

Figure 12-93   Drag Entity Selection

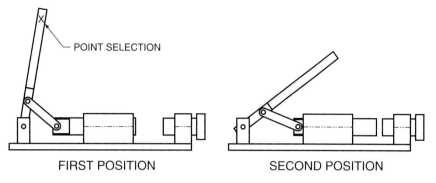

POINT SELECTION

FIRST POSITION   SECOND POSITION

Figure 12-94   Drag Positions

Figure 12–94 represents a side view of the assembly. You can utilize any orientation to include a user-defined viewpoint. Position the handle approximately 10 degrees from vertical.

STEP 4:   On the Drag dialog box, select the SNAPSHOT icon (Figure 12–93).

Snapshots can be used to restore a mechanism's position and to create animations. You will use the snapshots created in this segment to animate the mechanism in the last segment of this tutorial.

STEP 5:   Drag the handle to the second position shown in Figure 12–94, then create a second snapshot.

STEP 6:   Close the Drag dialog box.

RUNNING A MECHANISM'S MOTION

Motion, as defined by the degrees of freedom within a mechanism, can be animated. Within this segment of the tutorial, you will define the motion of the assembly through the use of a servo motor. Ensure that you are in the Mechanism application before performing the following steps (Application >> Mechanism).

Note: Servo motors were formerly known as drivers in previous releases of Pro/ENGINEER.

STEP 1:   Select the Define Servo Motors icon on the mechanism toolbar.

STEP 2:   For the Driven Entity, select the pick icon, then on the workscreen pick the pin joint shown in Figure 12–95.

STEP 3:   Select the Profile tab on the Servo Motor Definition dialog box, then select VELOCITY as the specification (Figure 12–96).

STEP 4:   Change the Magnitude option to COSINE, then enter the values shown in Figure 12–96.

STEP 5:   Select the GRAPH option to observe the graph of your mechanism.

STEP 6:   Close the Graph windows.

STEP 7:   On the Servo Motor Definition dialog box, select the Joint Axis Setting icon.

STEP 8:   Pick the Component Zero Reference plane shown in Figure 12–97.

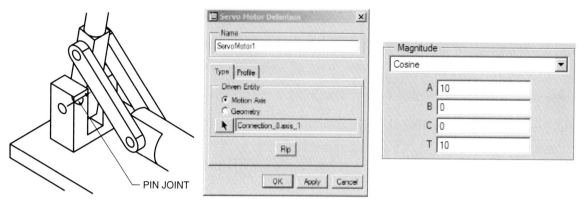

Figure 12–95 Driver Creation

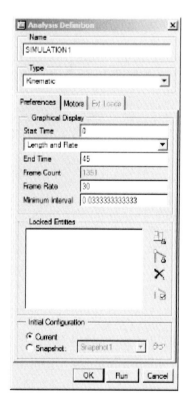

Figure 12–96 Servo Motor Definition Dialog Box

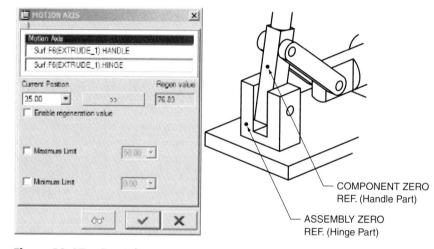

Figure 12–97 Zero Reference Selection

This step along with the next step will define the starting zero point for the mechanism animation.

STEP 9: **Pick the Assembly Zero Reference plane shown in Figure 12–97.**

INSTRUCTIONAL POINT Notice on the workscreen the arrow representing the connection's joint. Using the right-hand rule with your thumb pointing in the direction of the arrow, your fingers will point in the direction of driver rotation. Within the next step, you might have to enter a negative 35.00 value for the initial angle.

STEP 10: Enter 35.00 as the Joint Axis Position initial Value.

STEP 11: Select the continue checkmark to exit the Motion Axis dialog box.

STEP 12: Select OK to exit the Servo Motors Definition dialog box.

STEP 13: Close the Servo Motors dialog box.

STEP 14: Select the Define Analysis icon on the mechanism toolbar.

After you select the Define Analysis icon, Pro/ENGINEER will launch the Analysis Definition dialog box. This dialog box is used to establish multiple motion definitions for a mechanism.

STEP 15: Enter SIMULATION1 as the name for the analysis (Figure 12–98).

STEP 16: On the Analysis Definition dialog box, select KINEMATIC as the analysis type.

STEP 17: On the dialog box, select LENGTH AND RATE and the time definition and enter the values shown in Figure 12–98.

STEP 18: Enter the End Time (45) and Frame Rate (30) values shown in Figure 12–98.

STEP 19: Select RUN on the Analysis dialog box.

With any luck, after selecting the Run option, your mechanism should animate according to the defined degrees of freedom and the set servo motor. If you have unexpected results in your animation, try adjusting the servo motor's profile values or its initial angle value.

STEP 20: Select the Replay Analysis icon on the mechanism toolbar.

Figure 12–98   Motion Definition Dialog Box

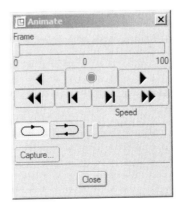

Figure 12–99   Animate Dialog Box

STEP 21:   ▶   Select the Play icon on the Playbacks dialog box.

STEP 22:   Use options on the Animate dialog box to run the results of your motion study (Figure 12–99).

STEP 23:   Close out of the Animate and Playbacks dialog boxes and save your assembly.

ANIMATING A MECHANISM

Mechanisms can be animated by using Pro/ENGINEER's Animation mode. Within this segment of the tutorial, you will use the snapshots created previously to animate the nut-cracker assembly.

STEP   1:   Using Pro/ENGINEER's menu bar, select APPLICATIONS >> ANIMATION.

When you select the Animation option, Pro/ENGINEER will reveal the Animation toolbar (Figure 12–100) and, at the bottom of the screen, a timeline.

STEP   2:   Select the Animation icon on the toolbar, then select NEW on the Animation dialog box.

STEP   3:   Use the Animation dialog box's Rename option to name the new animation *NUTCRACKER.*

STEP   4:   Close the Animation dialog box.

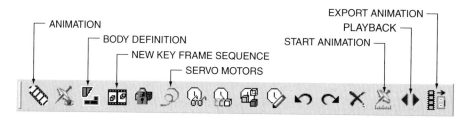

Figure 12–100   Animation Dialog box

STEP   5:   Double-pick the timeline at the bottom of the workscreen and set the time domain values shown in Figure 12–101.

You should set *Length and Frame Count* as the time domain setting with an end time of 20 seconds and a frame count of 201 frames.

STEP   6:   Select OK to create the time domain.

After selecting OK, notice how the timeline at the bottom of the screen now displays a scale from 0 to 20.

STEP   7:   Select the New Key Frame Sequence icon on the Animation toolbar.

Multiple key frame sequences can be created for an animation. Within this tutorial, two will be used. The first sequence will utilize the two snapshots created previously to animate the handle and piston linkage. The second sequence will animate the adjuster.

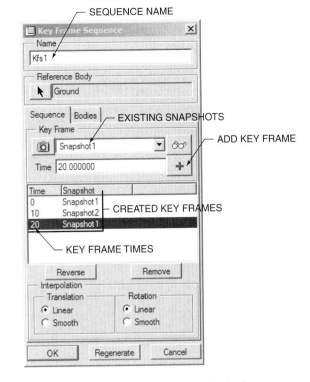

Figure 12–102   Key Frame Sequence Dialog Box

Figure 12–101   Time Domain

STEP 8: On the dialog box, use the Add Key Frame icon to create the three key frames shown on the Key Frame Sequence dialog box (Figure 12–102).

Two snapshots should currently exist. Use the Add Key Frame icon to create the three key frames shown in Figure 12–102 (snapshot1 at 0 second, snapshot2 at 10 seconds, and snapshot1 at 20 seconds). To perform this, select an existing snapshot from the drop-down list, enter a specific time value, and then select the Add Key Frame icon.

STEP 9: Select OK when your Key Frame Sequence dialog box matches Figure 12–102.

STEP 10: Select the Start Animation icon on the Animation toolbar (Figure 12–100).

Your handle and piston linkage should animate.

STEP 11: Select the New Key Frame Sequence icon on the Animation toolbar.

The next key frame sequence will animate the adjuster part.

STEP 12: Select the Snapshot icon on the Key Frame Sequence dialog box.

The Snapshot option will launch the Drag dialog box (Figure 12–103). This dialog box is also accessible directly from the Animation dialog box.

STEP 13: Use the Point Drag option to drag the adjuster part to the SNAPSHOT3 position shown in Figure 12–103.

Your snapshot numbers may be different from those represented in this tutorial. You can approximate the exact location for each shot.

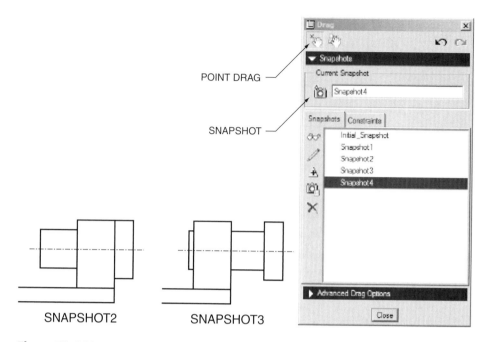

POINT DRAG

SNAPSHOT

SNAPSHOT2   SNAPSHOT3

Figure 12–103   Snapshot Creation

STEP 14:   Select the Snapshot icon to create snapshot3.

STEP 15:   Use the Point Drag option to drag the adjuster part to the SNAPSHOT4 position shown in Figure 12–103.

STEP 16:   Select the Snapshot icon to create snapshot4.

STEP 17:   Close the Drag dialog box.

STEP 18:   Modify SNAPSHOT4 (see Figure 12–104) to have a value of 10 seconds.

STEP 19:   Select OK to close the Key Frame Sequence dialog box.

Key Frame Sequences can be modified with the Animation >> Key Frame Sequence option on Pro/ENGINEER's menu bar. Your timeline should look similar to Figure 12–105. Key frames on the timeline are represented by the triangle symbol. They can be manipulated by dragging with the mouse.

Figure 12–104   Key Frame Sequence Dialog box

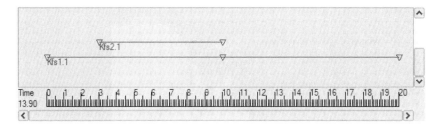

Figure 12-105   Animation Timeline

STEP 20:   Use your mouse to drag the second key frame sequence to the position shown in Figure 12–105.

STEP 21:   Run the animation by selecting the Start Animation icon.

STEP 22:   Playback the created animation by selecting the Playback icon.

STEP 23:   Save your assembly file.

PROBLEMS

1. This problem will have you model the assembly shown in Figure 12–106. This assembly consists of five components: one support, one cam, one guide, and two pins. Model each component in Part mode, then place them in the assembly using traditional Pro/ENGINEER constraints.

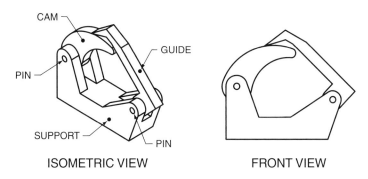

Figure 12-106   Assembly

Detail drawings of the support and cam parts are shown in Figures 12–107 through 12–109. One of the problems of this exercise is for you to design a guide part that will stay mated with the top surface of the cam (see Figure 12–106). As shown in Figure 12–108, when constructing the cam part, create a datum plane (DTM1) that will lie through the central axis of the cam and at an angle to one of Pro/ENGINEER's default datum planes. The angular dimension formed will be used within the assembly to vary the rotational position of the cam. The design intent for the assembly requires the angle shown in the cam drawing to be variable between °10 and 45 degrees (Figure 12–110). Within Figure 12–109, the dimensions for the guide part have been omitted. You must create a guide part that will allow for the °10 through 45 degrees of rotation of the cam, plus any necessary part clearances. Notice in Figures 12–106 and 12–110 how the point of the guide part is constrained to the top of the cam using a Point On Surface constraint. In addition, you must design and place the two pin components. After the assembly is complete, use a modify dimension option to change the angular value defining the datum plane.

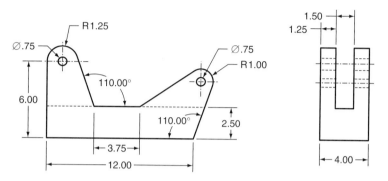

Figure 12–107   Support Part

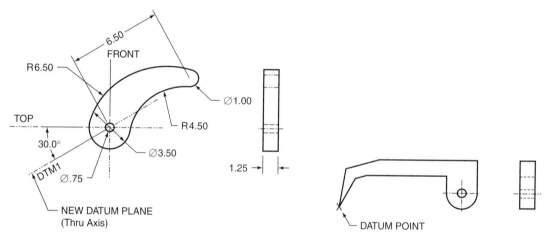

Figure 12–108   Cam Part

Figure 12–109   Guide Part

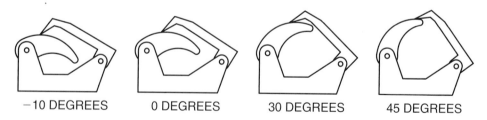

Figure 12–110   Degrees of Movement

2.   Use bottom-up assembly design techniques to create the assembly model shown in Figure 12–111. Model each component in Part mode, then assemble them within Assembly mode. Detailed drawings for the *key* and *plate* parts are shown in Figure 12–112. The drawing for the *arm* part can be found in the Problems section of Chapter 5 (Figure 5–87). The drawings for the *retainer, body,* and *shaft* parts can be found in the Problems section of Chapter 6 (Figures 6–58, 6–59, and 6–60).

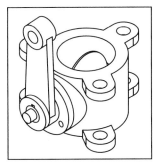

UNEXPLODED
ASSEMBLY VIEW

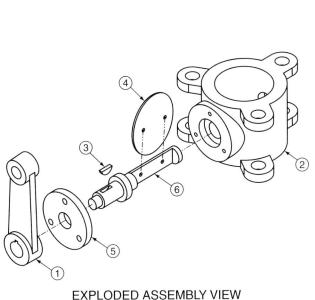

EXPLODED ASSEMBLY VIEW

ITEM	COMPONENT
1	ARM
2	BODY
3	KEY
4	PLATE
5	RETAINER
6	SHAFT

Figure 12–111   Exploded and Unexploded Model

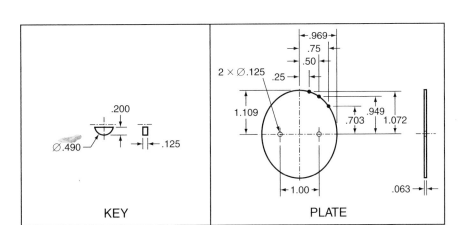

KEY   PLATE

Figure 12–112   Key and Plate Parts

3. Use top-down assembly design techniques to create the assembly model shown in Figure 12–111. Model each component in Assembly mode in the order listed in the Bill of Materials shown in the illustration. Detailed drawings for the *key* and *plate* parts are shown in Figure 12–112. The drawing for the *arm* part can be found in the Problems Section of Chapter 5 (Figure 5–87). The drawings for the *retainer*, *body*, and *shaft* parts can be found in the Problems section of Chapter 6 (Figures 6–58, 6–59, and 6–60).

4. Use bottom-up mechanism design techniques to create the assembly model shown in Figure 12–111. Model each component in Part mode, then assemble them as a mechanism within Assembly mode. Detailed drawings for the *key* and *plate* parts are shown in Figure 12–112. The drawing for the *arm* part can be found in the Problems section of Chapter 5 (Figure 5–87). The drawings for the *retainer, body*, and *shaft* parts can be found in the Problems section of Chapter 6 (Figures 6–58, 6–59, and 6–60). Animate your mechanism and export as a movie file.

QUESTIONS AND DISCUSSION

1. Describe the basic principles behind bottom-up assembly design.

2. Describe the basic principles behind top-down assembly design.

3. Describe the difference between the Align constraint option and the Mate constraint option.

4. How does the Offset Mate constraint differ from the Offset Align constraint?

5. What two constraint options can be used to place the axis of a shaft coaxial with the axis of a hole?

6. Describe four uses of a skeleton model.

7. How do dimension symbols within Assembly mode differ from dimension symbols within Part mode?

8. Describe uses of Pro/ENGINEER's Layout mode.

9. List and describe five mechanism joints.

13

SURFACE MODELING

Introduction

Pro/ENGINEER's surface creation tools are useful for modeling parts with complex curves and surfaces. While most of Pro/ENGINEER's solid modeling tools are ideal for creating components with planar surfaces and smooth curves, these tools are not always best for complex shapes. Creating a surface model requires a different strategy from that for a solid model. Even though many Pro/ENGINEER surface creation tools are similar to solid creation options (e.g., Extrude and Revolve), additional tools such as Boundaries and Merge are available. The key to creating a surface model is to define the skeleton of the part with appropriate datum options. Once this skeleton has been created, the surface skin can be added. Upon finishing this chapter, you will be able to:

- Create extruded, revolved, swept, and blended surface features.
- Create a flat surface feature.
- Create a round between two surface features.
- Utilize datum curves to create surface features.
- Merge two quilts.
- Create trimmed features on a surface.
- Create a surface from boundaries.
- Create a solid feature from a merged quilt.

DEFINITIONS

Boundaries	A technique for creating a surface feature by defining the surface's boundaries.
Merge	The process of joining two surface quilts. Quilts that intersect and quilts that share common boundaries can be joined.
Quilt	One or more surface features.
Solidify	The process of converting a surface quilt to a solid feature.

INTRODUCTION TO SURFACES

A surface is a geometric feature with no defined thickness. Surface features are often confused with thin features. Thin features are actually thin-walled solids. The wall of a thin feature has a defined thickness. The wall of a surface feature does not have a defined thickness. Surface tools are used to create geometric shapes with complex contours and

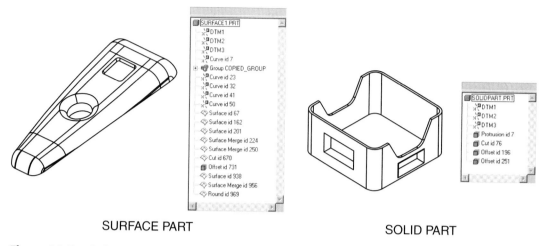

SURFACE PART   SOLID PART

Figure 13–1   Surface Versus Solid Part

undulations. Figure 13–1 shows a comparison of two parts. In the illustration, the solid part's features and geometric shape make an ideal model for using solid creation tools. The surface part, on the other hand, would be better served with Pro/ENGINEER's surface creation tools. On first observation, the solid part might appear to have surface modeling characteristics. It was actually created as a thin extruded protrusion. The indentations were created with Tweak menu options. One characteristic of a part created as a surface is the number of features required to construct the model. Figure 13–1 reveals the model trees for both parts. A typical surface part is composed of multiple datum curves and surfaces.

A surface model is composed of patches of surface features. The surface part shown in Figure 13–1 is composed of multiple surface features. The main body surface and the end features were created with the Boundaries option, while the hole was created with a combination of the Extrude, Merge, and Round commands. The indented square feature was created with the Offset option. Within Pro/ENGINEER, a quilt is a combination of one or more surface features. When surfaces are merged to form a completely enclosed quilt, the solid option Use Quilt (found under the Protrusion and Cut commands) can be used to convert the surface feature into a solid feature.

Surface features (quilts) are created in Part mode with the Insert menu option or with commands on Pro/ENGINEER's toolbar. Feature creation tools that utilize a dashboard, such as Revolve and Extrude, have options for creating either solid features or surface features (Figure 13–2).

SURFACE OPTIONS

The Surface option on feature dashboards and surface commands under the Insert menu are used to create basic surface features. Most of the options available are similar to options found for the creation of solids. Figure 13–3 shows four basic surface creation options: Extrude, Revolve, Sweep, and Blend. When a surface feature is created, with the exception

SURFACE FEATURE
SOLID FEATURE

Figure 13–2   Surface Feature Option

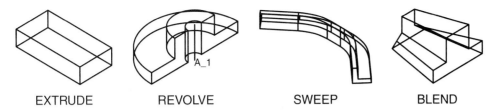

Figure 13–3   Surface Creation Options

of a shaded model, it is displayed in a wireframe format. The following is a description of basic surface creation tools.

EXTRUDE

The Extrude tool can create a surface feature by extruding a sketched section. This option is found on the extrude dashboard. When it is selected, an extra option is provided for either capping the ends of the extrusion or leaving the ends open.

REVOLVE

The Revolve tool can create a surface feature by revolving a sketched section around an axis of revolution. It is available on the revolve dashboard. Like a solid revolved protrusion, the axis of revolution is a sketched centerline. An option exists for either capping the ends of the revolution or leaving the ends option. Ends are available for capping with a less than 360 degree revolution.

SWEEP

The Sweep option creates a surface feature by protruding a section along a sketched or selected trajectory. It is available under the Insert >> Sweep menu. As with extruded and revolved surfaces, an additional option is available for either capping the ends of the sweep or leaving the ends open.

BLEND

The Blend option creates a surface feature by protruding between two or more sections. This option functions identical to the solid blend command. Options are available for creating parallel, rotational, and general blends. In addition, the ends of the blend can be capped or left uncapped.

FILL

The Fill tool (Edit >> Fill) is used to create a flat surface feature (Figure 13–4). This option functions similarly to the extrude option, but without the depth parameter. As in the extrude option, the section of the feature is sketched.

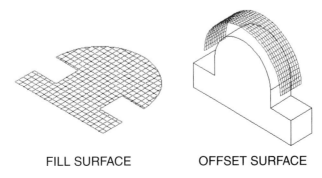

FILL SURFACE   OFFSET SURFACE

Figure 13–4   Flat and Offset Options

OFFSET

The Offset tool (Edit >> Offset) creates a new surface feature by offsetting from a solid or quilt (Figure 13–4). The user specifies the offset distance and surface to offset from. An option exists for adding side surfaces.

COPY

The Copy tool (Edit >> Copy) creates a surface feature on top of one or more selected surfaces. This option is useful for creating surface features out of existing solid features. The created surface can be exported as an IGES (International Graphics Exchange Standard) file to create a new surface model.

ROUND

The Round tool is available for creating a rounded transition at the intersection of two surface features. Like solid rounds, multiple rounds sets can be utilized. Other options such as Variable radius and Full Round are also available.

SURFACE OPERATIONS

As defined previously in this chapter, a quilt is a combination of one or more surface features. Quilt options are available to manipulate and modify existing quilts. These commands are accessible under the Edit menu. The following options are available:

MERGE

The Merge command is used to join two or more quilts. This option can be used to combine two adjacent surfaces or it can be used to join two intersecting surfaces. Figure 13–5 shows an example of two surfaces joined with the Merge option. Within the figure, three possible solutions to the merge are shown.

TRIM

The Trim tool is used to trim a surface feature (trimmed quilt) with an existing surface feature (trimming object). The trimming object is used as a cutting tool and is removed from the model. As an example, if an extruded surface feature is utilized as a trimming object while passing through a second surface feature (the trimmed quilt), surfaces within the extrusion are removed without the addition of surfaces defined by the trimming object. This command functions in a similar manner to setting the surface and cut options on a feature creation dashboard.

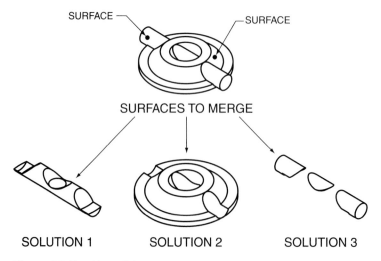

Figure 13–5   Merge Solutions

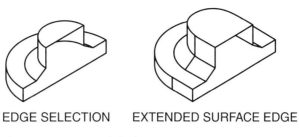

EDGE SELECTION   EXTENDED SURFACE EDGE

Figure 13–6   Extended Edge

EXTEND

The Extend option is used to extend a selected surface edge. Figure 13–6 shows an example of an edge that has been extended. This extended surface was created with the Same Surface option. Other options include Approximate Surface and Along Direction.

MIRROR

The Mirror tool is used to mirror selected surfaces over a picked mirror plane. This option is available under the Edit menu and works similarly to the Copy command found on the Feature Operations menu.

DRAFT

The Draft option creates a drafted surface from an existing planar quilt surface (Figure 13–7). This option works identically to the Draft command found under the Tweak menu. Refer to Chapter 5 for more information on creating a draft feature.

DRAFT OFFSET

The Draft Offset tool is actually a suboption under the Offset command. It creates a new surface by offsetting within the boundary of a defined sketched section (Figure 13–8). The following options exist:

- The offset can be defined normal to the offsetting surface or normal to the sketch.
- The offset can be straight or beveled (Figure 13–8).
- A draft angle can be entered, creating a beveled side around the offset.
- The new surface can extrude away from the offsetting surface or into the offsetting surface.

ADVANCED SURFACE OPTIONS

Advanced surface options are available by either selecting the command on the toolbar or through the Insert menu. Most advanced surfacing tools are suboptions under their respective solid command. The following is a description of some of the available options.

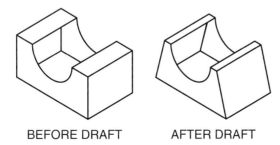

BEFORE DRAFT   AFTER DRAFT

Figure 13–7   Drafted Surfaces

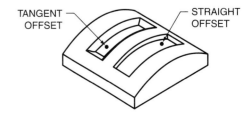

TANGENT OFFSET   STRAIGHT OFFSET

Figure 13–8   Area and Draft Offset

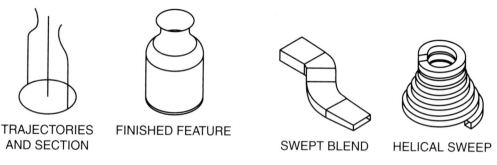

TRAJECTORIES   FINISHED FEATURE
AND SECTION   SWEPT BLEND   HELICAL SWEEP

Figure 13–9   Variable Section Sweep   **Figure 13–10**   Swept Blend and Helical Sweep

VARIABLE-SECTION SWEEP

The Variable-Section Sweep tool has an option for creating a solid or a surface feature. This tool is used to sweep a section along multiple trajectories. Figure 13–9 shows an example of a surface part created with this option. Refer to Chapter 11 for more information on creating a variable-section sweep feature.

SWEPT BLEND

A swept blend is a combination of a sweep and a blend (Figure 13–10). It is created by sweeping one or more sections along a user-defined trajectory. The trajectory can be selected on the workscreen or sketched. This option functions identical to the same option for creating solid features.

HELICAL SWEEP

The Helical Sweep option creates a surface feature by revolving a sketched section around an axis and along a user-defined trajectory (Figure 13–10). Common features created with this option are threads and springs. This option is identical to the Helical Sweep option used for the creation of solids. Refer to Chapter 11 for more information.

BOUNDARY BLEND

The Boundary Blend tool is used to create a quilt by defining its boundaries. The feature's surface can be defined by selecting reference entities in one or two directions. Figure 13–11 shows examples of quilt surfaces created in one and two directions.

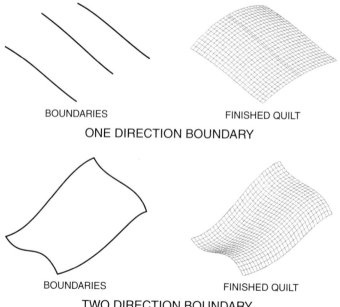

BOUNDARIES   FINISHED QUILT
ONE DIRECTION BOUNDARY

BOUNDARIES   FINISHED QUILT
TWO DIRECTION BOUNDARY

Figure 13–11   Boundaries Option

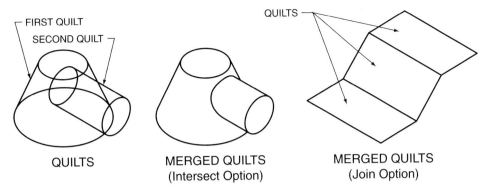

FIRST QUILT
SECOND QUILT
QUILTS

QUILTS

MERGED QUILTS
(Intersect Option)

QUILTS

MERGED QUILTS
(Join Option)

Figure 13–12   Surface Merge

MERGING QUILTS

The Merge option is used to join two or more quilts. Two options are available: Intersect and Join. The Intersect option merges two intersecting quilts (Figure 13–12). The intersection edge of the two quilts is utilized as a knife-edge for trimming each quilt. The Join option merges two adjacent quilts. Each quilt has to be joined on one or more edges. Used in conjunction with the Solidify tool, quilts can be converted to solid features.

This guide will demonstrate the creation of the merged intersecting quilts shown in Figure 13–12. Perform the following steps to merge two intersecting quilts:

STEP 1:   Using the Control key, preselect the two quilts to merge.

The Merge option combines two quilts. The first quilt selected is the primary quilt and the second quilt is the additional quilt.

STEP 2:   Select EDIT >> MERGE on the menu bar.

After you select the Merge option, Pro/ENGINEER will reveal the merge dashboard (Figure 13–13). This dialog is used to select quilts to merge and to select the type of merge to execute.

STEP 3:   On the dashboard, use the two material direction options to select the quilt sides that will create the correctly merged feature (Figure 13–14).

STEP 4:   On the Surface Merge dialog box, select the Build Feature icon to create the merged surface.

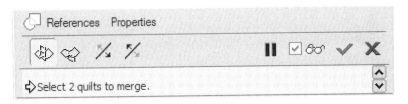

Figure 13–13   Merge Dashboard

PRIMARY QUILT: Side 1
ADDITIONAL QUILT: Side 1

PRIMARY QUILT: Side 1
ADDITIONAL QUILT: Side 2

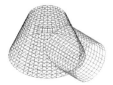

PRIMARY QUILT: Side 2
ADDITIONAL QUILT: Side 1

PRIMARY QUILT: Side 2
ADDITIONAL QUILT: Side 2

Figure 13–14   Merged Quilts Side Options

BOUNDARY SURFACE FEATURES

A surface feature can be created by selecting bounding edges of the feature through the Boundary Blend tool. Four options are available. The following is a description of each:

BOUNDARY BLEND TOOL

The Boundary Blend Tool creates a quilted surface by defining the surface's external boundaries (Figures 13–15). Selectable entities include datum curves and datum points. Selected entities can lie in either one or two directions. The illustration of the blended surface shown in the figure represents a quilt defined by selecting entities lying in two directions.

CONIC SURFACE

The Conic Surface option creates a quilted surface between two selected boundaries. The surface is formed by a third controlling curve (Figure 13–15). Two options are available for this controlling curve: Shoulder Curve and Tangent Curve. With the Shoulder Curve option, the surface feature passes through the control curve. With the Tangent Curve option, the surface feature does not pass through the curve. This option is available under the Insert >> Advanced menu.

APPROXIMATE BLEND

An approximate blend is a suboption under the Boundary Blend tool. Within a boundary blend's definition, additional fitting curves can be selected for the manipulation of the surface's form and shape. Additional curves do not form the boundary of the quilt.

N-SIDED SURFACE

The N-Sided Surface command is used to create a quilted surface from more than four bounding entities (Figure 13–15). Boundary entities, such as datum curves, can not be tangent. This option is available under the Insert >> Advanced menu.

CREATING A BOUNDARY BLEND SURFACE

This guide will utilize the Boundary Blend tool to create the quilted surface feature shown in Figure 13–16. As shown in the figure, the boundaries forming the quilt consist of datum

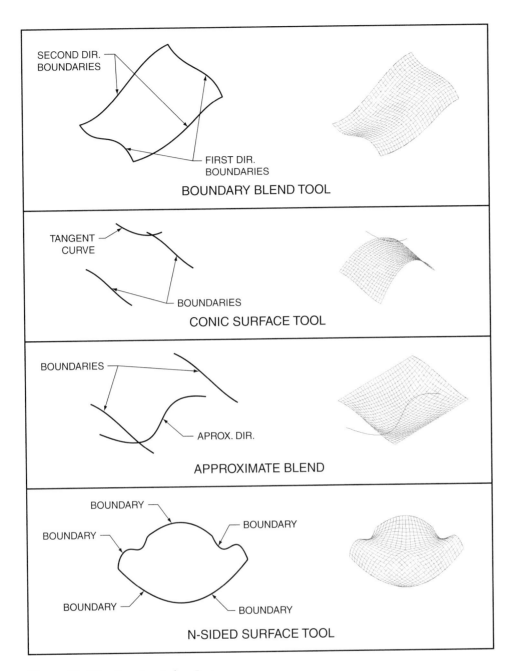

Figure 13–15   Boundary Surface Features

curves running in two directions. Datum curves defining a surface boundary in this manner must be aligned at their respective endpoints. Perform the following steps to create this feature:

STEP 1:   Select the Boundary Blend icon on the toolbar.

This option is also available under the Insert menu.

STEP 2:   Using the Control key, pick curves running in the first direction.

The first edit box, shown in Figure 13–16, is used to define boundaries in the first direction. This box is selected by default upon entering the Boundary Blend's dashboard. The second edit box is used in a later step to pick

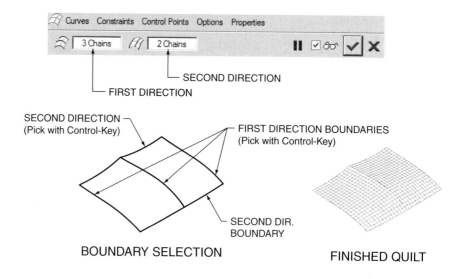

Figure 13-16   Blended Surface Feature

bounding curves in the second direction. Note, Pro/ENGINEER only requires a boundary definition in one direction.

Multiple entities can be defined as boundaries for the feature. The following rules apply:

- Curves, edges, datum points, and vertices can be used as bounding entities.

- Entities must be selected in consecutive order.

- For boundaries defined in two directions, the bounding entities must form a closed loop. As shown in Figure 13–16, the vertices of the bounding datum curves are connected to form a closed loop.

INSTRUCTIONAL NOTE   Boundaries for a surface can be selected in one direction or two directions. For two-directional boundaries, proceed to the next step of this guide. For one-directional boundaries, skip to Step 5.

STEP 3:   **On the dashboard, select the edit box for defining curves in the second direction (Figure 13–16).**

Second-direction curves can also be defined under the Cross Curves menu.

STEP 4:   **Using the Control key, pick curve entities that define the second direction of the surface feature.**

STEP 5:   ✓ **Build the feature.**

SOLIDIFYING QUILTS

The Edit menu's Solidify and Thicken tools can be used to create positive or negative space features from a merged quilt. For merged quilts that form a completely enclosed, watertight boundary, the Solidify tool is used to construct a formed solid (or a cut in a solid). For all

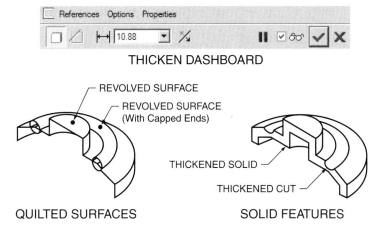

Figure 13–17   Solidified Surface Features

merged quilts, the Thicken tool can be used to thicken the surface, thus forming a thin solid feature (or a cut into a second solid).

Figure 13–17 shows an example of how quilts can be converted to positive and negative space solid features. The primary revolved feature is first constructed with the Revolve Surface tool. It is then solidified with the Thicken tool. The torus shaped feature was also constructed with the revolved surface tool. Since the ends of this feature are capped, it is watertight, which allows the Solidify tool to turn it into a solid cut.

Perform the following steps to create a solid from a quilt:

STEP 1: **On the model tree or on the workscreen, select a single quilt to solidify.**

Surfaces used to create a solid must be joined with the Merge option before a solid feature can be created from them.

STEP 2: **Under the Edit menu, select either the SOLIDIFY tool or the THICKEN tool.**

The Solidify tool will create a feature solid throughout its geometry. The Thicken tool will create a feature with a defined wall thickness.

STEP 3: **On the dashboard, select an appropriate material side (for thin features only).**

STEP 4: **For the Thicken tool, enter a thickness for the wall of the solid (Figure 13–17).**

STEP 5: **Select the Build Feature icon.**

SUMMARY

Pro/ENGINEER's surfacing tools are useful for the creation of complex geometric shapes. Solid creation tools are useful for creating hard mechanical components, but lack the flexibility to create free forming surfaces. While standard Pro/ENGINEER options such as Extrude, Revolve, Sweep, and Variable-Section Sweep are available for creating surfaces and solids, additional surface options like Merge and Boundaries are also available. Solids can be converted from surface quilts through the Use-Quilt option.

TUTORIAL

SURFACE TUTORIAL 1

The objective of this tutorial is to familiarize you with the basic techniques behind creating surface features in Pro/ENGINEER.

This tutorial exercise will provide instruction on how to model the part shown in Figure 13–18. Within this tutorial, the following topics will be covered:

- Creating an extruded surface feature.
- Creating datum curves.
- Creating an approximate boundary surface.
- Merging quilts.
- Creating a solid from a quilt.

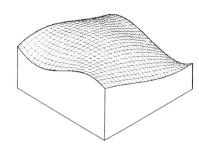

Figure 13–18   Finished Model

CREATE THE BASE EXTRUDED SURFACE FEATURE

The first segment of this tutorial will create an extruded surface feature (Figure 13–19). The Extrude tool with the Surface Feature option selected works basically the same as the Extrude options with Solid as the selection. For Extruded surfaces, an option is available for capping the end of the extrusion.

STEP   1:   ▢ **Start a New Part file.**

Use the File >> New option to create a new Part object file named *SURFACE1*. (Use the Default Template File)

STEP   2:   ⌇ **Select the Sketch option on the dashboard.**

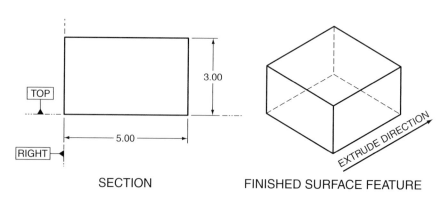

SECTION   FINISHED SURFACE FEATURE

Figure 13–19   Extruded Surface Feature

Step 3: Select datum plane FRONT as the sketching plane for the feature, accept the default direction of extrusion, then orient datum plane RIGHT toward the right of the workscreen.

Step 4: Sketch the Section shown in Figure 13–19.

Use the following options when creating the section:

- Specify datum planes RIGHT and TOP as references.
- Use the Rectangle option to sketch the section.
- Use the dimensional sizes shown in Figure 13–19.

Step 5: ✓ Select the Continue icon to exit the sketching environment.

Step 6: Select the Extrude icon on the toolbar.

Step 7: Select the Surface option on the Extrude dashboard.

Step 8: On the dashboard, enter 5.00 as the Blind extrude distance.

Step 9: Under the dashboard's Options menu, select CAPPED ENDS as an attribute for the surface feature.

With extruded surface features, the ends of the feature are left open by default. The Capped Ends option will close these ends.

Notice the difference in appearance between a surface model and a solid model. In No Hidden display mode, you cannot see the surface mesh of the surface feature. Because of this, the model can appear ambiguous. Try using the Shade display mode.

Step 10: ✓ Build the surface feature.

Step 11: Save your model.

CREATING DATUM CURVES

This segment of the tutorial will create the five datum curves shown in Figure 13–20. Datum curves and points are important features for the creation of surfaces. Within this tutorial, the four connected spline datum curves will be used as boundaries for the creation

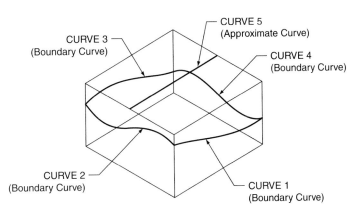

Figure 13–20 Datum Curves

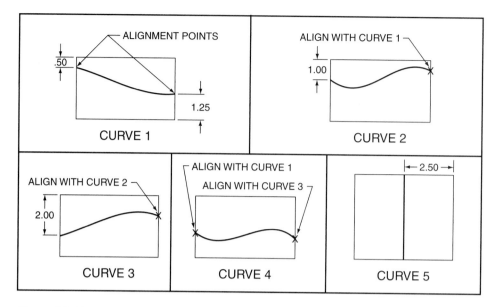

Figure 13–21   Curve Regenerated Sections

of a surface quilt. The fifth datum curve will be used as an approximate curve for manipulating the undulation of the surface.

Use the Sketch Tool to create each datum curve. Use appropriate existing planar surfaces as sketching planes. Each curve is a separate feature. If necessary, use the Pick From List mouse option to select hidden sketching planes. While sketching, the vertices of each curve must align. When specifying references for curves 3, 4, and 5, dynamically rotate the model to better select the vertices of existing curves. Within the sketching environment, use the Spline tool to create curves 1 through 4. The sections for each curve are shown in Figure 13–21. Approximate the exact spline shape for each curve.

CREATING A BOUNDARIES SURFACE

This segment of the tutorial will create the boundary blend surface shown in Figure 13–22. Surface features created with the Boundary Blend tool require the selection of boundary entities that define the exterior of the surface. With the Blended Surface, these boundaries must form a closed loop. With this surface feature, an influencing curve will be selected that will allow for extra control in the variation of the feature's surface.

STEP 1:   Select the Boundary Blend icon on the toolbar.

This option is also available under the Insert menu.

STEP 2:   Using the Control key, pick curves running in the first direction.

The first edit box shown in Figure 13–22 is used to define boundaries in the first direction. This box is selected by default upon entering the boundary blend's dashboard. The second edit box is used in a later step to pick bounding curves in the second direction. Note, Pro/ENGINEER requires a boundary definition in only one direction.

STEP 3:   On the dashboard, select the edit box for defining curves in the second direction (Figure 13–22).

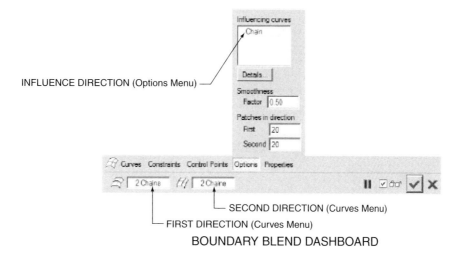

Figure 13–22   Curve Selection and Finished Surface

Second-direction curves can also be defined under the Cross Curves menu.

STEP 4: **Using the Control key, pick curve entities that define the second direction of the surface feature.**

STEP 5: **On the dashboard, select the Options menu, then select the CLICK HERE edit box for Influencing Curves (Figure 12–22).**

STEP 6: **Pick the approximate blend curve shown in Figure 12–22.**

The Influence curve option is used to select an additional reference curve. This additional curve will allow for a deviation in the surface of the quilt.

STEP 7: **On the workscreen, select the Influence Direction curve shown in Figure 13–22.**

STEP 8: **Also under the Options menu, enter 0.500 as the Smoothness parameter (Figure 13–22).**

STEP 9: **Enter 20 as the number of patches in the U and V Directions.**

The Smoothness parameter from Step 8 must be a value between 0 and 1. The Number of Patches parameter is used to specify the number of patches to create in each direction of the quilt. The more patches, the closer the surface will match the directional curves.

STEP 10: ✔ **Build the feature.**

If you get a failed feature, try increasing the number of patches for the fitting curve.

MERGING QUILTS

Your model currently consists of two surface features: one extrusion and one boundary blend. On the model tree, notice these two features and the five defined datum curves. Surface models typically have more features than solid models. This segment of the tutorial will merge the two surfaces. The intent of this tutorial is to create a final solid model of the design. After merging, the final quilt will be turned into a solid with the Solidify tool.

STEP 1: On the model tree, use the Control key to pick the two surface features (Extrude and Boundary Blend).

STEP 2: Select the Merge icon on the toolbar.

After selecting the Merge tool, Pro/ENGINEER will launch the surface merge dashboard (Figure 13–23). When you are picking quilts, the first surface is considered the primary quilt and the second surface is considered the additional quilt.

STEP 3: Under the Options menu, select the JOIN option.

Two Merge types are available: Intersect and Join. The Intersect option will merge two quilts that overlap. The Join option will merge two quilts that have common boundaries.

STEP 4: Shade (Render) your model to better visualize the surface feature.

STEP 5: Use the Material Direction option to select the side of the quilt to trim.

Select the Material Direction option and observe changes to your model. The intersecting quilts have two available final definitions. Your final merged quilt should resemble Figure 13–23.

STEP 6: Select the Build Feature icon to create the surface feature.

On your model tree, notice how a surface merge feature has been added. With the Merge option, the two parent features of the merge remain separate definable features in the order of regeneration.

STEP 7: Save your model file.

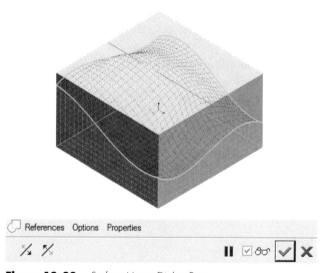

Figure 13–23   Surface Merge Dialog Box

CREATING A SOLID FROM A QUILT

Surface features have no defined thickness. How often does a design surface not have at least a minimum surface thickness? Merged surfaces can be converted to solids with the Solidify and Thicken tools, both which can be found under the Edit menu. This segment of the tutorial will create a solid feature from the existing merged quilts.

STEP 1: **On the model tree, select the last merged surface.**

STEP 2: **Select EDIT >> SOLIDIFY on the menu bar.**

When you created the boundary surface earlier in this tutorial, you defined the number of patches to equal 20 in both directions. If you had accepted the default value of 5, the model would not have been closed well enough to use the Solidify tool. The Solidify tool requires "watertight" surface quilts with no holes, openings, or gaps.

STEP 3: **On the solidify dashboard, select the Build Feature icon.**

No additional settings are needed on the dashboard. Your model should appear as shown in Figure 13–24. You can use solid modeling tools (protrusions, cuts, etc.) to build upon a model that has been solidified. On your model tree, notice the number of features it took to build this part.

STEP 4: **Save your model.**

MODELING POINT Notice the absence of datum curves in Figure 13–24. Datum curves are not merged with surfaces and are not affected by the Solidify option. The datums in this example were placed on a layer and the display of the layer hidden. See Chapter 2 for information on creating layers.

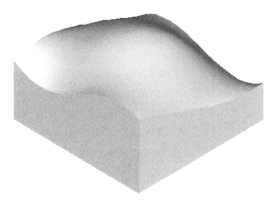

Figure 13–24 Finished Model

SURFACE TUTORIAL 2

This tutorial will expand on the concepts covered in the first tutorial. This project will focus on modeling the hair dryer shell shown in Figure 13–25. Like most surface models, the first segment of this tutorial will involve the creation of datum curves to control the creation of surface features. Within this tutorial, the following topics will be covered:

- Creating datum curves.
- Creating surfaces from boundaries.
- Defining boundary conditions.
- Merging surfaces.
- Filleting surface features.
- Trimming a surface.
- Creating a draft offset.
- Converting a surface to a solid.

Figure 13–25   Finished Model

CREATING DATUM CURVES

This segment of the tutorial will establish the object's modeling environment and create the datum curves necessary to define the part.

STEP 1:   Start Pro/ENGINEER and create a new Part file named *SURFACE2*.

STEP 2:   Use the Sketched Tool to create the first datum curve.

The curves used to construct this part are shown in Figure 13–26. With the exception of one mirrored curve, each will be created with the sketch option.

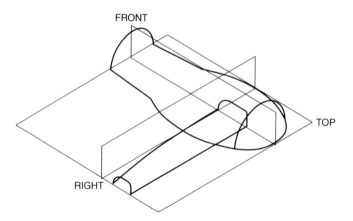

Figure 13–26   Datum Curves

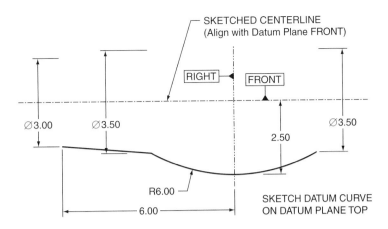

Figure 13-27   The First Datum Curve

STEP 3:   **Select datum plane TOP as the sketch plane for the datum curve.**

STEP 4:   **Accept the Default Direction and orient datum plane RIGHT toward the right of the workscreen (see Figure 13–27's orientation).**

STEP 5:   **Create the sketch shown in Figure 13–27.**

Sketch the section using one line and one arc. Create the entities oriented toward the datum planes as shown. Sketch the horizontal centerline aligned with datum plane FRONT and use the dimensioning scheme shown.

NOTE: The center of the arc entity is aligned with datum plane RIGHT but not with datum plane FRONT.

STEP 6:   **Exit the sketching environment.**

STEP 7:   **Use the COPY >> MIRROR option and the DEPENDENT suboption to mirror the first datum curve about datum plane FRONT.**

The Copy command can be found under Edit >> Feature Operations. The body of the part is symmetrical about datum plane FRONT. The Dependent option will help to keep the model symmetrical. Your first two datum curves should appear as shown in Figure 13–28.

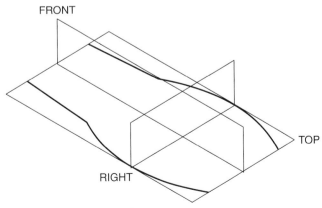

Figure 13-28   Datum Curves

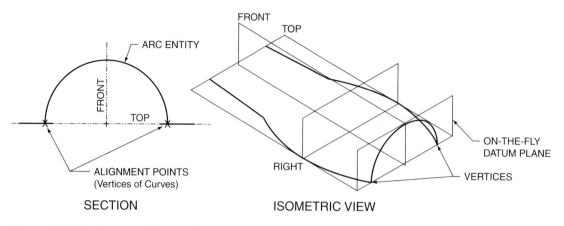

Figure 13–29 Section and Isometric View

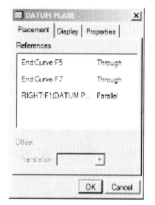

Figure 13–30 Datum Plane Creation

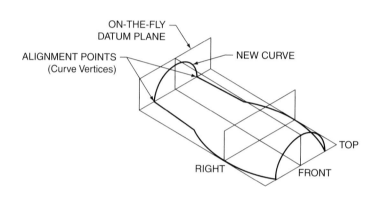

Figure 13–31 New Datum Curve

STEP 8: Setup the sketching plane for the datum curve feature shown in Figure 13–29.

Use the Sketch-Tool to create the curve feature. Sketch the feature on an on-the-fly datum plane that runs through the vertices at the end of the two existing datum curves (see Figures 13–29 and 13–30) and is constrained parallel to datum plane RIGHT.

STEP 9: **Sketch the Curve Feature.**

Select the end of the two existing datum curves as references within the sketching environment. Use the Pick From List option if necessary. When creating the section, use the Arc option and align the ends of the arc with the two referenced vertices as shown in Figure 13–29.

STEP 10: **Exit the sketching environment.**

STEP 11: **Create the Sketched Curve Feature shown in Figure 13–31.**

Use the same procedure from the previous datum curve to create this datum curve. Sketch the curve on an on-the-fly datum plane running through the two vertices and parallel to datum plane RIGHT.

STEP 12: **Create the Sketched Curve Feature shown in Figure 13–32.**

Sketch the curve on datum plane "TOP". Within the sketching environment,

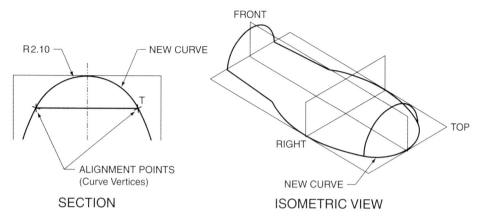

Figure 13–32   New Tangent Datum Curve

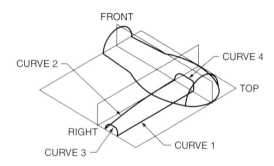

Figure 13–33   Datum Curves

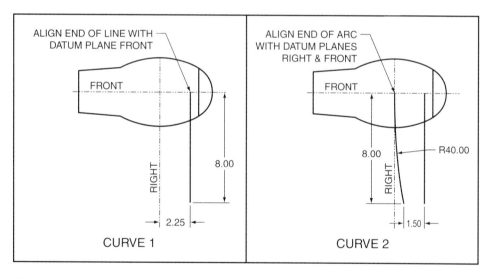

Figure 13–34   Datum Curves 1 and 2

reference the ends of the existing datum curves. One end of the curve should be constrained tangent to existing part curves.

STEP 13: 〰 **Use the Sketch Tool to create CURVE 1 (Figures 13–33 and 13–34).**

Sketch CURVE 1 as shown in Figure 13–34. Sketch the entity as a vertical line with one end aligned with datum plane FRONT.

STEP 14: 〰 **Use the Sketch Tool to create CURVE 2 (Figures 13–33 and 13–34).**

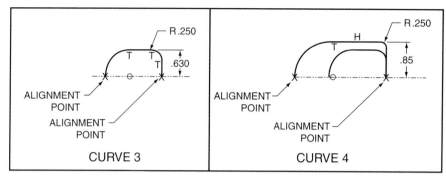

Figure 13–35   Datum Curves 3 and 4

Sketch CURVE 2 as shown in Figure 13–34. Sketch the entity as an arc with one end aligned at the intersection of datum planes RIGHT and FRONT.

STEP 15:   Use the Sketch Tool to create CURVE 3 (Figures 13–33 and 13–35).

Sketch CURVE 3 as shown in Figure 13–35. Sketch the section on an on-the-fly datum plane that runs through the ends of CURVE 1 and CURVE 2 and that is normal to datum plane TOP. Sketch the entities of the section aligned with the ends of CURVE 1 and CURVE 2.

STEP 16:   Use the Sketch Tool to create CURVE 4 (Figures 13–33 and 13–35).

Sketch CURVE 4 as shown in Figure 13–35. Sketch the section on an on-the-fly datum plane that runs through the ends of CURVE 1 and CURVE 2. Sketch the entities of the section aligned with the ends of CURVE 1 and CURVE 2.

INSTRUCTIONAL POINT   The primary surface quilts of this part will each be constructed with the Boundary Blend tool. Two of the boundaries surfaces will be constructed in two directions. Datum curves forming the boundaries with this type of surface must be aligned at their respective intersecting vertices. The previous segment of this tutorial specified alignments when necessary. You should double-check to verify that all alignments are correct.

CREATING SURFACES FROM BOUNDARIES

The previous segment of this tutorial created sketched curves for use in defining the shape of the part. This segment of the tutorial will use these curves to create surface features by defining each surface quilt's boundary. The four surfaces to be defined are shown in Figure 13–36.

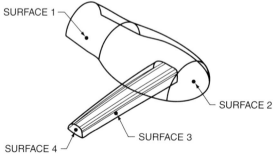

Figure 13–36   Surface Creation

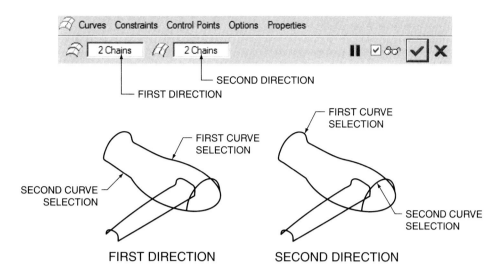

Figure 13-37   First Surface Creation

Step 1: Select the Boundary Blend icon on the toolbar.

The first surface to be created will be SURFACE 1 shown in Figure 13–36. The Boundary Blend tool is used to define a surface between two or more bounding entities. Boundaries can be defined in one or two directions. When a boundary is defined in two directions, the boundary entities must be connected at each entity's vertex.

Step 2: Using the Control key, pick the two datum curves in the FIRST DIRECTION shown in Figure 13–37.

Step 3: On the dashboard, select the second-direction edit box shown in Figure 13–37.

Step 4: Using the Control key, pick the two datum curves in the SECOND DIRECTION shown in Figure 13–37.

Step 5: Under the dashboard's Constraints menu, set the first direction's boundary type to NORMAL (Figure 13–38).

The Edge Alignment option allows surface boundaries to lie tangent or normal to a boundary entity. In this feature, your surface will lie normal to the sketching plane of the two bounding entities (datum curves in this case).

Step 6: Build the feature.

Step 7: Select the Boundary Blend icon on the toolbar.

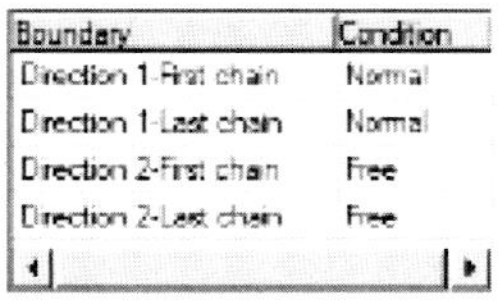

Boundary	Condition
Direction 1-First chain	Normal
Direction 1-Last chain	Normal
Direction 2-First chain	Free
Direction 2-Last chain	Free

Figure 13-38   Edge Alignment Settings

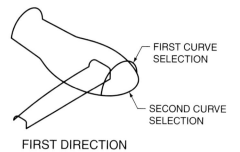

Figure 13–39   SURFACE 2 Creation

The next surface to be created will be SURFACE 2 shown in Figure 13–36. This surface feature will be created by defining the two boundaries shown in Figure 13–39. In addition, an edge alignment condition will be defined that makes this surface tangent to SURFACE 1.

STEP 8: **Using the Control key, pick the first curve shown in Figure 13–39.**

Within this tutorial, it is important that you pick the datum curves in the order shown in the figure. In a later step, the first datum curve will have an edge condition set that makes it tangent to the first surface feature.

STEP 9: **Using the Control key, pick the second curve (also first direction) shown in Figure 13–39.**

STEP 10: **Under the Constraints menu, set the first direction edge alignment to TANGENT (Figure 13–40).**

The following boundary conditions are available:

- **Free** No tangent condition is established between surfaces.
- **Tangent** Surface is created tangent across the boundary.
- **Normal** The surface is created normal across the boundary.
- **Curvature** The surface has curvature continuity across the boundary.

STEP 11: **Ensure that the Entity 1 surface is set to the surface shown in Figure 13–40.**

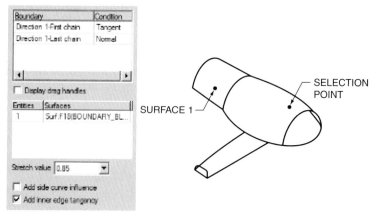

Figure 13–40   Edge Alignment Settings

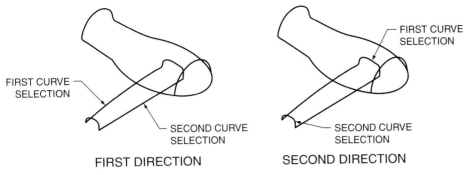

FIRST CURVE
SELECTION

FIRST CURVE
SELECTION

SECOND CURVE
SELECTION

SECOND CURVE
SELECTION

FIRST DIRECTION   SECOND DIRECTION

Figure 13–41   Surface Creation

If the Entity 1 surface is not set, select the Surfaces collector box then pick the surface shown in Figure 13–40.

STEP 12: **Under the Edge Alignment menu, set the second boundary under the first direction edge alignment to NORMAL and the Stretch Value to 0.85 (Figure 13–40).**

The Stretch value acts as a tension or pulling force on the tangent edge constraint. Try changing this value to see the affect it has on the surfaces curvature.

STEP 13: **Build the feature.**

The next section of this segment will create the surface feature for the handle of the part. You will create the quilt using the Boundaries option.

STEP 14: **Use the Boundary Blend tool to create the two-directional boundary feature surface shown in Figure 13–41.**

As shown in Figure 13–41, create the surface for the handle using the Boundary Blend tool. Create the boundaries feature in the two directions shown. Define the edge alignment for both first direction boundaries as Normal.

NOTE: In a later segment of this tutorial, this surface will be merged with the surface defining the main body of the part.

STEP 15: **Save your object file.**

CREATING A FLAT SURFACE

The next steps of this tutorial will use the Fill tool to create the surface feature located at the end of the handle (Figure 13–42).

STEP 1: **Select the Sketch tool on the toolbar.**

STEP 2: **Use the Datum Plane tool to create the on-the-fly datum plane for sketching (Figure 13–43).**

Use the Through and Normal constraint options to create the three datum plane constraint options shown in Figure 13–43. Remember to use the Control key to pick each constraint.

STEP 3: **Orient and enter the sketching environment.**

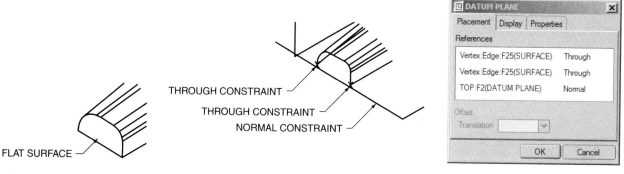

Figure 13–42   Flat Surface Option

Figure 13–43   Make Datum Constraint Options

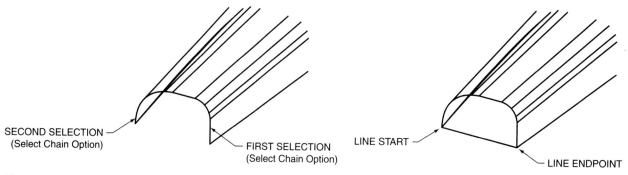

Figure 13–44   The Use Edge Option

Figure 13–45   Line Creation

STEP 4:   Use the USE EDGE option and the CHAIN suboption to project the entities shown in Figure 13–44.

The Use Edge option will project existing feature edges onto the sketching plane as sketcher entities. The Chain option will select a chain of entities between two picked entities.

STEP 5:   Use the Line tool to create the line shown in Figure 13–45.

STEP 6:   Exit the sketching environment.

If you get the error message "Section must be closed for this feature," the Use Edge option probably did not completely define the required loop. Observe the workscreen to find the problem area.

STEP 7:   Select EDIT >> FILL on the menu bar.

The Fill tool is used to create a two-dimensional surface feature. The construction technique is similar to the Extrude option, only without the depth definition.

STEP 8:   Build the feature.

STEP 9:   Save your part.

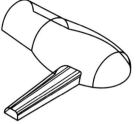

Figure 13–46
Merged Surfaces

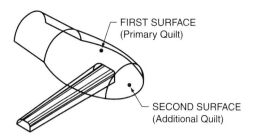

FIRST SURFACE
(Primary Quilt)

SECOND SURFACE
(Additional Quilt)

Figure 13–47
Surface Merge Dashboard

MERGING SURFACES

At this point in the modeling process, your model consists of four surface features. You will use the Merge option to join these surfaces to create one quilted feature. The finished merged feature is shown in Figure 13–46.

STEP 1: Using the Control key, pick the two surfaces shown in Figure 13–47.

STEP 2: Select the Merge icon on the toolbar.

After you select the Merge option, Pro/ENGINEER will launch the Surface Merge dashboard. This dashboard is used to select quilts that either intersect or share a common edge. Quilts that share a common edge can be merged with the Join option. Two quilts that intersect can be merged at their intersection.

STEP 3: On the Surface Merge dashboard under the Options menu, select the JOIN option.

The Join option is used to merge two quilts that share adjacent boundaries.

STEP 4: Select the Build Feature option.

After creating the merged feature, notice the Surface Merge feature on the model tree. The Merge option creates a separate feature from its parent surfaces.

STEP 5: Select RENDER on the toolbar as the models display model.

It is often easier to view merge operations with the model not shaded.

STEP 6: Using the Control key, pick the two surfaces shown in Figure 13–48.

STEP 7: Select the Merge icon on the toolbar.

STEP 8: Under the Options menu, select the INTERSECT option.

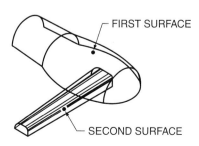

FIRST SURFACE

SECOND SURFACE

Figure 13–48   Merge Selections

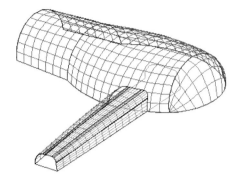

Figure 13–49   Side Options

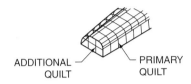

Figure 13–50   Merged Quilts

The Intersect option will merge two quilts that intersect. After you select two quilts, with the Intersect option, there are generally four possible ways to trim the surfaces. The dashboard provides two Material Direction options to set appropriate trim sides.

STEP 9:   On the dashboard, use the two Material Direction options to match the quilt shown in Figure 13–49.

STEP 10:   Select the Build Feature option.

STEP 11:   Use the Merge tool to join the handle's flat end quilt with the previously defined merged quilt (Figure 13–50).

STEP 12:   Save your model.

CREATING ADDITIONAL SURFACES

This segment of the tutorial will create additional surface features (Figure 13–51). The first feature created will be a round on the surface of the first quilt. The second surface feature will be a revolved quilt forming the trigger finger location.

STEP 1:   Select the Round icon on the toolbar.

The surface Round command works identically to the solid Round command.

STEP 2:   Pick the edge shown in Figure 13–52, then enter 2.00 as the radius.

STEP 3:   Build the feature.

Round features are automatically merged.

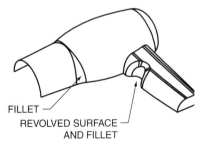

FILLET
REVOLVED SURFACE
AND FILLET

Figure 13–51   Additional Surfaces

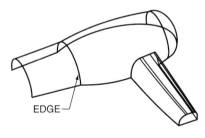

EDGE

Figure 13–52   Surface Selection

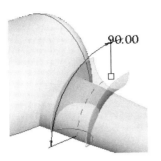

Figure 13-53 Revolved Surface

STEP 4: **Set up and create the sketch for the revolved surface shown in Figure 13–53.**

Next, you will create the revolved surface shown in Figure 13–53. After creation, you will merge this surface with the remaining part surfaces. Referring to Figure 13–54, use the following options when creating the sketch:

- Use datum plane TOP as the sketching plane.
- Orient datum plane RIGHT toward the bottom on the sketcher environment.
- Specify the edges of the handle and the round as references for aligning the sketched arc (Sketch >> References).
- Use the Arc Center/Ends option to sketch the section.
- Sketch the feature's section as shown in Figure 13–54.

STEP 5: **Select the Revolve icon on the toolbar.**

STEP 6: **On the dashboard, select Surface as the type of feature to create.**

- Revolved features require a centerline as shown.
- Use a 90 degree rotation angle.

STEP 7: **Use the Merge tool to merge the previously created revolved surface.**

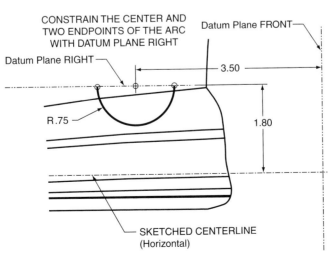

Figure 13-54 Surface Section

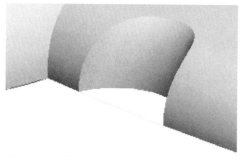

Figure 13-55   Merged Surfaces

Figure 13-56   Filleted Surface

Use the Merge tool and the Intersect option to merge the revolved surface with any intersecting surfaces. The final merge is shown in Figure 13–55.

STEP 8:   Use the Round tool to create the 0.350 radius fillet shown in Figure 13–56.

CREATING A DRAFT OFFSET

This segment of the tutorial will create the Draft Offset feature shown in Figure 13–57. Draft is an option under the Offset command (Edit >> Offset). The Draft Offset option creates a new surface by offsetting from an existing surface between the bounds of a sketched section. The sides of the new surface can be beveled similarly to a drafted surface or they can be defined normal to the sketching plane. The new surface can protrude into or away from the selected parent surface. In this tutorial, the surface will protrude into the part.

STEP 1:   Select the Sketch option on the dashboard.

You will create a sketch that defines the boundary for the offsetting surface.

STEP 2:   **Select datum plane TOP as the sketching plane and orient datum plane RIGHT toward the right of the sketching environment.**

Draft offset features are created by projecting a section onto a receiving surface. In this example, the section will be sketched on datum plane TOP.

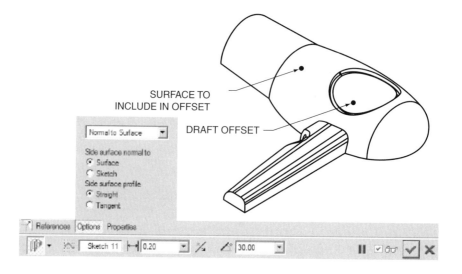

Figure 13-57   Draft Offset

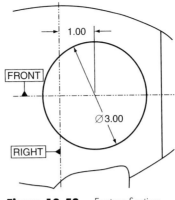

Figure 13–58　Feature Section

Step 3: Create the section shown in Figure 13–58.

Step 4: ✔ When the section is complete, exit the sketching environment.

Step 5: On the model tree, select the last merged surface feature.

Step 6: Select EDIT >> OFFSET on the menu bar.

Step 7: On the dashboard, enter an offset value of 0.20.

Step 8: On the dashboard, select WITH DRAFT as the offset type (Figure 13–57).

The With Draft option allows for a beveled (or drafted) offset surface.

Step 9: Pick your last sketch as the curve boundary for the offset.

Step 10: Under the Options menu on the dashboard, select the following options.

- Normal to Surface.
- Surface as the Side Surface Normal to setting.
- Straight as the Side Surface Profile setting.

Note: With the Tangent side surface profile option, the surfaces of the feature will be blended with round, tangent edges. With the Straight option, sharp edges remain.

Step 11: Dynamically rotate your model to observe the direction of offset.

Your offset should protrude toward the inside of the surface shell.

Step 12: If necessary, use the Material Direction option to change the direction of extrusion.

Observe the workscreen and make sure you select a material direction that will protrude the feature into the part.

Step 13: Enter a draft angle value of 30.00 degrees.

Step 14: ✔ Build the feature.

TRIMMING A SURFACE

This segment of the tutorial will create the four trimmed features shown in Figure 13–59. The first feature will be created with the Surface Trim command. You will pattern this feature to create the remaining three features.

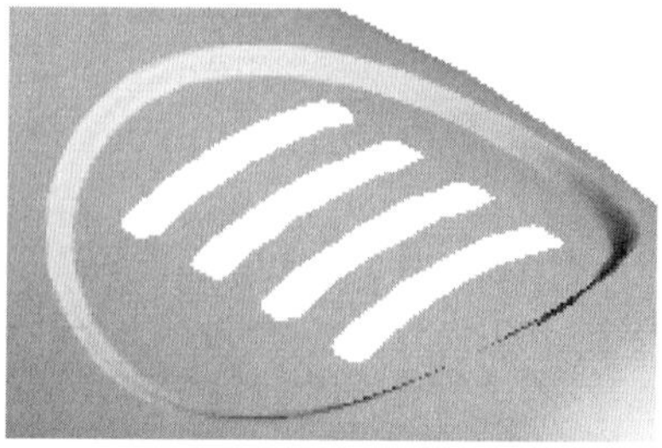

Figure 13–59 Trimmed Features

Step 1: 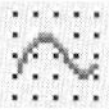 **Select the Sketch tool.**

Step 2: **Pick datum plane TOP as the sketching plane and orient datum plane RIGHT toward the right.**

Step 3: **SKETCH the section shown in Figure 13–60.**

Step 4:  **When the section is complete, exit the sketching environment.**

Step 5: 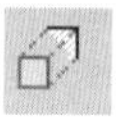 **After completing the sketch, select the Extrude tool.**

When used in combination, the Extrude tool's Surface and Cut suboptions can be utilized to trim existing surface features.

Step 6:  **On the Extrude dashboard, select the Surface option.**

Step 7:  **On the dashboard, select the Cut option.**

Step 8: **On the dashboard, select the Quilt collector (Figure 13–60).**

The Quilt collector is used to select intersecting surfaces for trimming. This edit box should currently display No Items to indicate that no surfaces have been selected.

Step 9: **On the workscreen, pick the Quilt To Intersect location shown in Figure 13–60.**

Since you currently have a completely merged quilt, you could pick anywhere on the model.

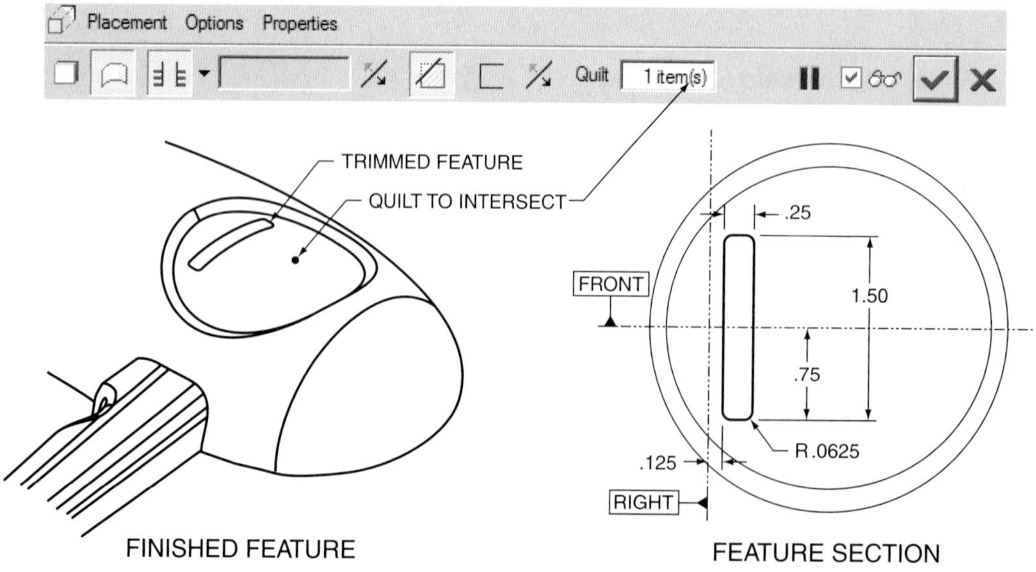

Figure 13–60 Trimmed Feature Construction

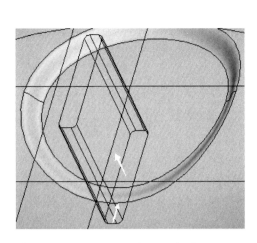

Figure 13–61   Material Removal Side

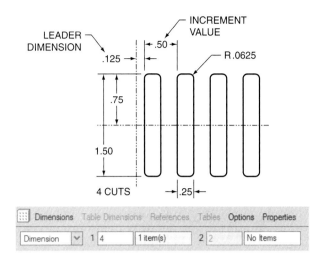

Figure 13–62   Patterned Feature

STEP 10:  Select Through All as the depth option.

STEP 11: Use the first Material Direction option to define an extrusion direction that will intersect the current surface features.

You might need to dynamically rotate your model to better visualize the direction of rotation.

STEP 12: Use the second Material Direction option to ensure that the material removal side for the cut is toward the inside of the section (Figure 13–61).

STEP 13: Build the feature.

Attributes and definitions used to create the trimmed feature can be redefined on the dashboard. The next several steps will pattern this feature.

STEP 14: Preselect the last trimmed surface feature then select the Pattern tool to create the pattern shown in Figures 13–59 and 13–62.

Define a one-directional linear pattern. Select the Leader dimension shown in Figure 13–62. Increment four instances of the pattern a distance of 0.50 inches.

STEP 15: Save your part file.

CONVERTING A SURFACE TO A SOLID

This segment of the tutorial will convert the part from a surface model to a solid model. In this tutorial, the part will be defined as a thin solid. Once a surface quilt is converted to a solid, normal solid commands can be used to manipulate the feature.

STEP 1: On the workscreen, pick the model.

STEP 2: Select EDIT >> THICKEN on the menu bar.

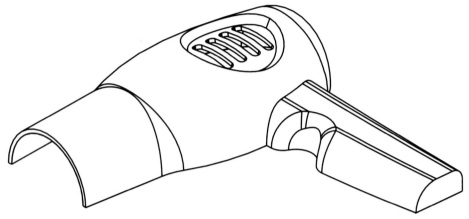

Figure 13–63   Thicken Dashboard

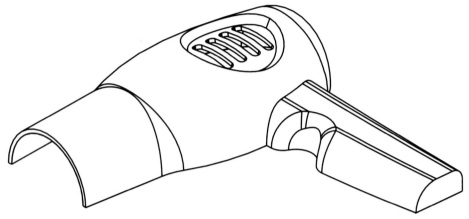

Figure 13–64   Finished Part

(The Thicken option is used to add solid material to the surface of an existing quilt.) For enclosed surfaces, the Solidify option is available for adding solid material throughout the interior of the quilt.

STEP 3:   **On the dashboard, enter 0.125 as the thicken depth (Figure 13–63).**

STEP 4:   **Select a Material Side that will create solid material toward the inside of the existing surface features.**

STEP 5:   **Select the Build Feature option.**

STEP 6:   **Save your part model.**

Your final part should appear as shown in Figure 13–64.

PROBLEMS

1. Model the plastic soap dish shown in Figure 13–65. In this problem, do not model the lid of the dish removable from the base.

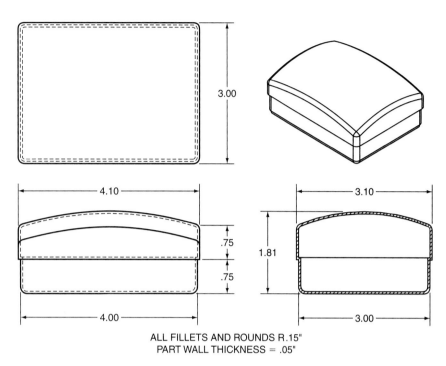

ALL FILLETS AND ROUNDS R.15"
PART WALL THICKNESS = .05"

Figure 13–65 Problem 1

2. Model the plastic bottle shown in Figure 13–66. The figure includes orthographic views of the part and a view with dimensions for the part's datum curves. Create the part using Pro/ENGINEER's surfacing tools. Use the Edit >> Thicken command to convert the surfaces to a solid feature.

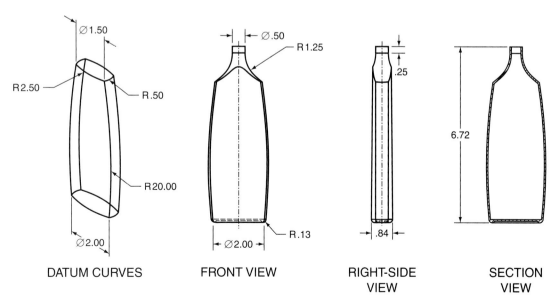

DATUM CURVES   FRONT VIEW   RIGHT-SIDE VIEW   SECTION VIEW

Figure 13–66 Problem 2

3. Model the part shown in Figure 13–67.

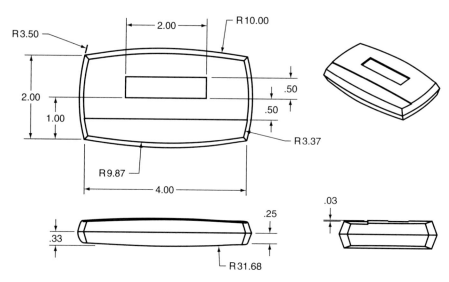

Figure 13–67   Problem 3

4. Model the part shown in Figure 13–68.

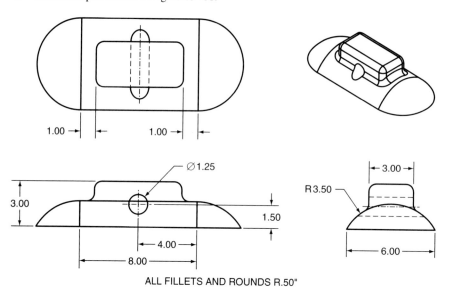

ALL FILLETS AND ROUNDS R.50"

Figure 13–68   Problem 4

QUESTIONS AND DISCUSSION

1. How does a surface feature differ from a thin solid feature?

2. Compare the Thicken tool to the Solidify tool.

3. Explain when two surfaces can be combined by using the Merge tool's Join suboption.

4. Explain when two surfaces can be combined by using the Merge tool's Intersect suboption.

APPENDIX

A

SUPPLEMENTAL FILES

The following model files are used in selected guides throughout this textbook. They can be downloaded from the following Web address: http://www.mhhe.com/kelley.

CHAPTER 2

interface.prt	Used in the Interface Tutorial

CHAPTER 4

datum1.prt	Used in the Datum Tutorial

CHAPTER 5

draft1.prt	Used in the "Creating a Neutral Plane No Split Draft" guide
draft2.prt	Used in the "Creating a Neutral Plane Split Draft" guide
draft3.prt	Used in the "Creating a Neutral Curve Draft" guide
pattern1.prt	Used in the "Creating a Linear Pattern" guide
rib.prt	Used in the "Creating a Rib" guide

CHAPTER 7

copy_mirror.prt	Used to practice the Copy-Mirror option
copy_new_ref.prt	Used to practice the Copy-New Reference option
copy_rotate.prt	Used to practice the Copy-Move-Rotate option
copy_translate.prt	Used to practice the Copy-Move-Translate option
generic_bolt.prt	Used in the "Creating a Family Table" guide
group_pattern.prt	Used in the "Patterning a Group" guide
pattern1.gph	Used in the "Placing a User-Defined Feature" guide
udf_part.prt	Used in the "Creating a User-Defined Feature" guide

CHAPTER 8

dtl_no_tol.dtl	Drawing Setup File with tolerances not displayed
dtl_yes_tol.dtl	Drawing Setup File with tolerances displayed
detail_view.prt	Part used to practice the establishment of a drawing

CHAPTER 9

align.drw	Drawing used in the "Aligned Section Views" guide
align.prt	Part used in the "Aligned Section Views" guide
auxiliary.drw	Drawing used in the "Auxiliary Views" guide
auxiliary.prt	Part used in the "Auxiliary Views" guide
broken_out.drw	Drawing used in the "Broken Out Section" guide

broken_out.prt	Part used in the "Broken Out Section" guide
offset.drw	Drawing used in the "Offset Sections" guide
offset.prt	Part used in the "Offset Sections" guide
revolve_sect.drw	Drawing used in the "Revolved Sections" guide
revolve_sect.prt	Part used in the "Revolved Sections" guide
section.drw	Drawing used to practice the creation of section views
section.prt	Part used to practice the creation of section views
section1.prt	Used in "Advanced Drawing Tutorial 1"

CHAPTER 11

var_sec_swp_traj.prt	Used in the "Creating a Variable Section Sweep" guide

CHAPTER 13

surface1.prt	Used in the "Merging Quilts" guide
surface2.prt	Used in the "Creating a Blended Surface from Boundaries" guide
surface3.prt	Used in the "Use Quilt" guide

MISCELLANEOUS

start.prt

config.pro

Option/Possible Values	Description
ANGULAR_TOL *ANGULAR_TOL_0.00*	Used to set angular tolerance values.
ALLOW_ANATOMIC_FEATURES *YES* *NO*	Used to set the display of menu options to include Slot, Shaft, Neck, and Flange.
BELL *YES* *NO*	Used to turn on or off Pro/ENGINEER's message bell.
DEFAULT_DEC_PLACES	Used to set the default number of decimal places.
DEFAULT_DRAW_SCALE	Used to set a default initial drawing scale factor.
DEF_LAY *LAYER_GEOM_FEAT features*	Used to create a default layer and to automatically add items to the layer.
DRAWING_SETUP_FILE C:\dtl\dtl_no_tol.dtl	Used to establish a default drawing setup file.
FONTS_SIZE *SMALL* *MEDIUM* *LARGE*	Used to set the size of menu fonts.
LINEAR_TOL *LINEAR_TOL_0.000*	Used to set linear tolerance values.
ORIENTATION *ISOMETRIC* *TRIMETRIC* *USER-DEFAULT*	Used to set the default orientation of models.
PEN#_LINE_WEIGHT *PEN1_LINE_WEIGHT 2*	Used to set the line weight for a specific pen number. Possible settings can range from 1 to 16. Each increment equals a width value of 0.005 in.
PRO_CROSSHATCH_DIR	Used to specify the directory where hatch patterns are saved.
PRO_DTL_SETUP_DIR	Used to set a default drawing setup file.
PRO_FORMAT_DIR	Used to set the default directory for locating drawing formats.
PRO_GROUP_DIR	Used to set the default directory for locating user-defined features.
PRO_MATERIAL_DIR	Used to set the default directory for locating material files.
PRO_PLOT_CONFIG_DIR	Used to specify the directory where Pro/ENGINEER searches for plotter configuration files (.pcf files).
SAVE_MODEL_DISPLAY	Used to set the shading display in a graphical simplified representation. Possible options include *wireframe*, *no_display, shading_low, shading medium, shading_high,* and *shading_lod*. The default is *wireframe*.
SKETCHER_DISPLAY_CONSTRAINTS *YES* *NO*	Used to control the display of constraints in the sketcher environment.

Option/Possible Values	Description

SKETCHER_DISPLAY_DIMENSIONS
  YES
  NO
	Used to control the display of dimensions in the sketcher environment.

SKETCHER_DISPLAY_GRID
  YES
  NO
	Used to control the display of grids in the sketcher environment.

SKETCHER_DISPLAY_VERTICES
  YES
  NO
	Used to control the display of vertices in a sketcher environment.

SKETCHER_INTENT_*MANAGER*
  YES
  NO
	Used to activate Intent Manager in a sketcher environment.

SPIN_CENTER_DISPLAY
  YES
  NO
	Used to control the display of the spin center symbol.

SYSTEM_BACKGROUND_COLOR *100 100 100*	Used to set the background color of Pro/ENGINEER's workscreen.
SYSTEM_DIMMED_MENU_COLOR	Used to set the color of dimmed entities and menu options.

TANGENT_EDGE_DISPLAY
  NO
  SOLID
  CENTERLINE
  PHANTOM
  DIMMED
	Used to set the default tangent edge display style.

TOL_DISPLAY
  YES
  NO
	Used to set the default display of tolerances.

TOL_MODE
  NOMINAL
  LIMITS
  PLUSMINUS
  PLUSMINUSSYM
	Used to set the default display mode of tolerances.

TOLERANCE_STANDARD
  ANSI
  ISO
	Used to select a default tolerance standard.

INDEX